PRACTICAL IMMUNOASSAY

CLINICAL AND BIOCHEMICAL ANALYSIS

A series of monographs and textbooks

EDITOR

Morton K. Schwartz

Chairman, Department of Biochemistry
Memorial Sloan-Kettering Cancer Center
New York, New York

1. Colorimetric and Fluorimetric Analysis of Organic Compounds and Drugs, *M. Pesez and J. Bartos*

2. Normal Values in Clinical Chemistry: Statistical Analysis of Laboratory Data, *Horace F. Martin, Benjamin J. Gudzinowicz, and Herbert Fanger*

3. Continuous Flow Analysis: Theory and Practice, *William B. Furman*

4. Handbook of Enzymatic Methods of Analysis, *George G. Guilbault*

5. Handbook of Radioimmunoassay, *edited by Guy E. Abraham*

6. The Hemoglobinopathies, *Titus H. J. Huisman and J. H. P. Jonxis*

7. Automated Immunoanalysis (in two parts), *edited by Robert F. Ritchie*

8. Computers in the Clinical Laboratory: An Introduction, *E. Clifford Toren, Jr. and Arthur A. Eggert*

9. The Chromatography of Hemoglobin, *Walter A. Schroeder and Titus H. J. Huisman*

10. Nonisotopic Alternatives to Radioimmunoassay: Principles and Applications, *edited by Lawrence A. Kaplan and Amadeo J. Pesce*

11. Biochemical Markers for Cancer, *edited by T. Ming Chu*

12. Clinical and Biochemical Luminescence, *edited by Larry J. Kricka and Timothy J. N. Carter*

13. Advanced Interpretation of Clinical Laboratory Data, *edited by Camille Heusghem, Adelin Albert, and Ellis S. Benson*

14. Practical Immunoassay: The State of the Art, *edited by Wilfrid R. Butt*

ADDITIONAL VOLUMES IN PREPARATION

19059

PRACTICAL IMMUNOASSAY

The State of the Art

Edited by
WILFRID R. BUTT
*Birmingham and Midland Hospital for Women
Birmingham, England*

MARCEL DEKKER, INC.   New York and Basel

LIBRARY OF CONGRESS CATALOGING IN PUBLICATION DATA
Main entry under title:

Practical immunoassay.

   (Clinical and biochemical analysis ; 14)
   Includes index.
   1. Hormones--Analysis. 2. Immunoassay. 3. Diagnosis,
Laboratory. I. Butt, Wilfrid R. II. Series. [DNLM:
1. Immunoassay--Methods. W1 CL654 v.14 / QW 570 P895]
RB48.5.P73   1984   616.07'56   84-1884
ISBN 0-8247-7094-3

COPYRIGHT © 1984 BY MARCEL DEKKER, INC. ALL RIGHTS RESERVED.

Neither this book nor any part may be reproduced or transmitted in any form
or by any means, electronic or mechanical, including photocopying, micro-
filming, and recording, or by any information storage and retrieval system,
without permission in writing from the publisher.

MARCEL DEKKER, INC.
270 Madison Avenue, New York, New York 10016

Current printing (last digit):
10 9 8 7 6 5 4 3 2 1

PRINTED IN THE UNITED STATES OF AMERICA

Preface

Immunoassays have become indispensable in laboratories used by many disciplines: in endocrinology it is hard to imagine the days when these techniques were not in common use. Few, if any, hormones could be measured satisfactorily in serum by other means, and the availability of radioimmunoassays has led to appreciable advances in our knowledge of endocrine function.

There are many ways in which antibodies can be used as assay reagents and there has been considerable progress in techniques over the last few years. In this book some of the more recent work has been collected together and the methods described by authors who have had considerable experience in developing and applying them. I am grateful to these people for their cooperation, undertaken in addition to their other heavy commitments.

In any laboratory it is difficult to decide when to change a well-established method for a new one. The decision will usually depend on changing needs and facilities and it is hoped that the information collected together here will help those who have to make the choice. For this reason it is expected that the readership will largely be made up of those who are engaged in directing, teaching, developing, or performing immunoassays. Some of the material is based on that presented at a successful meeting held in Birmingham, England, in 1980 on the same subject, and this and other material is here brought up to date.

I thank the publishers for encouraging the preparation of this book, for their continuous help throughout the production, and for their patience and understanding.

WILFRID R. BUTT

Contributors

WILFRID R. BUTT   Professor and Director, Department of Clinical Endocrinology, Birmingham and Midland Hospital for Women, Central Birmingham Health Authority, Edgbaston, Birmingham, England

ANTHONY K. CAMPBELL   Senior Lecturer, Department of Medical Biochemistry, Welsh National School of Medicine, Cardiff, Wales

SALIFU DAKUBU   Research Assistant, Department of Molecular Endocrinology, Middlesex Hospital Medical School, London, England

ROGER EKINS   Professor, Department of Molecular Endocrinology, Middlesex Hospital Medical School, London, England

THOMAS JACKSON   Research Assistant, Department of Molecular Endocrinology, Middlesex Hospital Medical School, London, England

ROYSTON JEFFERIS   Senior Lecturer, Department of Immunology, University of Birmingham School of Medicine, Edgbaston, Birmingham, England

RHYS JOHN   Principal Biochemist, Department of Medical Biochemistry, University Hospital of Wales, Cardiff, Wales

HEATHER A. KEMP   Research Assistant, Department of Medical Biochemistry, Welsh National School of Medicine, Cardiff, Wales

NOEL R. LING   Reader, Department of Immunology, University of Birmingham School of Medicine, Edgbaston, Birmingham, England

NICHOLAS J. MARSHALL   Senior Lecturer, Department of Molecular Endocrinology, Middlesex Hospital Medical School, London, England

FRANK McCAPRA   Professor, Department of Chemistry, School of Chemistry and Molecular Sciences, University of Sussex, Falmer, England

MICHAEL J. O'SULLIVAN[*] Biochemist, The Blond McIndoe Centre for Transplantation Biology, Queen Victoria Hospital, East Grinstead, Sussex, England

DAVID E. PERRY Senior Biochemist, Protein Reference Unit, Department of Chemical Pathology, Westminster Medical School, London, England

R. P. CHANNING RODGERS[†] I.C.P. Fellow, Unité de Medicine Experimentale, International Institute of Cellular and Molecular Pathology, Faculte de Medicine, Université Catholique de Louvain, Brussels, Belgium

IAN WEEKS Senior Research Scientist, Department of Medical Biochemistry, Welsh National School of Medicine, Cardiff, Wales

JOHN T. WHICHER Consultant Chemical Pathologist, Department of Chemical Pathology, Bristol Royal Infirmary, Bristol, England

J. STUART WOODHEAD Reader in Endocrine Biochemistry, Department of Medical Biochemistry, Welsh National School of Medicine, Cardiff, Wales

---

Present affiliation:
[*] Department of Tritium Products, Amersham International plc, Cardiff Laboratories, Cardiff, Wales
[†] Department of Laboratory Medicine, University of California School of Medicine, San Francisco, California.

Contents

Preface	iii
Contributors	v

1. **INTRODUCTION AND OVERVIEW** — 1
   Wilfrid R. Butt

I.	Introduction	1
II.	Requirements for satisfactory immunoassays	2
III.	Current trends	13
	References	14

2. **PROBLEMS OF IODINATION** — 19
   Wilfrid R. Butt

I.	Introduction	19
II.	Methods of iodination	20
III.	Assessment and storage of iodinated products	31
IV.	Conclusions	32
	References	33

3. **ENZYME IMMUNOASSAY** — 37
   Michael J. O'Sullivan

I.	Introduction	37
II.	Assay principles	38
III.	Choice of enzyme	42
IV.	Preparation of enzyme labels	45
V.	Methodology	50
VI.	Applications	54
VII.	Characteristics	55
VIII.	Conclusions	60
	References	60

4. HIGH SENSITIVITY, PULSED-LIGHT, TIME-
   RESOLVED FLUOROIMMUNOASSAY   71
   Salifu Dakubu, Roger Ekins, Thomas Jackson,
   and Nicholas J. Marshall

   I.   Introduction   71
   II. Time-resolved fluorescence: General principles   84
   III. Present applications and future prospects   95
        References   99

5. IMMUNOASSAYS USING CHEMILUMINESCENT LABELS   103
   Ian Weeks, Anthony K. Campbell, J. Stuart Woodhead,
   and Frank McCapra

   I.   Introduction   103
   II. Chemiluminescence   104
   III. Immunoassays using chemiluminescent probes   108
   IV. Conclusion   113
        References   114

6. NEPHELOMETRIC METHODS   117
   John T. Whicher and David E. Perry

   I.   Introduction   117
   II.   The physics of light scattering   118
   III. Optical requirements for nephelometry   120
   IV.   Formation of light-scattering complexes by
         antigen-antibody reaction   121
   V.   End-point nephelometric assays   126
   VI.   Kinetic nephelometric assays   135
   VII. Commercial immunonephelometric analysis systems   143
   VIII. Practicalities of immunonephelometry   155
   IX.   Applications   167
   X.   Conclusion   170
         References   170

7. LABELED ANTIBODY IMMUNOASSAYS   179
   Heather A. Kemp, Rhys John, and J. Stuart Woodhead

   I.   Introduction   179
   II. Methodology   183
   III. Conclusion   193
        References   195

Contents

8. **MONOCLONAL ANTIBODIES** — 199
   Noel R. Ling and Royston Jefferis

   I. Introduction — 199
   II. Principles of production — 199
   III. Applications of monoclonal antibodies — 208
   References — 212

9. **FREE HORMONES IN BLOOD: THEIR PHYSIOLOGICAL SIGNIFICANCE AND MEASUREMENT** — 217
   Roger Ekins

   I. Introduction — 217
   II. Concepts of hormone delivery: The role of serum-binding proteins — 219
   III. Measurement of free hormones in blood — 227
   IV. Standardization and quality control — 246
   V. Summary and conclusions — 248
   References — 249

10. **DATA ANALYSIS AND QUALITY CONTROL OF ASSAYS: A PRACTICAL PRIMER** — 253
    R. P. Channing Rodgers

    I. Introduction — 253
    II. The language and goals of assay data processing and quality control — 255
    III. Interlude 1: Counting statistics and common sense — 257
    IV. Calibration curve-fitting — 262
    V. Interlude 2: Which of these curve-fitting methods is the best? — 276
    VI. Errors and assays — 278
    VII. Interlude 3: Which of these assays is superior? — 284
    VIII. Assay design — 287
    IX. Interlude 4: What does this Scatchard plot mean? — 290
    X. Conclusions and recommendations — 292
    References — 295

Index — 309

PRACTICAL IMMUNOASSAY

1
Introduction and Overview

WILFRID R. BUTT
Birmingham and Midland Hospital for Women, Birmingham, England

I. INTRODUCTION

The ideas of using natural proteins such as thyroid-binding globulin (Ekins, 1960) or antibodies (Yalow and Berson, 1960) as assay reagents were described at about the same time. The methods using natural binding proteins (receptors) may give closer estimates of the biological activity of a hormone, but antibodies, when specific and of high titer, offer the advantages that they need not be freshly prepared and they can be widely distributed and used over extended periods.

Radioactive tracers provide convenient and highly sensitive end points for immunoassays, but they produce an environmental hazard and many attempts have been made to devise nonradioactive end points instead. Even so, the isotopes ^{125}I and ^{3}H continue to be used and will probably remain in use for some time to come. Tritium-labeled compounds are widely available commercially and no detailed description of their preparation is required here. The iodination of antigens, however, is required in many laboratories and some of the methods available are described in Chapter 2. Much progress has also been made with methods employing nonradioactive labels, enzymes, chemiluminescent and fluorescent compounds, and these are described in succeeding chapters.

Although the emphasis here is on radioimmunoassays and related techniques there are other methods of great value such as nephelometry, which is described in detail in Chapter 6. For methods of complement fixation, immunoprecipitation and electrophoresis in gels, etc., the reader is referred to other reviews (Weir, 1978; Oudin, 1980).

II. REQUIREMENTS FOR SATISFACTORY IMMUNOASSAYS

There are a number of basic requirements for the setting up of a satisfactory immunoassay. The first of these concerns the substance to be analyzed, the analyte. A purified preparation is required for raising antiserum and to use as tracer in the assay. Provided the assay developed is specific, the standards used do not necessarily have to be of the same degree of purity but they must be stable and they must perform identically to the pure substance in the assay. The second requirement is the antibody itself and until recently this has been raised by immunizing rabbits, guinea pigs, sheep, or other convenient species. Nowadays, however, much research is in progress on the raising of monoclonal antibodies and in certain respects these are likely to revolutionize methods of immunoassay: they form the subject of Chapter 8. Lane and Koprowski (1982) have drawn attention to some of the differences between molecular recognition by conventional antibodies and monoclonals. They refer to numerous examples of monoclonal antibodies which recognize cross-reacting sites on molecules in which conventional serology had failed to detect homology, and point out that this may not be appreciated by many who turn to monoclonal reagents for immunochemistry. Cross-reactions may be given by two protein antigens which share a small and precise detail of their surface topology: such a determinant may not be detected by conventional antibodies which could be directed against other specific structures in the molecules. Another type of cross-reaction occurs with two protein antigens having dissimilar structures but which have epitopes that interact with the same antibody molecule. Their affinities may be lower than in the first type of cross-reaction but this means that reaction with a monoclonal antibody cannot of itself be interpreted as proving molecular identity.

    The tracer used in the assay is usually a labeled antigen but in the immunoradiometric type of assay (IRMA) the antibody is labeled and here monoclonal antibodies may be particularly valuable (Chapter 7). Assays that depend on hemagglutination or particle agglutination (e.g., Latex) make use of a carrier, the red cell, or the solid particle as the index of the reaction. These methods are very convenient to carry out but they are less sensitive than radioimmunoassays and related techniques. They are suitable when sensitivity is not a problem and when the result is required rapidly, as in routine tests for early pregnancy which depend on the detection and measurement of human chorionic gonadotrophin (CG).

    Even if highly satisfactory antigens, antibodies, and labeling methods are available, results obtained in different laboratories may vary greatly unless careful attention is paid to the protocols used. Therefore, efficient internal and external quality control procedures are necessary and several national and international quality assessment schemes have now been running for some years. The mathematical treatment of results has raised some controversial points and some guidelines are outlined in Chapter 10. Perhaps too much has sometimes been read into the significance of too

Introduction and Overview

few results and there has been confusion over the definition of such apparently straightforward concepts as accuracy and sensitivity (the minimum detectable amount, as the English meaning of sensitivity would infer, or a definition in terms of the slope of the dose-response curve as preferred by certain international organizations). With proteins it is difficult to define the "correct value," as there is no independent physicochemical method such as mass-spectrometry-gas-liquid chromatography as there is for smaller molecules like steroids. In the history of external quality assessment formerly accepted values have been changed in the light of subsequent results.

Since the majority of assays are performed on blood samples, and the standards are prepared from other sources (e.g., the pituitary), the provision of blood which is free of the analyte has sometimes been necessary, and this has proved difficult. Bias may be introduced if no such "filler" is used. Another source of error in immunoassays is the methods used for separating the labeled antigen which is bound to antibody from the labeled antigen which is "free." Examples of the methods employed are mentioned in subsequent chapters but some general remarks are included in this chapter. Other topics which call for further comment here are the preparation of the antigen, the raising of polyclonal antibodies, and the separation of free and bound fractions.

A. Preparation of Antibodies

Small molecules such as steroids, drugs, thyroid hormones, and peptide hormones are available in pure form, but in order to render them immunogenic they must be linked to a larger molecule. Bovine serum albumin (BSA) has been the most commonly used carrier but many others have been suggested (Erlanger, 1980). Alternatives include ovalbumin, fibrinogen, thyroglobulin, and keyhole limpet hemocyanin. It is not entirely clear whether the choice of carrier affects significantly the antibody response: certainly synthetic polypeptide carriers such as poly(L-lysine) or poly(L-glutamic acid) do not appear to have any great advantage over BSA (Jaffe et al., 1970; Walker et al., 1973).

Experimentation is required to ascertain the optimum position in the antigen molecule to which to link the carrier, bearing in mind the need to leave biologically important groups in the molecule accessible for antibody response. The antibody specificity is directed primarily at that part of the hapten molecule furthest removed from the linkage with the carrier. Good results may therefore be obtained by linking the carrier at position 3 in those steroids such as testosterone, progesterone, corticosteroids, etc., which have a common ring A structure, since this leaves specific groupings in ring D exposed. Similarly, positions 6 and 11 in the steroid molecule are useful for linkage since important groupings in both rings A and D are available for antibody reactions. Fuller details will be found in the review by Pratt (1978).

1. Preparation of Steroid-Protein Conjugates

The methods of linking steroids to carrier have been reviewed by Erlanger (1980). Two commonly used steroid derivatives are hemisuccinates and 0-carboxylmethyl oximes which may be linked to protein by the mixed anhydride reaction.

Hemisuccinates: Steroids with a single primary hydroxyl group or an unhindered secondary hydroxyl group may be converted to hemisuccinates by reaction with succinic anhydride:

$$\text{Steroid.OH} + (CH_2)_2\begin{matrix}CO\\ \\CO\end{matrix}O \longrightarrow \text{Steroid.COO.}(CH_2)_2.\text{COOH}$$

(Succinic anhydride)   (Steroid hemisuccinate)

The steroid (3 mmol) dissolved in dry pyridine (10 ml) is mixed with succinic anhydride (10 mmol) and boiled for 4-5 hr. The reaction mixture is evaporated to dryness, redissolved in chloroform, washed three times with water, and dried with sodium sulphate. The solvent is removed by distillation and the product recrystallized from ethanol.

(O-carboxylmethyl) oximes: This method is suitable for alkali-stable steroids containing a single oxo group:

Steroid=O + $NH_2.CO.CH_2COOH.HCl$

(0-carboxymethyl)hydroxylamine hydrochloride) $\longrightarrow$

Steroid=N.O.CH_2COOH

(Steroid carboxymethyl oxime)

A mixture of steroid (2 mmol) and the carboxylmethyl reagent (4 mmol) in 2 M NaOH (2 ml) and ethanol (25 ml) is refluxed for 3 hr. Then the ethanol is removed by distillation and the residue is dissolved in water and the pH adjusted to 8.5. The aqueous solution is washed twice with ethyl acetate or diethyl ether and then acidified to yield a precipitate of the oxime. The produce is recrystallized from ethanol.

The purity of the products may be examined by thin-layer chromatography and by melting point determinations. The addition of tritium-labeled steroid at the beginning of the reaction is useful in the later determination of the number of steroid residues in the protein conjugate.

The Mixed Anhydride Reaction: The steroid hemisuccinate or oxime (in the acid form) (3 mmol) and tri-N-butylamine (3 mmol) are dissolved in dioxan (30 ml) and after cooling to $10^{\circ}C$ isobutylchlorocarbonate (3 mmol) is added. The reaction is allowed to proceed for 20 min at $4^{\circ}C$ and the mixture is then added to a stirred solution of BSA (0.06 mmol) in 1:1 (v/v) water: dioxan (200 ml) followed by M-NaOH (5 ml). The reaction mixture is maintained at $4^{\circ}C$ at pH 8 for approximately 4 hr.

Introduction and Overview

The mixture is then dialyzed against frequent changes of water for 15 hr and then the pH is adjusted to 4.5 with M-HCl. The product is collected by centrifugation and redissolved in water with the addition of a small quantity of $NaHCO_3$. The dialysis may be replaced by gel filtration in which case the reaction mixture is transferred to a G-25 Sephadex column (50 x 2 cm) and eluted with water. The steroid-protein complex appears in the retention volume. The reaction may be represented as:

$$\text{Steroid-bridge-COOH} + (CH_3)_2 \cdot CH \cdot CH_2 \cdot COOCl$$
$$\text{(Steroid derivative)} \qquad \text{(isobutylchlorocarbonate)}$$
$$\downarrow$$
$$\text{Steroid-bridge-COO} \cdot CO \cdot CH_2 \cdot CH(CH_3)_2$$
$$\text{(mixed-anhydride)}$$

$$+ \; NH_2\text{-Protein}$$
$$\downarrow$$
$$\text{Steroid-bridge-CONH-Protein}$$

If a tritium-labeled tracer is included the number of steroid molecules per molecule of conjugate can be calculated from the relative specific activity of the hapten and the conjugate. If not, optical density in the ultraviolet of the conjugate at its absorption maximum (about 250 nm) may be compared with the absorption of the steroid derivative and the protein separately. In general, best results are obtained when the molar ratio of the hapten to carrier is between 8 and 25 but the length of the bridge (Bermudez et al., 1975) and the nature of the hapten influence antibody production.

2. Antibodies to Proteins and Glycoproteins

<u>Specificity and Dosage</u>: Specificity has been a problem in raising antibodies to some closely related protein hormones and glycoproteins. The glycoproteins are an interesting group of substances which include both the gonadotrophins, follicle-stimulating hormone (FSH) and luteinizing hormone (LH), and also thyrotrophin (TSH). They consist of α and β subunits, the α-subunit being chemically identical in these three hormones and also in the placental gonadotrophin hCG, but the β-subunit is hormone specific. It is not surprising, therefore, that before these hormones were available in pure form, the antibodies produced showed considerable cross-reactions. When methods had been developed for the purification of each however, the antibodies which were subsequently produced were highly specific (Parlow and Shome, 1974; Torjesen et al., 1974; Lynch and Shirley, 1975; Prentice and Ryan, 1975; Suginami et al., 1978; Marana et al., 1979).

In parallel there have been advances in the immunization procedure although there is still no foolproof method. The dosage of immunogen is important: it was early shown that the dosage not only modulates the classes of immunoglobulin formed but also affects the capacity of the animal to produce antibodies. Tolerance, which may be induced with most proteins of low or moderate molecular weight, is the refractory state of an animal to react to a material which is normally immunogenic (Kabat, 1978). High zone tolerance is induced by dosages which are much greater than the optimum (these may be repeated doses of 10 mg for example) while low zone tolerance is induced with suboptimum amounts (1 μg or lower). The use of potent adjuvants tends to make tolerance less likely. Full practical details of immunization procedures are included in the review of Hurn and Chantler (1980).

The Multisite Intradermal Immunization Method: In the earlier methods of immunization when milligram quantities of immunogen were used, it was tempting to use relatively crude material. Vaitukaitis et al. (1971) showed that only microgram quantities were required for raising satisfactory antisera to the α and β-subunits of hCG when the immunogens were administered intradermally at many sites. Total doses of about 50 μg were given initially followed by 20 μg as booster injections. Lynch and Shirley (1975) extended this work to include the purified native hormones FSH, LH, CG, and TSH. An area 20 x 30 cm of fur was shaved from the backs of rabbits (New Zealand Whites) approximately 20 weeks old. The immunogens were dissolved in saline and mixed with 2 vol Freund's complete adjuvant by repeated ejection from a hypodermic syringe until a drop of emulsion was completely contained when placed on water. Each rabbit received 1.5 ml emulsion to 30-70 sites on the back, the maximum amount of antigen used being 100 μg per rabbit. Booster injections were given as for the primary immunization but the quantity used was only 10-20% of that used initially.

Samples of blood were taken at intervals until an acceptable titer was obtained. The booster injections were given when the titer had dropped by at least 50% of the maximum titer reached. Typical results are shown in Fig. 1. High titers following the primary injections were produced by 14-20 weeks and within 2-3 weeks of the secondary immunization. The association constants which ranged from 10^{10} to 10^{11} mol/liter and the high specificities made these antisera ideal for use in radioimmunoassay.

B. Separation Methods

In radioimmunoassays the double antibody method has become a standard against which other methods can be compared. Of the solid phase methods, covalent binding of the antibody or antigen to finely divided particles such as Sepharose or cellulose (Wide, 1981) has proved reliable, and dextran-coated charcoal for the separation of small molecules such as the steroids has been widely used (Odell, 1980). The charcoal method requires careful control of temperature and timing to avoid imprecision, and there is at present a trend away from the method to other solid phases such as magnet-

Introduction and Overview

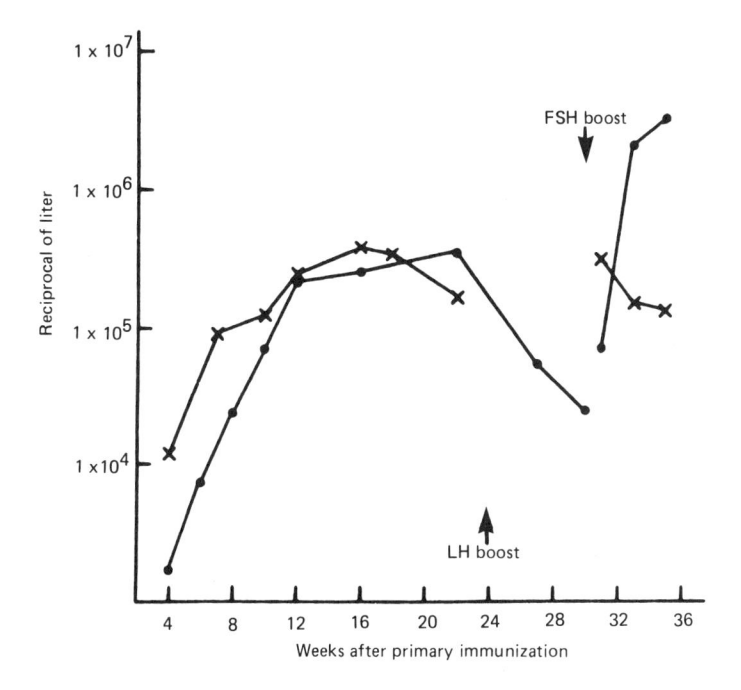

FIGURE 1   Average titers of FSH (●———●) and LH (X———X) antisera taken at various intervals after immunization. (Reproduced by permission from Lynch and Shirley, J. Endocrin., 1975.)

ized particles and to methods that do not require any separation. Errors arising at the separation stage contribute to the imprecision of an assay and to bias. A separation system which is satisfactory for standard materials or for purified antigens may be unsatisfactory when used in the presence of plasma or serum, and this may necessitate the use of antigen-free serum in all tubes.

The choice of method from the many that have been described will depend largely on the type of application required. Some methods may be highly efficient but not be generally applicable if large numbers of assays are required. Included here are those methods such as electrophoresis or gel filtration which depend on the charge of the molecule or on molecular size differences between the free and bound fractions. The double-antibody method, as other methods, requires careful optimization and this is not difficult but requires titration of the second antibody with the carrier serum to be used. The method requires fairly long incubation times to achieve adequate precipitation and relatively large volumes of the precipitated antiserum. To overcome these problems accelerated precipitation methods or linking of the second antibody to a solid phase have been proposed.

1. Accelerated Precipitation Methods

Polyethylene glycol (PEG) has proved effective in precipitating molecules according to their molecular weights and it is not as susceptible to interference from serum proteins as are many other methods. Furthermore, it does not interfere with the specificity of the antibody reaction (Creighton et al., 1973; Chard, 1980; Edwards, 1983). In several of the methods described PEG 6000 is used in 0.05 M phosphate buffer pH 7 at a concentration of 4%. Edwards (1983) found that in assays for T_3 and T_4 the second antibody (a donkey antisheep serum) did not interfere with the primary reactions with the first antibody (anti-T_3 and anti-T_4 sera, respectively) when precipitation was aided by 4% PEG. In some other assays however, (e.g., prolactin and TSH) there was interference with the first antibody reaction so that this was allowed to proceed for about 24 hr at room temperature before the addition of the second antibody-PEG.

In the methods for T_3 and T_4, first antibodies and antigens (total 200 µl) were mixed with the second antibody (50 µl) and incubated for 2 hr (T_4) or overnight (T_3) at room temperature. Then 4% PEG (500 µl) was added, mixed on a vortex and centrifuged, the supernatants were aspirated, and the precipitates counted.

For TSH and prolactin the antigens and first antibodies (450 µl) were mixed and allowed to incubate at room temperature overnight. Then the second antibody (50 µl, sheep antirabbit serum at optimal concentration), with normal rabbit serum carrier at 1:100 dilution, was added and, after mixing, left at room temperature for 2 hr. Then 4% PEG (1 ml) was added, mixed, and centrifuged. The supernatants were aspirated and the precipitates counted.

The accelerated second antibody precipitation method has proved to be robust and widely used in hospital laboratories routinely. It is not restricted to applications in radioimmunoassays and examples of its use in nephelometric assays will be found in Chapter 6.

2. Double Antibody Solid Phase Methods (DASP)

These methods were originally described using the second antibody linked covalently to cellulose particles (Midgley et al., 1969; Den-Hollander et al., 1972). They have the advantage that they require less carrier serum and therefore much less second antibody than liquid phase methods and the reaction is rapid. They require centrifuging however, and a recent trend has been the introduction of solid-phase magnetizable particles in which the separation is easily and efficiently achieved without the need for centrifugation (Hersh and Yaverbaum, 1975; Nye et al., 1976). The second antibody coupled to magnetic particles (DAMP: double antibody magnetic particles) provides a reliable method of separation, and since an excess of DAMP reagent may be used, accurate pipetting of the DAMP reagents is not critical.

Black iron oxide (Fe_3O_4) has proved to be a satisfactory magnetic component due to its small particle size (100Å-200Å) combined with a good magnetic response. Hersh and Yaverbaum (1975) prepared silane-coated Fe_3O_4 particles which were coupled after diazotization to the antibody. Nye et al. (1976) coated finely divided Fe_3O_4 particles with Enzacryl (polymerized m-diaminobenzene). The diazotized free amino groups were coupled to antibody by incubating for 3 days at 4°C at pH 8.6. Other polymers based on cellulose, agarose etc., have been employed. Cellulose has many advantages, being inexpensive and of low density and high surface capacity and it provides a stable reagent (Forrest and Rattle, 1983).

The magnetic particles can be coupled covalently to the antibodies by conventional methods. The cyanogen bromide method is suitable for cellulose particles: cellulose is activated by treatment with the reagent for a few minutes at pH 10-11

$$\begin{array}{c} \diagdown CH-OH \\ | \\ \diagup CH-OH \end{array} + CN.Br \longrightarrow \begin{array}{c} \diagdown CH-O \\ | \diagdown \\ C = NH \\ | \diagup \\ \diagup CH-O \end{array}$$

Antibody may then be covalently coupled by incubation with the activated cellulose at 4°C and pH 8.6 using either the whole antiserum or a precipitated γ-globulin fraction

$$\begin{array}{c} \diagdown CHO \\ | \diagdown \\ C = NH + H_2N\text{-protein} \\ | \diagup \\ \diagup CHO \end{array} \longrightarrow \begin{array}{c} \diagdown CHO \\ | \diagdown \\ C = N\text{-protein} \\ | \diagup \\ \diagup CHO \end{array}$$

The appropriate coupling ratio of antibody to magnetic particle is important and depends on the titer of the antiserum: it can range from 0.005 to 2.0 ml/g. Forrest and Rattle (1983) recommend 0.5 - 1.0 mg of solid phase per tube and at this level found very low nonspecific binding (< 0.5%). Particles could be stored for more than 2.5 years at 2-4°C, pH 7.4 without loss of immunoreactivity.

In assays using DAMP, the first antibody reaction is first set up and after a suitable incubation about 1 mg of DAMP is added and after 5-20 min incubation the particles are separated magnetically. Good agreement has been found between this method and the conventional liquid phase separation methods in assays for prolactin (Forrest and Rattle, 1983) and for a variety of other substances (Kamel et al., 1979; Shine et al., 1981). A method based on that of Robinson and co-workers (1973) has proved suitable for assays of reproductive hormones (J. Williams, 1982, unpublished): precipitated magnetic iron oxide was coated with α-cellulose and coupled to the second antibody by the cyanogen bromide method.

Magnetic particles can also be used for linking the primary antibody but best results appear to have been obtained when applied to hapten assays. The preparative methods are similar to those for second antibody.

Magnetic adsorbent particles have been prepared by entrapping charcoal and the ferric oxide particles within a polyacrylamide matrix (Dawes and Gardner, 1978). These particles are suitable for adsorbing steroids or small polypeptide hormones. Al-Dugaili et al. (1979) pointed out several advantages possessed by these particles over conventional charcoal particles: they did not require prior coating with dextran, there was not so much variation with temperature and time, and no centrifuging was required. The correlation with conventional methods for steroid hormones was excellent and the coefficient of variation was 0.5% compared with 0.8% using ordinary charcoal.

3. Sucrose Layering Separation

A novel way of avoiding centrifugation is the sucrose layering separation method (Wright and Hunter, 1983). The first stage of the method is between antigen and solid-phase antibody: diluent is then added and the solid phase allowed to settle. Then a 10% sucrose solution is layered beneath the incubate through a stainless steel tube. In 15 min the solid-phase particles carrying the bound fraction settle, leaving the free fraction in the top layer. The settling process effectively washes the particles as they pass through the sucrose layer. The free fraction and most of the sucrose layer is removed by suction leaving the bound fraction in the tube ready for counting. The precision of the method is good and the nonspecific binding low: it can be reduced still further by including a second sucrose wash.

Several solid phases for use in this method, were investigated by Wright and Hunter (1983). To avoid the use of hazardous cyanogen bromide an alternative method was suggested in which active dialdehyde groups were introduced into Sephacryl gels (Pharmacia) by mild periodate oxidation: the dialdehydes are then coupled to the amino groups of the proteins (Wright and Hunter, 1982). The activated gels proved to be stable and could be stored for a considerable time before coupling to antibody. They have similar properties to the Sepharoses and are suitable for use in the sucrose method. The nonspecific binding was low both in radioimmunoassay and IRMA procedures.

The procedure may be illustrated by the flow chart in Fig. 2.

4. Methods Requiring No Separation

Radioimmunoassays would be considerably simplified if the separation of free and bound fractions could be avoided all together. Edwards et al. (1975) partitioned free and bound steroid fractions between an ammonium sulphate solution and a toluene-based scintillant. In assays for cortisol, Morris (1978) carried out the reaction in scintillation vials: the details of this method as used in this laboratory are as follows:

The serum sample (50 μl) is diluted with distilled water (1 ml) and the solution is incubated at 60°C for 30 min to denature the cortisol-binding globulin and so avoid the need for solvent extraction of cortisol. After cooling, the denatured serum (0.1 ml) is transferred into a scintillation vial

Introduction and Overview

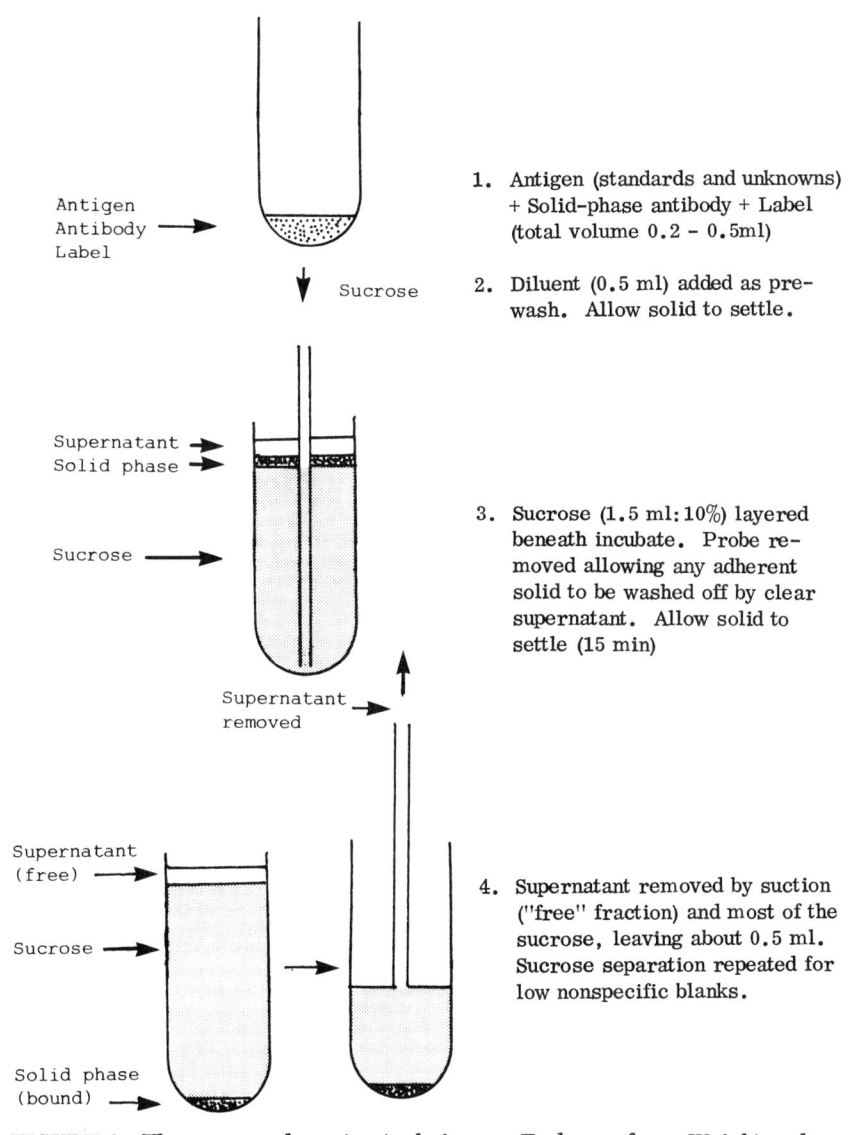

FIGURE 2 The sucrose layering technique. (Redrawn from Wright and Hunter, 1983.) Reproduced by permission.

1. Antigen (standards and unknowns) + Solid-phase antibody + Label (total volume 0.2 - 0.5ml)

2. Diluent (0.5 ml) added as pre-wash. Allow solid to settle.

3. Sucrose (1.5 ml: 10%) layered beneath incubate. Probe removed allowing any adherent solid to be washed off by clear supernatant. Allow solid to settle (15 min)

4. Supernatant removed by suction ("free" fraction) and most of the sucrose, leaving about 0.5 ml. Sucrose separation repeated for low nonspecific blanks.

and a mixture (0.2 ml) of labeled cortisol ([1,2-^{3}H]-cortisol) and antiserum (rabbit antiserum to cortisol 21-monosuccinate:BSA) is added. This mixture is prepared by evaporating 50 μl labeled cortisol in ethanol (50 μCi/ml) and adding 20 ml buffer (0.05 M Tris, pH 8, containing 0.1% BSA and 0.1% bovine γ-globulin) followed by suitably diluted antiserum (200 μl).

After incubation at room temperature for 30 min saturated ammonium sulfate (0.2 ml) is added and mixed. The scintillation fluid (PPO-POPOP in toluene) (2 ml) is now added and shaken for 1 min. The ammonium sulfate ensures that the free steroid is extracted into the scintillant: no separation is required and counting can be performed directly. It is important to count in the cold to avoid dissociation of the antigen-antibody complex in the presence of the ammonium sulfate.

A variant of this is the so-called ligand differentiation immunoassay (LIDIA) (Barnard et al., 1983). This depends on the separation of bound and free ligand by introducing a change in the partition characteristics of one of the components after binding to antibody. The method has been developed for several steroids using tritiated labels and it involves three steps.

1. Competitive binding of labeled and unlabeled antigen to specific antibody
2. Differential hydrolysis (or conjugation) of unbound ligand
3. Partition of the reactants into an organic and an aqueous phase

Water-soluble steroid conjugates such as the glucuronides and sulfates offer examples of this assay. Labeled and unlabeled steroid first reacts with a limited amount of the antibody.

$$(Ag\text{-cong}) + (Ag^*\text{-cong}) + (Ab)$$
$$\longrightarrow (Ab\text{-}Ag^*\text{-cong}) + (Ag^*\text{-cong})$$
$$ \text{Bound} \text{Free}$$

where Ag-cong = unlabeled conjugate and Ag^*-cong = labeled conjugate.

Free steroid is now hydrolyzed by enzyme (β-glucuronidase or aryl sulfatase), the bound steroid being protected against hydrolysis.

$$(Ag^*\text{-cong}) + \text{enzyme} \longrightarrow (Ag^*) + \text{free conjugate}$$
$$\text{Free}$$

The partition is achieved by adding the scintillant when the hydrolyzed free steroid is taken up in the solvent. The bound ligand remains hydrophilic and is not taken up in the organic phase and does not interfere with the scintillation process.

$$(Ag^*) + \text{scintillation fluid} \longrightarrow Ag^*$$
$$\text{(Toluene based)} \text{Organic phase}$$

An example is the assay of urinary estriol-16α-glucuronide (E_3-16α-G). For samples from women of reproductive age urine (100 µl) is diluted to 1 ml with assay buffer (0.1 M phosphate containing 0.9% sodium chloride, 0.1% BSA, and 0.1% sodium azide, pH 7.5). The

Introduction and Overview

diluted sample (100 µl) and the standard (100 µl with concentrations ranging from 31.25 to 2000 pg/100 µl) are added in duplicate to the assay tubes. Antiserum to E_3-16 α-G-6-BSA (100 µl) at a dilution giving about 60% binding and tritiated antigen (100 µl: 20,000 dpm) is added and the mixture incubated for 20 min at room temperature. A solution of β-glucuronidase (100 µl: 2.5 mg/ml: 128 units) is added to each tube and the contents incubated at room temperature for 15 min. Then scintillation fluid (1 ml: 5 g PPO/liter toluene: ethanol 99:1 v/v) is added. The tubes are stoppered, shaken, and the radioactivity determined.

The assay was evaluated by comparison with a conventional radio-immunoassay. It compared well in sensitivity (7.3 nmol/liter) and precision and the correlation of results was good (Barnard et al., 1983).

An example of an assay where conjugation rather than hydrolysis is used for the change of phase is afforded by the assay of plasma testosterone (Makawiti et al., 1983).

The plasma sample (50 µl from women, 20 µl from men) is extracted with diethyl ether (400 µl) and the dried extract is reconstituted in assay buffer (100 µl phosphate as for E_3-16α-G). The tracer, ^{3}H testosterone (100 µl: 20,000 dpm) is added followed by diluted antiserum (50 µl). The contents of the tube are mixed and incubated at room temperature for 45 min. The competitive binding reaction (T)+(T*)+(Ab) results in free and bound fractions ((T*-Ab)+(T*)) as in usual assays. The bound fraction is protected against conversion but the free is converted to a glucuronide (T*-glucuronide) by adding a preparation of UDP-glucuronyl transferase (20 µl: 400 µg protein) and UDP-glucuronic acid (300 µg) and incubating the mixture for 90 min.

Then scintillation fluid (1 ml) is added, the tubes are stoppered and shaken, and the radioactivity counted. The T*-glucuronide is retained in the aqueous phase and does not interfere in the scintillation process. Again, the correlation of results by this method and by radioimmunoassay was good.

III. CURRENT TRENDS

The need for large screening programs for substances such as TSH, α-fetoprotein, and carcinoembryonic antigen has necessitated the development of completely automated methods. There are a number of advantages in using IRMA techniques for this purpose and eventually these methods may incorporate nonradioactive indicators (Chapter 7).

Traditionally, clinical assays have been performed either on blood or urine but for certain hormones there are advantages in using saliva. As the sampling is noninvasive, studies requiring repeated sampling present no problems, and in studies on children the ethical objection to taking blood samples is removed. Saliva is particularly suitable for the investigation of steroid hormones, and the commonly estimated cortisol as an index of adrenal function and progesterone for the detection of ovulatory changes are obvious examples. Samples of saliva are best handled by freezing overnight, thawing, and centrifuging. The method of assay is then the same as for serum samples with minor adjustments to make the assays more sensitive. Furthermore, the fraction of the total circulating hormone which appears in the saliva represents the "free" or biologically active fraction in serum and is therefore the fraction of most value in clinical diagnosis. A full description of "free" hormones is contained in Chapter 9.

International organizations have supported a number of studies connected with the detection of hormonal changes during the menstrual cycle which would enable a subject to detect ovulatory changes reliably. This would be a considerable help to those who for religious or other reasons wish to use natural methods of family planning. Branch and colleagues (1982) measured estrone, estradiol, and estriol glucuronides by radioimmunoassays not requiring extraction of the urine samples. Each conjugate provided similar information but quantitatively the estrone glucuronide was the most important. These assays employed tritium tracers: clearly nonextraction methods as well as nonradioactive end points are required; solid phase methods with end points depending on a color change, such as enzymic reactions, or on chemiluminescence (Barnard et al., 1981) offer attractive possibilities.

REFERENCES

Al-Dugaili, E. A. S., Forrest, C. G., Edwards, C. R. W., and Landon, J. (1979). Evaluation and application of magnetizable charcoal for separation in radioimmunoassays. Clin. Chem. 25: 1402.

Barnard, G., Collins, W. P., Kohen, F., and Lindner, H. R. (1981). The measurement of urinary estriol-16α-glucuronide by a solid phase chemiluminescence immunoassay. J. Steroid Biochem. 14: 941.

Barnard, G. J. R., Matson, C. M., Makawiti, D. W., Kilpatrick, M. J., and Collins, W. P. (1983). A novel liquid phase separation system for automated immunoassay. In Immunoassays for Clinical Chemistry: A Workshop Meeting, W. M. Hunter and J. E. T. Corrie (Eds.), Churchill Livingstone, Edinburgh, p. 231.

Bermudez, J. A., Coronado, V., Mijares, A., Leon, C., Velazquez, A., Noble, P., and Mateos, J. L. (1975). Stereochemical approach to increase the specificity of steroid antibodies. J. Steroid Biochem. 6: 283.

Branch, C. M., Collins, P. O., and Collins, W. P. (1983). Ovulation prediction: Changes in the concentrations of urinary estrone-3-glucuronide, estradiol-17β-glucuronide and estriol-16α-glucuronide during conceptional cycles. J. Steroid Biochem. 16: 345.

Chard, T. (1980). Ammonium sulphate and polyethylene glycol as reagents to separate antigen from antigen-antibody complexes. Methods Enzymol. 70: 280.

Creighton, W. D., Lambert, P. H., and Meischer, P. A. (1973). Detection of antibodies and soluble antigen-antibody complexes by precipitation with polyethylene glycol. J. Immunol. 111: 1219.
Dawes, C., and Gardner, J. (1978). Radioimmunoassay of digoxin employing charcoal entrapped in magnetic polyacrylamide. Clin. Chim. Acta 86: 353.
Den-Hollander, F. C., Schuurs, A. H. W. M., and van Hell, H. (1972). Radioimmunoassays for human gonadotrophins and insulin employing a "double-antibody solid-phase" technique. J. Immunol. Methods 1: 247.
Edwards, C. R. W., Taylor, A. A., Baum, C. K., and Kurtz, A. B. (1975). Evaluation of two new separation techniques for steroid radioimmunoassays. In Steroid Immunoassay, E. H. D. Cameron, S. G. Hillier, and K. Griffiths (eds.), Alpha Omega Publishing, Cardiff, p. 229.
Edwards, R. (1983). The development and use of polyethylene glycol assisted second antibody precipitation as a separation technique in radioimmunoassays. In Immunoassays for Clinical Chemistry, W. M. Hunter and J. E. T. Corrie (eds.), Churchill Livingstone, Edinburgh, p. 139.
Ekins, R. P. (1960). The estimation of thyroxine in human plasma by an electrophoretic technique. Clin. Chim. Acta 5: 453.
Erlanger, B. F. (1980). The preparation of antigen hapten-carrier conjugates: A survey. Methods Enzymol. 70: 85.
Forrest, G. C., and Rattle, S. J. (1983). Magnetic particle radioimmunoassay. In Immunoassays for Clinical Chemistry, W. M. Hunter and J. E. T. Corrie (eds.), Churchill Livingstone, Edinburgh, p. 147.
Hersh, L. S., and Yaverbaum, S. (1975). Magnetic solid phase radioimmunoassay. Clin. Chim. Acta 63: 69.
Hurn, B. A. C., and Chantler, S. M. (1980). Production of reagent antibodies. Methods Enzymol. 70: 104.
Jaffe, B. M., Newton, W. T., and McGuigan, J. E. (1970). The effect of carriers on the production of antibodies to the gastrin tetrapeptide. Immunochemistry 7: 715.
Kabat, E. A. (1978). Dimensions and specificities of recognition sites on lectins and antibodies. J. Supramolec. Stud. 8: 79.
Kamel, R. S., Landon, J., and Smith, D. S. (1979). Novel ^{125}I-labelled nortriptyline derivatives and their use in liquid phase or magnetisable solid phase second antibody radioimmunoassay. Clin. Chem. 25: 1997.
Lane, D., and Koprowski, H. (1982). Molecular recognition and the future of monoclonal antibodies. Nature 296: 200.
Lynch, S. S., and Shirley, A. (1975). Production of specific antisera to follicle-stimulating hormone and other hormones. J. Endocrinol. 65: 127.

Makawiti, D. W., Barnard, G. J. R., Matson, C. M., and Collins, W. P. (1982). A novel pseudohomogeneous radioimmunoassay for the measurement of plasma testosterone. J. Steroid Biochem. 18: 619.

Marana, R., Suginami, H., Robertson, D. M., and Diczfalusy, E. (1979). Influence of purity of the iodinated tracer on the specificity of the radioimmunoassay of human follicle-stimulating hormone. Acta Endocrinol. 92: 585.

Midgley, A. R. Jr., Rebar, R. W., and Niswender, G. D. (1969). Radioimmunoassays employing double antibody techniques. Acta Endocrinol. Suppl. 142: 247.

Morris, R. (1978). A simple and economical method for the radioimmunoassay of cortisol in serum. Ann. Clin. Biochem. 15: 178.

Nye, L., Forrest, G. C., Greenwood, M., Gardner, J. S., Jay, R., Roberts, J. R., and Landon, J. (1976). Solid phase magnetic particle radioimmunoassay. Clin. Chim. Acta 69: 387.

Odell, W. D. (1980). Use of charcoal to separate antibody complexes from free ligand in radioimmunoassay. Methods Enzymol. 70: 274.

Oudin, J. (1980). Immunochemical analysis by antigen-antibody precipitation in gels. Methods Enzymol. 70: 166.

Parlow, A. F., and Shome, B. (1974). Specific, homologous radioimmunoassay (RIA) of highly purified subunits of human pituitary follicle stimulating hormone (hFSH). J. Clin Endocrinol. Metab. 39: 195.

Pratt, J. J. (1978). Steroid immunoassay in clinical chemistry. Clin. Chem. 24: 1869.

Prentice, L. G., and Ryan, R. J. (1975). LH and its subunits in human pituitary, serum and urine. J. Clin. Endocrinol. Metab. 40: 303.

Robinson, P. J., Dunhill, P., and Lilly, M. D. (1973). The properties of magnetic supports in relation to immunobiological enzyme reactors. Biotechnol. Bioeng. 15: 503.

Shine, B., de Beer, F. C., and Pepys, M. B. (1981). Solid phase radioimmunoassays for human C-reactive protein. Clin. Chim. Acta 117: 13.

Suginami, H., Robertson, D. M., and Diczfalusy, E. (1978). Influence of the purity of the iodinated tracer on the specificity of the radioimmunoassay of human luteinizing hormone. Acta Endocrinol. 89: 506.

Torjesen, P. A., Sand, T., Norman, N., Trygstad, O., and Foss, I. (1974). Isolation of LH, FSH and TSH from human pituitaries after the removal of HGH. Acta Endocrinol. 77: 485.

Vaitukaitis, J., Robbins, J-B., Nieschlag, E., and Ross, G. T. (1971). A method for producing specific antisera with small doses of immunogen. J. Clin. Endocrinol. Metab. 33: 988.

Walker, C. S., Clark, S. J., and Wotiz, H. H. (1973). Factors involved in the production of specific antibodies to estriol and estradiol. Steroids 21: 259.

Weir, D. M. (1978). Handbook of Experimental Immunology, 3rd Ed. Blackwell Scientific Publications, Oxford.
Wide, L. (1981). Use of particulate immunosorbents in radioimmunoassay. Methods Enzymol. 73: 203.
Wright, J. F., and Hunter, W. M. (1982). A convenient replacement for cyanogen bromide-activated solid phases in immunoradiometric assays. J. Immunol. Methods 48: 311.
Wright, J. F., and Hunter, W. M. (1983). The sucrose layering separation: a non-centrifugation system. In Immunoassays for Clinical Chemistry, W. M. Hunter and J. E. T. Corrie (eds.), Churchill Livingstone, Edinburgh, p. 170.
Yalow, R. S. and Berson, S. A. (1960). Immunoassay of endogenous plasma insulin in man. J. Clin. Invest. 39: 1157.

from about 15 µCi/µg in immunoglobulin (molecular weight 150,000) to about 100 µCi/µg for human growth hormone (hGH; molecular weight 22,000), to about 1500 µCi/µg for a peptide of the size of luteinizing-hormone-releasing hormone (LH-RH, molecular weight 1360).

Ideally, an antigen will be labeled without loss of affinity for the antibody. This is not always possible and Bolton and Hunter (1973) described an antiserum to hGH which was affected by even a single atom of iodine introduced into the molecule.

Iodinations are performed with very small quantities of antigen in dilute solution while unprotected by carrier protein. Losses by adsorption may therefore occur on the surface of the iodination vessel and on chromatographic columns used in the procedure unless precautions are taken to prevent these. Minimum losses would be expected with those methods completed rapidly and after iodination a carrier protein must be added and the columns must be prewashed with protein and potassium iodide.

In order to avoid repeated iodinations, the product should be reasonably stable and the best conditions of storage have to be established for each antigen to avoid polymer formation, dissociation, etc. Freezing the solution is not always advisable as repeated thawing and freezing may lead to aggregation, as for instance with hGH. Iodine may be slowly released from the product so that after storage it may be necessary to rechromatograph before use.

II. METHODS OF IODINATION

Because of the problems associated with iodination it is not surprising that many different methods have been described, sometimes because of a special need, such as for a labeled ligand which is required for purposes other than RIA (e.g., as tracer in biological studies). In this chapter we will not survey the earlier iodination methods but describe those which are widely used nowadays. Undoubtedly the chloramine-T method is the best known and it will be discussed first with some of the variants which have been introduced. In the search for milder methods other chemical oxidants have been proposed which include chlorine, hypochlorite, N-bromosuccinimide, Iodo-Gen, and the enzyme lactoperoxidase. Finally, for proteins that do not contain tyrosine or histidine and for ligands that are not proteins, there are methods in which an iodine-labeled compound or a compound which can be iodinated is linked covalently to the molecule.

A. Chemical Methods

1. Chloramine-T

Basically the method is unchanged since it was first described by Greenwood and co-workers (1963). Modifications that have been introduced are con-

2
Problems of Iodination

WILFRID R. BUTT
Birmingham and Midland Hospital for Women, Birmingham, England

I. INTRODUCTION

The process of radioiodination is crucial for the satisfactory performance of radioimmunoassays (RIA), immunoradiometric assays (IRMA), and radioreceptor assays for proteins and for smaller molecules such as steroids to which an iodine-containing molecule can be attached. Poor iodination has often been blamed in turn on radiation damage, the radioiodine itself, the oxidant, the time of exposure to the oxidant, the reducing agent used to stop the reaction, and the method of separation of excess iodine from the labeled product. It is well to remember however, that heterogeneity, apparent after labeling, may have been present before and only became apparent after labeling.

With present methods, radiation damage is no longer a serious problem. The antigen is exposed to the radioiodine for the minimum possible time compatible with satisfactory incorporation of iodine and the reagent is used at the minimum concentration that gives satisfactory iodination. Too great an incorporation of iodine leads to rapid decay of the product and the greater the number of radioactive atoms incorporated per molecule the greater the number of radiation decay fragments which could show reduced immunoradioactivity. Although greater incorporation of iodine produces a molecule with greater specific activity, the performance of the labeled compound is not necessarily improved. The optimum specific activity is given by the minimum quantity of iodine compatible with the desired assay sensitivity and the availability of counting time. It is generally accepted that the average incorporation of one iodine atom per protein molecule is the most desirable. The specific activity of the protein, therefore, varies with its molecular weight, the theoretical specific activity ranging cerned with the use of smaller amounts of oxidant, the time of the reaction, and the use of different reducing agents to stop the reaction.

In aqueous solution chloramine-T (the sodium salt of the N-monochloro derivative of p-toluene sulfonamide) breaks down slowly to form hypochlorous acid. Under mildly alkaline conditions (pH 7.5) sodium iodide is oxidized by the hypochlorous acid, forming cationic iodine. Sonoda and Schlamowitz (1970) made a careful study of the reaction and showed that chloramine-T produced "active iodine," H_2OI^+. Maximum production of I_2 occurred with stoichiometric ratios of iodide (1 mol) to chloramine-T (0.5 mol). Excess reagent led to the conversion of iodine to higher states of oxidation, IO_3^- and IO_4^- and this occurred more rapidly at pH 6 than at pH 7-8. In the presence of protein such as gamma globulin, however, the "active iodine" also oxidized sulfydryl groups so that a 10-fold excess of the molar ratio was required. McConahey and Dixon (1980) also pointed out that proteins with high reducing capacity may be poorly iodinated under standard conditions and require increased amounts of chloramine-T: keyhole limpet hemocyanin is an example.

At pH 7.5 tyrosine is only slightly ionized to the anion, the pK of the phenolic side chain being greater than 10, but the iodination reaction proceeds sufficiently at this pH (Rudinger, 1971). The monoiodo derivative of tyrosine has a pK of 8.5 and it is thus more highly ionized at the pH of the reaction. The iodine atom substitutes in the ortho position to the hydroxyl group in the phenol ring (Fig. 1). Different tyrosine residues in a protein may show different degrees of reactivity depending on their accessibility. In insulin the tyrosine at residue 14 in the A chain is iodinated preferentially as also is the C-terminal residue in the β-chain of chymotrypsin.

A typical protocol for the iodination of a protein of molecular weight 30,000 is as follows.

Chloramine-T (5 mg/ml) is made up freshly before use in 0.05 M phosphate buffer (pH 7.5). Sodium iodide (Na^{125}I), obtained commercially, is diluted in 0.5 M phosphate buffer (pH 7.5) to 100 mCi/ml. Then 0.5 mCi

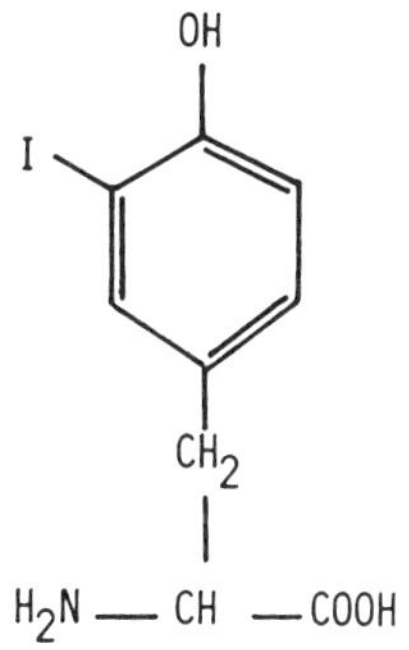

FIGURE 1 Monoiodotyrosine.

(5 µl) of the iodide is transferred to a polystyrene tube (4-ml volume). The protein for iodination is prepared in 0.05 M phosphate buffer at a concentration of 200 µg/ml and 10 µl (2 µg) is transferred to the iodination tube. Chloramine-T (10 µl: 50 µg) is added immediately and the mixture is agitated on a vortex mixer for 15 sec. The reaction is then stopped by adding a solution of sodium metabisulfite (200 µl: 200 µg) and potassium iodide (100 µl: 1 mg). The labeled protein is then separated from unreacted iodide by a simple chromatographic system, usually gel filtration on a column of Sephadex G-25.

There are many variations to this method designed to avoid some of the problems which have been encountered with the method. The oxidizing action of chloramine-T may damage some proteins and render the product unstable; therefore, the concentration of the reagent and the reaction time are reduced as much as possible. Brown and co-workers (1983) used only 16 µg in 10 µl for iodinations of 2-6 µg protein, while Edwards and colleagues (1983) reduced the amount still more to 6 µg in 10 µl for the iodination of 10 µg prolactin, and both achieved satisfactory specific activities. The amount of radioactive reagent is also kept to the minimum compatible with a satisfactory specific activity. The reducing reagent used to stop the reaction has often been suspected where damage to the protein has occurred, so several alternatives have been tried. Edwards et al. (1983) used only 6 µg sodium metabisulfite (10 µl) and obtained satisfactory results. Others have preferred to use alternative reducing agents such as L-cysteine (56 µg in 100 µl; Brown et al., 1983). They demonstrated by starch gel electrophoresis that sodium metabisulfite produces molecular heterogeneity following the iodination of calcitonin and this did not occur when L-cysteine was used instead (or cysteamine, 2-mercaptoethyl ammonium chloride). The damage with bisulfite may have resulted from cleavage of the 1-7 intradisulfide bonds by sulfite ions. They reported similar behavior during the iodination of insulin.

An example is now given of the method of calculating the specific activity of an iodinated antigen using as an illustration the iodination of the glycoprotein FSH. The hormone (1 µg in 5 µl) was iodinated using 195 µCi (5 µl) Na^{125}I. The total counts for 30 sec were 366,000. Free iodine was separated after the reaction on a column of Sephadex G-25 (20 cm x 1cm). Iodinated protein was eluted first (fractions 5-10) followed by the free iodide (fractions 11-20).

The counts/30 sec in the protein fraction were 219,533 and in the iodide fraction 102,577.

The counts in the protein = total counts - counts in the ^{125}I iodide peak.

Problems of Iodination

$$= 366{,}000 - 102{,}577 = 263{,}423.$$

Specific radioactivity
$$= \frac{\text{counts in protein} \times \mu Ci}{\text{total counts} \times \mu g\ \text{protein}}$$

$$= \frac{263{,}423 \times 195}{366{,}000 \times 1} = 140 \quad Ci/\mu g$$

Percentage incorporation of radioactivity into protein

$$= \frac{\text{counts in protein} \times 100}{\text{total counts}}$$

$$= \frac{263{,}423 \times 100}{366{,}000} = 72\%$$

Percentage of total protein recovered

$$= \frac{\text{counts in protein fraction} \times 100}{\text{counts in protein}}$$

$$= \frac{219{,}533 \times 100}{263{,}423}$$

$$= 83\%$$

2. Chlorine

This method makes use of the reaction between chloramine-T and sodium chloride whereby chlorine is released in the same way as iodine is released when chloramine-T reacts with sodium iodide. The chlorine then acts as oxidant for the iodination reaction and the chloramine-T does not come in contact with the antigen. The method has been simplified since it was first described (Butt, 1972, 1979).

The reaction is carried out in a small pointed polypropylene vial with a cap (Fig. 2). The protein antigen (2 µg in 10 µl phosphate buffer pH 7.5) is transferred to the vial and mixed with 0.4 mCi Na^{125}I (10 µl). Sodium chloride solution (50 µg: 5 µl) is pipetted onto a small piece of filter paper wedged into the top of the vial and chloramine-T (50 µg: 5 µl) is added. The lid is closed and chlorine is released and slowly enters the iodination solution. The reaction proceeds for 5 min (or as long as is found necessary by experiment with the protein being iodinated) with careful agitation of the solution for a few seconds every minute. The reaction is then

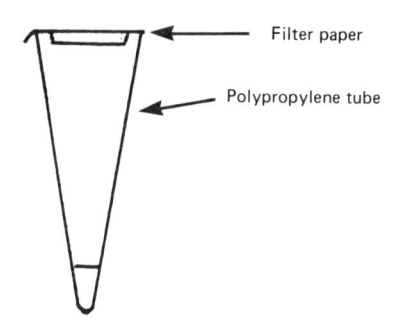

FIGURE 2 The chlorine gas method of iodination. In tube: antigen (2 μg in 10 μl), Na^{125}I (400 μCi in 10 μl). On lid/filter paper: NaCl (50 μg in 5 μl), Chloramine-T (50 μg in 5 μl), reaction stopped by sodium metabisulphite (200 μg in 50 μl). (From Butt, 1972.)

stopped by adding metabisulfite (200 μg: 50 μl) or other suitable reducing agent.
     This method may be used for a wide variety of protein hormones and also for hormones such as follicle-stimulating hormone (FSH) when they are to be used as tracers in biological experiments (Davies et al., 1978). Glycoprotein hormones are very sensitive to damage and the biological activity is soon destroyed if the terminal sialic acid residues are removed or amino acid residues are disrupted. Chaplin and co-workers (1970) reported that tryptophan, cysteine, and methionine in FSH are rapidly oxidized by chloramine-T. When the amount of chloramine-T was limited to 5 mols/mol tryptophan, 39% of the biological activity was destroyed; with a large excess of chloramine-T all biological activity was destroyed, but it was interesting to note that immunological activity was preserved. The biological activity was destroyed by reactions with chloramine-T and metabisulfite even in the absence of the radioiodine, indicating that the damage was not caused by the iodination (Butt, 1969; Butt et al., 1970). Even if immunological activity was not destroyed the modified molecular structure is less stable and the useful life of the labeled preparation is much reduced.

3. Sodium Hypochlorite

This procedure was designed to overcome the slight practical problems involved in the chlorine method described above (Redshaw and Lynch, 1974). Instead of the process generating chlorine gas, sodium hypochlorite itself was used as oxidant. A solution of this oxidant is made up freshly at an appropriate previously determined dilution for the iodination of 2.5 μg of antigen. Dilutions containing from 6.8 to 13.5 nmols were required for a number of gonadotrophins, prolactin, and thyroid-stimulating hormone

Problems of Iodination 25

(TSH). The iodination is performed in a small glass or polystyrene tube: the antigen (2.5 µg: 10 µl 0.05 M sodium phosphate pH 7.5) is added, followed by the appropriately diluted oxidant in 10 µl of the same buffer. Na^{125}I (500 µCi: 10 µl) is added and the mixture agitated for 30 sec. The reaction is terminated by adding the reducing agent; metabisulfite was used in the original method but milder reagents can be used.

4. N-Bromosuccinimide

This mild oxidant can be used for a variety of antigens. Iodinations are carried out by a similar procedure to that using chloramine-T (Reay, 1982). The reagent is freshly prepared before use in water as a 0.01 M solution and dilutions are made as required. For a typical iodination, 2 µg of the antigen (10 µl) is mixed with 200 µCi of Na^{125}I and N-bromosuccinimide (5 µl: 90 ng) is added. After 15 sec the reaction is stopped by adding KI/tyrosine mixture (20 µl: both 1 mmol) containing carrier protein (BSA 5 g/l). Recent work with this method has indicated that even this mixture of KI and tyrosine is not essential for stopping the reaction and simple dilution with buffer (1 ml) is satisfactory.

The hormones prolactin, hGH, FSH, LH, and TSH were iodinated to specific activities suitable for RIA by this method (Reay, 1982). The molar incorporation of iodine was more closely related to the amount of oxidant than to the amount of iodide used.

In a comparative study with other methods of iodination the iodinated products prepared by this method compared very favorably giving low nonspecific binding and satisfactory stability (Butt et al., 1983). The method is simple to perform and the reagent is stable and readily available.

5. Iodo-Gen

This reagent (1,3,4,6-tetrachloro-3α,6α-diphenyl glycoluril) is almost insoluble in water so the risk of the oxidant attacking the protein during iodination is small (Fraker and Speck, 1978). Dichloromethane is a suitable solvent but when in solution the reagent decomposes gradually in light: it is stable for several weeks in the dark, however.

The reagent is prepared at a concentration of 1 mg in 10 ml dichloromethane. Then 10 µl is transferred to a glass reaction vessel (75 mm x 12 mm) and left until the solvent has evaporated. The rate of reaction is dependent upon the area of the coating film of Iodo-Gen so a standard procedure should be used for reproducible results. Phosphate buffer pH 7.5 (0.05 M: 20 µl) is now added followed by the protein antigen (2 µg in 10 µl) and Na^{125}I (200 µCi). Incorporation of iodine is excellent and almost quantitative by 20 min at room temperature so that a large excess of Na^{125}I is not required. The reaction is more rapid at 37°C but the reagent then begins to decompose. The reaction can be stopped by simple dilution but in view of the slight solubility of the reagent it is preferable to add one of the usual reducing agents.

Satisfactory and repeatable iodinations have been reported for the glycoprotein hormones using this reagent (Ferguson et al., 1983). Less satisfactory results, however, have been obtained for polypeptide hormones (Butt, 1981).

B. Enzymic Methods: Lactoperoxidase

This enzyme catalyzes the oxidation of iodide in the presence of small amounts of hydrogen peroxide (for recent review see Morrison, 1980). When carefully used it provides another method whereby the biological activity of the iodinated product may be preserved and, as with other methods, there have been many variations in the amount of reagent used and the time of reaction. Hydrogen peroxide is added sequentially in small amounts but in order to avoid this, glucose oxidase has been used instead (Murphy, 1976; Tower et al., 1977). Small, controlled, and continuous amounts of H_2O_2 may be generated from the action of glucose oxidase on its substrates, glucose and oxygen. Murphy (1976) used 5 μM glucose, 3.6 mu glucose oxidase, and 3.6 mu lactoperoxidase at pH 7.2 for 10-40 min at 37°C for the iodination of erythropoietin and achieved satisfactory iodination without loss of biological activity.

During the procedure the enzyme itself may become iodinated and to avoid this, solid-phase preparations of the enzyme have been used (David and Reisfeld, 1974; Thorell and Larsson, 1974). However, when carbohydrate-based enzyme carriers are used they too may be iodinated to a considerable extent and Karonen et al. (1975) used instead a cross-linked copolymer of maleic anhydride and butanediol divinylether.

In the method of Karonen et al. (1975) this cross-linked carrier (100 mg) is suspended in an ice-cold solution of 15 mg lactoperoxidase in 0.05 M phosphate, pH 7.0 (7 ml). The mixture is incubated for 75 min in an ice bath with stirring and the pH is maintained between 8 and 9 by the gradual addition of 2 M NaOH. The carrier is then homogenized and washed twice with 0.05 M phosphate pH 7.0 (7 ml) and twice with 0.01 M borate pH 8.0 (7 ml) containing 1 M NaCl. The amount of lactoperoxidase linked can be calculated from the difference between the initial amount and that estimated in the washings. The carrier is suspended in 0.01 M sodium acetate, pH 5.0 (5 ml) and stored at 4°C or frozen at -70°C. Horseradish peroxidase can be linked and used in the same way.

The iodinations are performed in small plastic tubes. For amounts of antigen up to 10 μg, 1 mCi Na^{125}I in 100 μl 0.1 M sodium acetate (pH 5.0) and 20 μl of a 1:60 dilution of the solid-phase stock are mixed and the reaction is started by adding 1 μl H_2O_2 (0.3 g/liter). The mixture is incubated with mixing for 30 min during which time two 0.5 μl additions of H_2O_2 are made. For optimum iodinations it is better to have an excess of iodine but the amounts of H_2O_2 should be carefully controlled. The method was found to compare favorably with the chloramine-T method for LH, FSH, and angiotensin II.

C. Conjugation Labeling for Proteins and Polypeptides

Several conjugation labeling methods have been described and one of the best known is that using N-succinimidyl 3-(4-hydroxy 5-[^{125}I]iodophenyl) propionate (Bolton and Hunter, 1973), a reagent which is available commercially. The N-succinimidyl reagent condenses with free amino groups of peptides to form a conjugate in which a radioiodinated phenyl group is covalently linked via an amide bond to the protein (Fig. 3). The succinimidyl reagent is radioiodinated first using the chloramine-T reaction, and because the reagent is rapidly hydrolyzed the reaction must be carried out as rapidly as possible. The labeled material can be separated from the aqueous phase by an organic solvent and can be used directly after removal of the solvent or can be further purified, the monoiodinated and diiodinated compounds being separated by thin-layer chromatography.

The second stage of the procedure involves reaction of this labeled compound with the protein. The reaction is dependent on the concentration and therefore the protein is dissolved in the minimum volume, preferably in a narrow pointed conical tube. In a typical iodination the antigen (5 μg: 10 μl) in 0.1 M borate, pH 8.4 is reacted with 3-5 mol labeled ester/mol protein. The mixture is kept in an ice bath for 15-30 min after which any unreacted ester is conjugated to an excess of a low-molecular-weight amino-containing compound, such as glycine.

The labeled protein is then separated from excess reagent and from low-molecular weight iodinated conjugate by gel filtration, a protein carrier being used to minimize adsorption of the labeled compound.

This method was developed to overcome the problems of damage associated with exposure of proteins to oxidizing and reducing agents and it also offers the potential advantage over direct methods of not requiring a tyrosine residue for iodination. It may also be important where reaction with tyrosine destroys enzymic or biological activity. Fang and associates (1975) used the method to produce labeled enzyme preparations and an extensive review which lists many applications of this method has appeared (Langone, 1980).

Alternative reagents listed by Langone (1980) are the iodinated derivatives of methyl p-hydroxybenzimidate which is selective for amino groups and iodinated diazotized aniline which presumably couples with the phenolic tyrosine and therefore appears to lack one of the advantages of conjugation methods over direct methods of iodination.

The method is technically more complex to perform and requires more handling of radioactive materials with the associated hazards. Furthermore, because it is a two-stage reaction the yields are lower than with direct iodination methods.

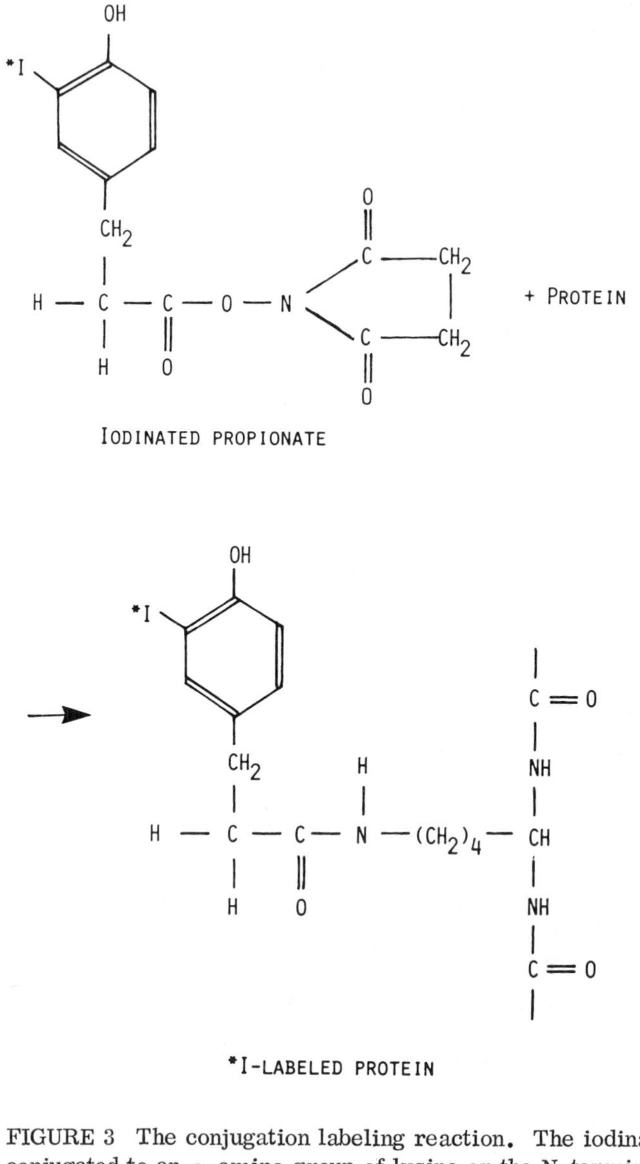

FIGURE 3 The conjugation labeling reaction. The iodinated propionate is conjugated to an ε-amino group of lysine or the N-terminus in the protein to be labeled.

D. Iodination of Haptens

Small molecules such as steroids are available as tritiated derivatives but because of the convenience of gamma counting there has been increasing use of iodination methods for such compounds. Usually a molecule which can be, or is already, labeled with iodine is coupled to the hapten but some molecules such as estrogens and thyroid hormones can be iodinated directly.

Externally labeled ligands in steroid immunoassays raise the problem of recognition of the bridge structure (Corrie and Hunter, 1981). The methods may be classified as homologous or heterologous assays according to whether or not the immunogen and tracer share an identical bridge structure. It is convenient to use the same bridge structure (hemisuccinate, carboxymethyl oxime, etc.) but the antibody may recognize the bridge as well as the hapten and if so the tracer usually has a higher affinity for the antibody than does the analyte, and therefore the assay is insensitive. To overcome this, the bridge may be attached at different sites on the molecule in the immunogen and the tracer (heterologous site assay) or the structure of the bridge may be varied, keeping the site of linkage unchanged (heterologous bridge).

An example was afforded by Nordblom et al. (1981). They compared the sensitivities of assays in which the same and different bridge structures were used in methods for androstenedione. A succinyl ester linkage ($-CO_2.CH_2.CH_2.CO.O$-steroid) and an ether linkage ($-NH.CO.CH_2.O$-steroid) both at position 19, were used. Assays were then set up with the antibody to the ester linkage with labels to the ether and the ester, and also with antibody to the ether linkage and with the same two labels. The heterologous bridge increased sensitivity considerably: with the ester antibody the sensitivity was 15 pg/tube with the ether label but 625 pg/tube with the ester label. Similarly with the ether antibody the sensitivity with the ether label was 425 pg/tube and 76 pg/tube with the ester.

An identical glucuronide bridge was used in the immunogen and in the radioligand in an assay for progesterone which had a sensitivity approaching that of an assay using tritiated progesterone (Corrie et al., 1981). The glucuronide bridge is only weakly antigenic and was only poorly recognized by the antibody. The glucuronide was prepared from 11α-hydroxyprogesterone to which tyramine was conjugated for labeling with iodine as the tracer and to which bovine serum albumin was conjugated to serve as the immunogen (Fig. 4, I and II). The tyramine derivative was labeled by the chloramine-T method using cysteine hydrochloride as the reducing agent. It was purified by extraction in ethyl acetate and by chromatography on a column of Sephadex LH-20. Greatly improved standard curves were obtained, however, when the radioiodinated glucuronide conjugate was used in a heterologous bridge

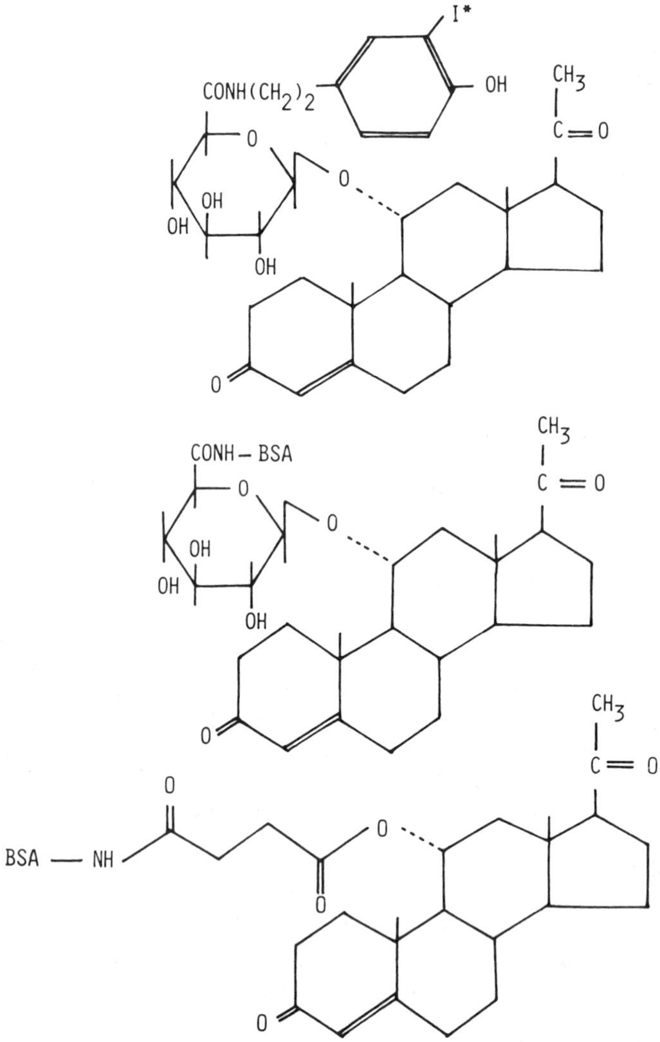

I

II

III

FIGURE 4 Iodinated tyramine conjugated to 11α-hydroxyprogesterone glucuronide (I) and bovine serum albumin conjugated to the glucuronide (II: for homologous bridge assay) and through a hemisuccinate (III: for heterologous bridge assay).

system with an antiserum raised against a bovine serum albumin conjugate of progesterone 11α-hemisuccinate (Corrie, 1983; Fig. 4, III).

Direct labeling of steroids has also been described. Hochberg (1979) synthesized 17β-[16α-^{125}I]-iodoestradiol by exchange of 16β-bromoestradiol with Na^{125}I. Na^{125}I (1 mCi) in 3 µl water at pH 8-10 was mixed with 100 µl acetonitrile in a small tapered tube and was evaporated to dryness under nitrogen. Next, 16β-bromoestradiol (10 µg) in 10 µl freshly distilled 2-butanone was added and the mixture placed at 66-68°C for 4-6 hr. The reaction was completed at room temperature overnight. The product was stable for several weeks.

III. ASSESSMENT AND STORAGE OF IODINATED PRODUCTS

The most important property of the iodinated product is its performance in radioimmunoassay. Bolton (1977) divided methods of assessment into (1) physicochemical, (2) binding of the tracer to excess antibody, and (3) the direct comparison of the immunoreactivity of labeled and unlabeled material. Changes in the physicochemical properties may be observed by altered electrophoretic mobility or reduced adsorption to charcoal, cellulose powder, or silica, etc. These methods do not measure directly changes in immunoreactivity but they may be useful guides to behavior.

Binding of the tracer to excess antibody is easily measured and is a useful guide, but no more, to tracer quality. A high level of binding may be given with material of lowered affinity and this loss only becomes apparent under the conditions used for the assay. The direct comparison of the immunoreactivity of the labeled and unlabeled material in standard curves for each, on the other hand, has the advantage that the tracer is tested under similar conditions to those used in the assay itself. Such tests however, are time-consuming. Tower et al. (1978) have used a rapid test in which three physicochemical properties are examined: adsorption to talc, exclusion from an anionic resin, and precipitation with trichloroacetic acid. The test gave patterns distinctive for monomeric iodinated hormones, aggregated material, and free iodine. Satisfactory results were obtained when there was greater than 90% binding to talc, less than 25% binding to resin, and greater than 90% precipitation with trichloroacetic acid.

After iodination, purification by gel filtration or ion exchange chromatography is used to remove excess iodine. Sephadex G-25 is suitable in most instances but more efficient separation systems such as Sephadex G-100 chromatography may be necessary for proteins such as prolactin and growth hormone which tend to form polymers. The purity of certain antigens, including LH is improved by chromatography on Ultrogel. Suginami et al. (1978) demonstrated that the performance of iodinated LH was improved after purification by adsorption on cellulose and even more so by chromatography on Ultrogel AcA (0.9 x 60 cm). Although more tedious than simple forms of chromatography, separation of subunit material was achieved more efficiently by this method. After storage of some labeled antigens repurification may be necessary immediately before use in an assay.

The stability during storage can be affected by many factors which include temperature, pH, the type and concentration of protein carrier, and the presence of protective agents such as preservatives. Commonly, iodinated preparations are stored frozen at -20°C, preferably in small amounts each sufficient for a complete assay. Repeated freezing and thawing is to be avoided. Hormones such as hGH tend to form polymers when frozen and are better stored at 4°C. Freeze-dried preparations of tracers may also be satisfactory and undoubtedly are the preparations of choice when they are to be distributed to other laboratories.

The type of protein carrier is important. Brown et al. (1983) noted a change in the elution profile of iodinated hGH when gelatin instead of bovine serum albumin was used as carrier. The concentration of carrier protein was also found to be important: there was a marked increase in the formation of high-molecular-weight components when the concentration was increased from 1.0 to 3.0 g/liter.

IV. CONCLUSIONS

A newcomer to radioimmunoassay will find a confusing selection of methods described for radioiodination. A considerable amount of experience is required to perform any of these methods satisfactorily and it is advisable therefore to retain a technique that performs well rather than to try every new method as it is described. Should a fault occur it is not always easy to discover the cause, but it will probably take less time and effort to attempt to do so than to set up a new method.

Some reasons have been mentioned why radioiodinated preparations are not ideal. In addition the half-life of ^{125}I is 60 days so that the shelf-life of the tracer is limited. There is also a health hazard connected with the labeling procedure and problems of disposal of the radioactive waste. To help with the latter problem Conroy and Fletcher (1981) have described a method whereby radioactive components may be recovered from large volumes of liquid waste by the use of a trichloroacetic acid flocculant. The other chapters in this book clearly indicate that there are many alternatives to radioactive tracers and the search for still more goes on (see also Schall and Tenoso, 1981). It is probably true to say, however, that radioiodine-labeled antigens and antibodies will continue to be usefully employed for some years to come and that a search for an ideal method of iodination is worthwhile.

REFERENCES

Bolton, A. E. (1977). Radioiodination Techniques. R.C.C. Review 18. The Radiochemical Centre, Amersham.

Bolton, A. E., and Hunter, W. M. (1973). The labelling of proteins to high specific radioactivities by conjugation to a ^{125}I-containing acylating agent. Biochem. J. 133: 529.

Brown, N.S., Abbott, S. R., and Corrie, J. E. T. (1983). Some observations on the preparation, purification and storage of radioiodinated protein hormones. In Immunoassays for Clinical Chemistry: A Workshop Meeting, W. M. Hunter and J. E. T. Corrie (eds.), Churchill Livingstone, Edinburgh, p. 267.

Butt, W. R. (1969). Chemistry of gonadotrophins in relation to their antigenic properties. Acta Endocrinol. Suppl. 142: 13.

Butt, W. R. (1972). The iodination of follicle-stimulating and other hormones for radioimmunoassay. J. Endocrinol. 55: 453.

Butt, W. R. (1979). Gonadotrophins. In Hormones in Blood, 3rd Ed., Vol. 1, C. H. Gray and V. H. T. James (eds.), Academic Press, London and New York, p. 412.

Butt, W. R. (1981). Techniques of radioiodination. Anal. Proc. 18: 100.

Butt, W. R., Lynch, S. S., Chaplin, M. F., Gray, C. J., and Kennedy, J. F. (1970). The effects of chemical modifications on the biological and radioimmunological activity of pituitary follicle stimulating hormone. In Gonadotrophins and Ovarian Development, W. R. Butt, A. C. Crooke, and M. Ryle (eds.). Churchill Livingstone, Edinburgh, p. 171.

Butt, W. R., Lynch, S. S., Reay, P., Robinson, W. and Sage, J. (1983). Comparison of iodination methods. In Immunoassays for Clinical Chemistry: A Workshop Meeting, W. M. Hunter and J. E. T. Corrie (eds.), Churchill Livingstone, Edinburgh, p. 286.

Chaplin, M. F., Gray, C. J., and Kennedy, J. F. (1970). Chemical studies on a pituitary FSH preparation. In Gonadotrophins and Ovarian Development, W. R. Butt, A. C. Crooke, and M. Ryle (eds.), Livingstone, Edinburgh, p. 77.

Conroy, P., and Fletcher, J. (1981). A technique for the removal of radioactive components from aqueous RIA waste. Clin. Chem. 27: 1073.

Corrie, J. E. T. (1983). ^{125}Iodinated tracers for steroid radioimmunoassay: The problem of bridge recognition. In Immunoassays for Clinical Chemistry: A Workshop Meeting, W. M. Hunter and J. E. T. Corrie (eds.), Churchill Livingstone, Edinburgh, p. 353.

Corrie, J. E. T., and Hunter, W. M. (1981). ^{125}Iodinated tracers for hapten-specific radioimmunoassays. Methods Enzymol. 73: 79.

Corrie, J. E. T., Hunter, W. M., and Macpherson, J. S. (1981). Progesterone with homologous-site ^{125}I-labelled radioligand. Clin. Chem. 27: 594.

David, G. S., and Reisfeld, R. A. (1974). Protein iodination with solid state lactoperoxidase. Biochemistry: 13: 1014.

Davies, A. G., Lawrence, N. R., and Lynch, S. S. (1978). Binding of ^{125}I-labelled FSH in the mouse testis in vivo. J. Reprod. Fertil. 53: 249.

Edwards, R., Lalloz, M., and Pull, P. I. (1983). Radioiodination of proteins by three procedures: Solid phase lactoperoxidase, chloramine-T and iodogen. In Immunoassays for Clinical Chemistry: A Workshop Meeting, W. M. Hunter and J. E. T. Corrie (eds.), Churchill Livingstone, Edinburgh, p. 277.

Fang, V. S., Cho, H. W., and Meltzer, H. Y. (1975). Labelling of creatine phosphokinase without loss of enzyme activity. Biochem. Biophys. Res. Commun. 65: 413.

Ferguson, K. M., Hayes, M. M., and Jeffcoate, S. L. (1983). Preparation of tracer LH and FSH for multicentre use. In Immunoassays for Clinical Chemistry: A Workshop Meeting, W. M. Hunter and J. E. T. Corrie (eds.), Churchill Livingstone, Edinburgh, p. 289.

Fraker, P. J., and Speck, J. C. (1978). Protein and cell membrane iodinations with a sparingly soluble chloramide 1,3,4,6-tetrachloro-3α, 6α-diphenyl glycoluril. Biochem. Biophys. Res. Commun. 80: 849.

Greenwood, F. C., Hunter, W. M., and Glover, J. S. (1963). The preparation of ^{131}I-labelled human growth hormone of high specific radioactivity. Biochem. J. 89: 114.

Hochberg, R. A. (1979). Iodine-125-labelled estradiol. A gamma-emitting analog of estradiol that binds to the estrogen receptor. Science 205: 1138.

Karonen, S.-L., Mörsky, P., Siren, M., and Seuderling, U. (1975). An enzymatic solid-phase method for trace iodination of proteins and peptides with ^{125}iodine. Anal. Biochem. 67: 1.

Langone, J. J. (1980). Radioiodination by use of the Bolton-Hunter and related reagents. Methods Enzymol. 70: 221.

McConahey, P. J., and Dixon, F. J. (1980). Radioiodination of proteins by the use of the chloramine-T method. Methods Enzymol. 70: 214.

Morrison, M. (1980). Lactoperoxidase-catalyzed iodinations as a tool for investigation of proteins. Methods Enzymol. 70: 210.

Murphy, M. J., Jr. (1976). ^{125}I-labelling of erythropoietin without loss of biological activity. Biochem. J. 159: 287.

Nordblom, G. D., Webb, R., Counsell, R. E., and England, B. G. (1981). A chemical approach to solving bridge phenomena in steroid radioimmunoassays. Steroids 38: 161.

Reay, P. (1982). Use of N-bromosuccinimide for the iodination of proteins for radioimmunoassay. Ann. Clin. Biochem. 19: 129.

Redshaw, M. R., and Lynch, S. S. (1974). An improved method for the preparation of iodinated antigens for radioimmunoassay. J. Endocrinol. 60: 527.

Rudinger, J. (1971). In Radioimmunoassay Methods, K. E. Kirkham and W. M. Hunter (eds.), Churchill Livingstone, Edinburgh, p. 104.

Schall, R. F. Jr., and Tenoso, H. J. (1981). Alternatives to radioimmunoassay: Labels and methods. Clin. Chem. 27: 1157.

Sonoda, S., and Schlamowitz, M. (1970). Studies of ^{125}I trace labelling of immunoglobulin G by chloramine-T. Immunochemistry 7: 885.

Suginami, H., Robertson, D. M., and Diczfalusy, E. (1978). Influence of the purity of the iodinated tracer on the specificity of the radioimmunoassay of human luteinizing hormone. Acta Endocrinol. 89: 506.

Thorell, J. I., and Larsson, I. (1974). Lactoperoxidase coupled to polyacrylamide for radio-iodination of proteins to high specific activity. Immunochemistry II: 203.

Tower, B. B., Clark, B. R., and Rubin, R. T. (1977). Preparation of ^{125}I polypeptide hormones for radioimmunoassay using glucose oxidase with lactoperoxidase. Life Sci. 21: 959.

Tower, B. B., Sigel, M. B., Rubin, R. T., Poland, R. E. and Vanderhaan, W. P. (1978). The talc-resin-TCA test: Rapid screening of radioiodinated polypeptide hormones for radioimmunoassay. Life Sci. 23: 2183.

3
Enzyme Immunoassay

MICHAEL J. O'SULLIVAN*
The Blond McIndoe Centre for Transplantation Biology,
Queen Victoria Hospital, East Grinstead, Sussex, England

I. INTRODUCTION

Radioimmunoassay (RIA) is a sensitive and specific technique which allows the quantitation of a wide variety of compounds. No purification step is usually required, even when complex biological fluids are used in the assay. RIA has proved extremely valuable in clinical chemistry and other fields but the method does have a number of limitations. The most useful radioisotopes have relatively short half-lives, the equipment required to quantitate their activity can be expensive, and potentially hazardous levels of radioactivity are used during labeling procedures. The requirement for a separation step during the assay is a particular disadvantage of RIA. This separation step increases the complexity of the assay and is a major obstacle to overcome when automating RIA.

      The properties of a variety of nonisotopic labels have been investigated in an attempt to retain the advantages of RIA without the drawbacks associated with radioisotopic labels. Nonisotopic labels have included bacteriophages (Haimovich et al., 1970), erythrocytes (Alder and Chi-Tan, 1971), stable free radicals (Leute et al., 1972), fluorescent molecules (Ullman et al., 1976), chemiluminescent precursors (Schroeder et al., 1976), latex particles (Cambiaso et al., 1977), metal ions (Cais et al., 1977), and enzymes.

      Enzymes can function as labels because their catalytic properties allow the detection and quantitation of extremely small quantities of labeled immune reactants. The first enzyme immunoassays (EIA) were reported independently by Engvall and Perlmann (1971), and Van Weemen and Schuurs (1971). However, enzyme labels had been used in immunocytochemistry prior to that date (Nakane et al., 1966).

---

*Present affiliation: Amersham International, Cardiff, Wales.

In this Chapter the principles and practice of EIA are discussed, the more recent developments in methodology described, and the merits of the technique are critically evaluated. Earlier work in the field has been reviewed (Wisdom, 1976; Scharpe et al., 1976; Schuurs and Van Weemen, 1977; Voller et al., 1978; O'Sullivan et al., 1979a).

II. ASSAY PRINCIPLES

The nomenclature of EIA has been rather confused. In this chapter the term EIA will embrace all immunoassays that utilize an enzyme label, or where the interaction between the antibody and the ligand is quantitated by its effect on the activity of an enzyme. EIA can be subdivided into heterogeneous and homogeneous assays. In heterogeneous assays, the interaction between the ligand and antibody does not affect the activity of the enzyme label. These assays are analogous to RIA and require a separation step. In homogeneous assays, the interaction between ligand and antibody alters the activity of an enzyme. These assays do not require a separation step and have no analogy in RIA.

A. Heterogeneous Enzyme Immunoassays

1. Competitive Enzyme Immunoassay

This technique involves competition between enzyme-labeled and unlabeled ligand for a limited quantity of ligand-specific antibody. Antibody-bound ligand is separated from unbound ligand and the enzyme activity in either phase measured. The activity in the bound phase is inversely proportional to the concentration of unlabeled ligand in the assay (Al-Bassam et al., 1978a). This type of EIA is analogous to the classic RIA method (Yalow and Berson, 1959).

2. Immunoenzymometric Assays

In an immunoenzymometric assay the ligand is mixed with excess enzyme-labeled antibody. This is followed by sufficient solid-phase ligand to bind any remaining ligand-free antibody. The amount of enzyme adsorbed to the solid phase is inversely proportional to the concentration of free ligand (Maiolini et al., 1975).

The double-antibody immunoenzymometric assay is a variation of this technique. Binding of the ligand-specific antibody to the solid phase is quantitated indirectly using an enzyme-labeled second antibody directed against immunoglobulins of the animal species in which the ligand-specific first antibody was raised (Gnemmi et al., 1978).

3. Two-Site Immunoenzymometric Assay

In this technique, ligand is mixed with an excess of ligand-specific antibody attached to a solid-phase. Binding of ligand to the solid phase is quantitated using an enzyme-labeled ligand-specific antibody. The amount of enzyme bound to the solid phase is proportional to the amount of ligand present in the assay (Maiolini et al., 1978). This procedure has been termed a "sandwich assay" and requires that the antigen has at least two sites to which antibody can bind. In the two-site double-antibody immunoenzymometric assay, binding of the ligand-specific antibody to the solid phase is quantitated using an enzyme-labeled second antibody (Belanger et al., 1973).

4. Enzyme-Linked Immunosorbent Assay (ELISA)

The ELISA assay is used to detect specific antibodies. Solid-phase ligand is mixed with the sample, the solid phase is then washed, and excess enzyme-labeled second antibody added. The amount of enzyme bound to the solid phase is proportional to the quantity of ligand-specific antibody in the sample (Engvall and Perlmann, 1972).

B. Homogeneous Enzyme Immunoassays

A number of novel EIAs have been developed which do not require a separation phase. These assays are based upon a variety of ingenious mechanisms.

1. Assays Based On Enzyme-Ligand Conjugates

This technique has usually been used to quantitate haptens, the key reagent being an enzyme-hapten conjugate. Binding of hapten-specific antibody to this conjugate inhibits the activity of the enzyme. The antibody is believed either to hinder access of substrate sterically to the active site of the enzyme (Rubenstein et al., 1972) or induce conformational changes in the enzyme which prevent the enzyme from acting upon its substrate (Rowley et al., 1975). Free hapten competes with the hapten-enzyme conjugate for a limited quantity of antibody and so reduces this inhibition. The enzyme activity of the conjugate is proportional to the concentration of free hapten in the sample or standards. The principle of this assay is illustrated in Fig. 1. This assay has been termed the enzyme multiplied immunoassay technique (EMIT).

The EMIT thyroxine assay is an exception to this mechanism. In this instance the thyroxine-malate dehydrogenase conjugate is enzymatically inactive. However, binding of thyroxine-specific antibody to the conjugate actually activates the enzyme. The enzyme activity of this label is inversely proportional to the concentration of unlabeled thyroxine in the assay (Ullman et al., 1975).

1 No unlabeled hapten present

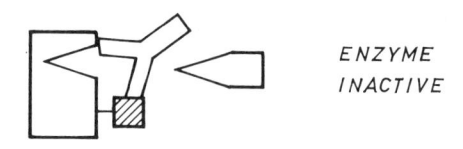

ENZYME
INACTIVE

2 Unlabeled hapten present

ENZYME
ACTIVE

KEY

active site

enzyme

hapten

hapten in sample

hapten-specific antibody

substrate

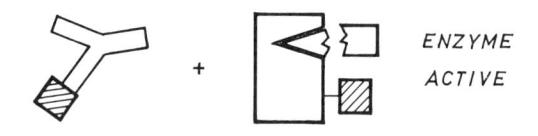

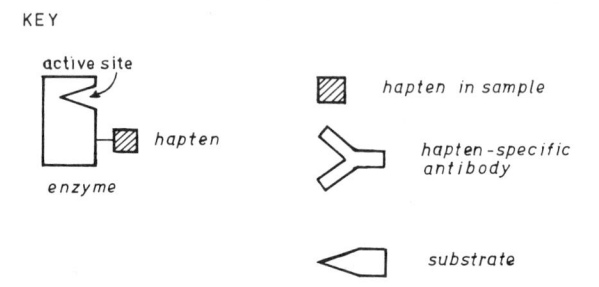

FIGURE 1 The principle of the EMIT assay.

It has proved difficult to quantitate high-molecular-weight antigens by the EMIT assay. It would appear that binding of antibody to antigen-enzyme conjugates does not sterically hinder access of small-substrate molecules to the enzyme's active site. A possible solution to this problem was indicated when Wei and Reibe (1977) found that human IgG inhibited the enzyme activity of a phospholipase-antihuman IgG conjugate, provided erythrocyte membrane phospholipids were used as substrate. Wei and Reibe postulated that binding of IgG was sufficient to prevent access of this high-molecular-weight substrate to the active site of the enzyme.

A homogeneous EIA for quantitating proteins has been reported (Gibbons et al., 1980). This assay is based upon a galactosidase-IgG conjugate and a synthetic high-molecular-weight substrate. Binding of IgG-specific antibodies to this conjugate inhibited the activity of the enzyme, provided the high-molecular-weight substrate was used to quantitate the activity of the enzyme. The assay was reported to have a sensitivity of 25 ng/ml of human IgG.

A homogenous EIA for polyvalent ligands and antibodies, which avoids the need for a labeled antigen, has been recently reported by Gibbons and co-workers (1981).

2. Assays Based On Hapten-Substrate Conjugates

The key reagent in this assay is the hapten-substrate conjugate. One example of such a conjugate consisted of gentamicin linked to an umbelliferone-galactoside derivative (Burd et al., 1977b). β-Galactosidase was able to hydrolyze this derivative, even when the derivative was conjugated to gentamicin. However, when bound by gentamicin-specific antibody, the derivative did not function as a substrate for the enzyme. This inhibition was relieved by unlabeled gantamicin; the amount of substrate hydrolyzed was proportional to the concentration of unlabeled gentamicin in the assay. The principle of the assay is illustrated in Fig. 2.

It is worth noting that the potential sensitivity of this type of assay is limited as the amplification property of the enzyme is not used. This effect is minimized as a highly fluorescent product is formed when the substrate is hydrolyzed by the enzyme. This type of assay has mainly been used to quantitate haptens, but has recently been adapted to the determination of high-molecular-weight compounds (Ngo and Lenhoff, 1981).

3. Assays Based On Ligand-Modulator Conjugates

These assays employ a modulator which acts to alter the activity of an indicator enzyme. The key reagent is a ligand-modulator conjugate: binding of ligand-specific antibody to this conjugate neutralizes the activity of the modulator.

Modulators such as cofactors (Carrico et al., 1976) or prosthetic groups (Ngo and Lenhoff, 1981) are necessary for enzyme activity. In this situation the presence of ligand-specific antibody results in inhibition of the enzyme. Unlabeled ligand, by competing for ligand-specific antibody, reduces this inhibition. Enzyme activity is proportional to the concentration of free ligand in the assay. Where the modulator acts to inhibit the enzyme, the reverse is true. Such modulators have included enzyme inhibitors (Finley et al., 1980) or inhibitory enzyme-specific antibodies (Ngo and Lenhoff, 1980).

4. Assays Based On Enzyme Channeling

These assays share features of both homogeneous and heterogeneous assays. The assay is based upon the finding that the overall rate of two consecutive reactions is enhanced when the enzymes catalyzing these reactions are co-immobilized on the same support. In the assay, a ligand-enzyme conjugate competes with free ligand for a limited quantity of ligand-specific antibody coimmobilized on a solid support with a second enzyme. The product of

1 No unlabeled hapten present

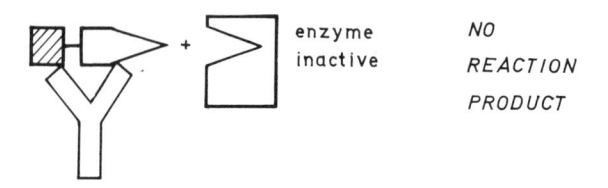

enzyme inactive

NO REACTION PRODUCT

2 Unlabeled hapten present

enzyme active

FLUORESCENT REACTION PRODUCT

KEY

unlabeled hapten

hapten-specific antibody

hapten-substrate conjugate

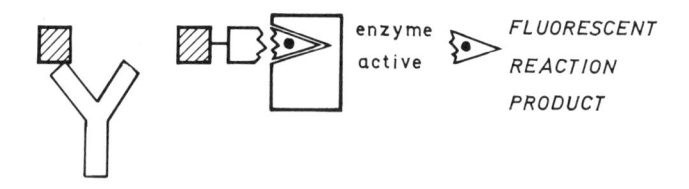

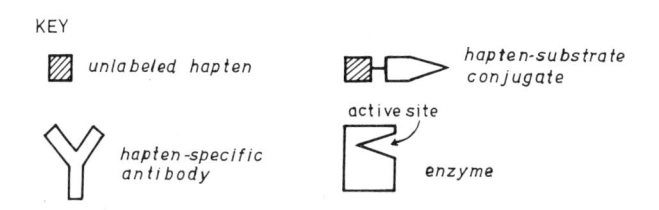

active site

enzyme

FIGURE 2 The principle of homogeneous assays based upon hapten-substrate conjugates.

the first enzyme acts as substrate for the second. When the ligand-enzyme conjugate is bound to the solid support, the two enzymes are brought into close proximity and the rate of the overall reaction is enhanced. This rate is inversely proportional to the concentration of unlabeled ligand in the assay. In practice the system would appear to be rather complicated, particularly as a third enzyme is required to minimize background reactions in the assay (Ngo and Lenhoff, 1981).

III. CHOICE OF ENZYME

The requirements of an immunoassay impose severe constraints upon any enzyme label. Ideally the enzyme should meet the requirements listed in Table 1. Most of the enzymes that have been used in EIA are tabulated in

TABLE 1 Criteria for Choice of an Enzyme Label
___

Availability of purified, low-cost enzyme preparations

High turnover number

Presence of residues through which the enzyme can be conjugated with minimal loss of activity

Stable enzyme conjugates

Enzyme absent from biological fluids

Assay method that is simple, sensitive, precise, and not affected by factors present in biological fluids

Enzyme, substrate, cofactors, etc., should not pose potential health hazards
___

Table 2. Few of these enzymes completely satisfy all the requirements in Table 1. In practice the choice of enzyme should be dictated by the requirements of the actual assay. In heterogeneous assays for instance, enzyme activity is usually measured in the washed bound phase. Any endogenous activity originally present in the sample is likely to have been removed before this stage of the assay. This means that the fifth criterion is not an absolute requirement, at least in heterogeneous assays.

In homogeneous assays the enzyme activity is measured in the presence of the biological sample. Thus endogenous activity must not be present in the sample or the sample must be pretreated to remove such activity (Finley et al., 1980). This was one reason for the replacement of lysozyme in EMIT assays. In addition, homogeneous assays based upon substrate-ligand conjugates were first developed using a carboxylic ester hydroxylase (Burd et al., 1977a). However, plasma contains endogenous esterase activity and this enzyme has now been replaced by β-galactosidase, which is not present in human plasma (O'Sullivan et al., 1978a).

The criteria listed in Table 1 can be regarded as the properties required for a general label. In homogeneous assays, the enzyme must also satisfy the requirements imposed by the particular assay mechanism. For example, acetylcholinesterase was used in an assay employing a ligand-inhibitor conjugate, because potent high-affinity inhibitors were available for this enzyme (Finley et al., 1980).

The activity of enzyme labels has been quantitated using a wide variety of techniques. These have included spectrometry (Al-Bassam et al., 1978b), fluorimetry (Ishikawa, 1973), turbidity (Rubenstein et al., 1972) scintillation counting (Van der Waart and Schuurs, 1976), thermometry (Mattiasson et al., 1978), enzyme electrodes (Mattiasson and Nilsonn, 1977)

TABLE 2 Enzymes Used as Labels in EIA

Enzyme	Enzyme commission number (E.C.)	Reference
Horseradish peroxidase[a]	EC. 1.11.1.17	Van Weemen and Schuurs, 1971
Alkaline phosphatase[a]	EC. 3.1.3.1	Engvall and Perlmann, 1972
β-D-galactosidase[a]	EC. 3.2.1.23	Lauer and Erlanger, 1974
Glucoamylase[a]	EC. 3.2.1.23	Ishikawa, 1973
Penicillinase[a]	EC. 3.5.2.6	Joshi et al., 1978
Acetylcholinesterase[a]	EC. 3.1.1.7	Van der Waart and Schuurs, 1976
Catalase[a]	EC. 1.11.1.6	Borrebaeck et al., 1978
Lysozyme[b]	EC. 3.2.1.17	Rubenstein et al., 1972
Malate dehydrogenase[b]	EC. 1.1.1.37	Ullman et al., 1975
β-D-galactosidase[b]	EC. 3.2.1.23	Wong et al., 1979
Glucose-6-phosphate dehydrogenase[b]	EC. 1.1.1.49	Kleine, 1978
Acetylcholinesterase[b]	EC. 3.1.1.7	Finley et al., 1980
Glucose oxidase[b]	EC. 1.1.3.4	Ngo and Lenhoff, 1981
Horseradish peroxidase[b]	EC. 1.11.1.17	Ngo and Lenhoff, 1980
Lactate dehydrogenase[b]	EC. 1.1.1.27	Carrico et al., 1976
Lipoamide dehydrogenase[b]	EC. 1.6.4.3	Carrico et al., 1976
Hexokinase[b]	EC. 2.7.1.1	Ngo and Lenhoff, 1981
Glucosephosphate isomerase[b]	EC. 5.3.1.9	Ngo and Lenhoff, 1981

[a] Heterogeneous EIA.
[b] Homogeneous EIA.

and chemiluminescence (Velan and Halmann, 1978). Spectrometry is sensitive, technically simple, and by far the most widely used procedure. Fluorimetry may be of value in situations requiring extreme sensitivity. The other techniques have not, as yet, been widely adopted.

The enzyme β-galactosidase is one which fulfills many of the requirements in Table 1. Its activity can be quantitated by spectrophotometric or fluorimetric assays. The spectrophotometric assay is straightforward (Craven et al., 1965), the reaction can be readily terminated, and the rate of the reaction remains linear and proportional to the amount of enzyme over a wide concentration range (O'Sullivan et al., 1975, unpublished observations).

The fluorimetric assay is extremely sensitive and an EIA with a sensitivity of 0.03 fm of ornithine-aminotransferase, which uses a fluorimetric detection system, has been reported by Hamaguchi and co-workers (1976).

IV. PREPARATION OF ENZYME LABELS

In this section, the techniques used to prepare enzyme labels will be outlined. The preparation of one enzyme-hapten conjugate and one enzyme-antibody conjugate will be discussed in more detail.

A. Enzyme-Hapten Conjugates

A number of factors need to be considered when preparing enzyme-hapten conjugates. These include the chemistry of the cross-linking reaction, the presence of residues on the hapten suitable for modification, the nature of the cross-link, the number of haptens coupled to each enzyme, the effect of the conjugation reaction on the enzyme activity, and, finally, the purity of the label.

Many of the reactions used to prepare enzyme-hapten labels and hapten-protein immunogens are similar (Erlanger, 1973). The mixed anhydride reaction is one of the more popular cross-linking techniques. This reaction is usually used to form a peptide bond between a carboxylic acid on the hapten and an amino group from the enzyme. We have used this technique to couple methotrexate (Marks et al., 1978), bile salts (Al-Bassam, 1979), and 6-methyl-prednisolone (Ngowi, 1979) to β-galactosidase. The procedure used to form a β-galactosidase-methotrexate conjugate is described below (Al-Bassam et al., 1979).

Methotrexate (6.8 mg, 15 μmol) and tri-n-butylamine (2.8 mg, 15 μmol) were sequentially dissolved in 0.1 ml dimethylformamide. The solution was cooled to 10°C then isobutyl chloroformate (2 mg, 15 μmol) was added. The solution was stirred for 30 min at 10°C before mixing with β-galactosidase (3 mg. 6 nmol) dissolved in 2 ml of 0.1 $\underline{M}$ sodium bicarbonate buffer, pH 9.5, containing 0.01 $\underline{M}$ $MgCl_2$. The reaction was maintained at 10°C for 4 hr then left overnight at 4°C. To separate compounds of low relative molecular mass from the label, the solution was chromatographed on a Sephadex G25 column (60 x 1 cm) and eluted in 0.05 $\underline{M}$ tris-acetate buffer, pH 7.5, containing 0.01 $\underline{M}$ $MgCl_2$, 0.01 $\underline{M}$ β-mercaptoethanol, and 0.1 $\underline{M}$ NaCl. Fractions containing enzyme activity were pooled and stored at 4°C in the above buffer which also contained 0.02% w/v NaN_3 and 0.1% w/v bovine serum albumin (BSA). The enzyme retained >90% of its original activity; 80% of the enzyme was bound by methotrexate-specific antibody.

Haptens occasionally contain no reactive residues suitable for cross-linking. In such a situation it may prove possible to introduce suitable residues into the hapten (Erlanger, 1973) or to synthesize the label using a more reactive analog of the hapten (Al-Bassam et al., 1978a). To develop specific and sensitive assays, it is considered advisable to avoid modifying sites on the hapten which distinguish the hapten from closely related analogs or inactive metabolites (Van Weemen and Schuurs, 1975).

Antibodies raised against immunogens tend also to recognize the bridge between the hapten and the carrier protein. If the same cross-link is present in the label, it is possible that the antibody will bind the label more avidly than the free hapten, resulting in an insensitive assay. This situation was encountered during the attempted development of an EIA for V116, a potent cytotoxic drug (O'Sullivan and Horten, 1979, unpublished data). Both the label and the immunogen were produced by the periodate oxidation reaction (Lauer and Erlanger, 1974). Unfortunately, free hapten did not displace the V116-galactosidase label from the antibody. In contrast, free methotrexate was able to displace label from antibody (Marks et al., 1978) even though the label and immunogen (Aherne et al., 1977) contained the same cross-link. The reason for the difference between these two assays may be that the periodate reaction modifies the structure of the hapten more than the carbodiimide reaction which was employed to synthesize the immunogen used to raise the methotrexate-specific antibodies. Possibly antibodies recognize the modified hapten, in both the immunogen and label more readily than the original hapten.

Noncoupled hapten is usually removed from the label by gel filtration, but it is often difficult to remove noncoupled enzyme. This may be accomplished during the actual assay when aspirating the supernatant from the antibody-bound phase (Schall et al., 1978). However, such an approach may decrease the precision of the assay, particularly if the noncoupled enzyme forms a high proportion of the total enzyme in the preparation. Alternatively the amount of free enzyme can be minimized by using an efficient cross-linking reaction. A galactosidase-methotrexate conjugate was synthesized, in which virtually all the enzyme was conjugated to immunoreactive methotrexate (Fig. 3). Unfortunately a relatively shallow standard curve was obtained using this label (Marks et al., 1978). Labels produced using a lower ratio of methotrexate to enzyme yielded more sensitive assays even though the preparations contained approximately 40% free enzyme (Al-Bassam et al., 1979). To develop sensitive EIAs using enzyme-hapten labels, it may prove necessary to employ labels with a low ratio of hapten to enzyme in the conjugates. In this situation the presence of a high proportion of noncoupled enzyme is likely to be unavoidable. Affinity chromatography was used to remove noncoupled enzyme during the development of a sensitive estradiol assay (Exley and Abuknesha, 1978).

The principles outlined in this section have been discussed in more detail elsewhere. The development of steroid immunoassays (Riad-Fahmy et al., 1981) and of drug immunoassays (Marks et al., 1980) have been reviewed.

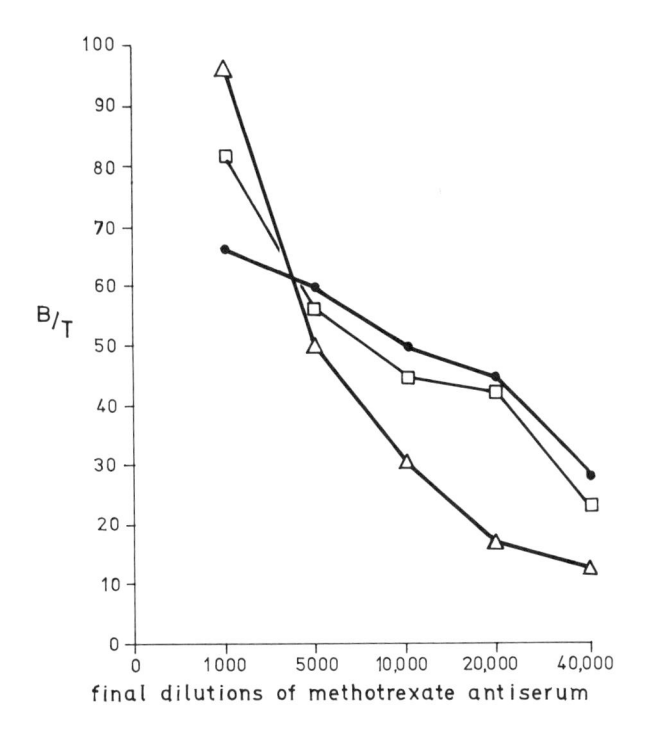

FIGURE 3 Binding of galactosidase-methotrexate conjugates by methotrexate-specific antibodies. B, enzyme activity bound to antibody; T, total enzyme activity. Ratio of methotrexate to β-galactosidase in coupling reaction: Δ = 5000 : 1; □ = 2500 : 1; ● = 1000 : 1.

B. Enzyme-Antibody Conjugates

Numerous techniques have been used to prepare protein-protein conjugates (Kennedy et al., 1976). However, relatively few of these techniques have been successfully adapted to the synthesis of enzyme-antibody conjugates. The more popular cross-linking techniques have been recently reviewed (O'Sullivan and Marks, 1981) and are listed in Table 3. The major problem in preparing enzyme-antibody conjugates is in the similar chemical composition of the two reactants. This means that it is difficult to minimize side reactions that lead to the formation of enzyme-enzyme and protein-protein conjugates rather than the desired enzyme-antibody conjugate.

Horseradish peroxidase, alkaline phosphatase, and β-galactosidase are the most popular labels. However, β-galactosidase has the advantage that it contains reactive sulfydryl residues not essential for enzyme activity (Craven et al., 1965). This enzyme has been coupled to antibody using $\underline{N}$,

TABLE 3 Techniques used to Prepare Enzyme-Antibody Conjugates

Cross-linking technique	Reference
The one-step glutaraldehyde method	Engvall and Perlmann, 1971
The two-step glutaraldehyde method	Avrameas and Ternynck, 1971
The periodate oxidation method	Nakane and Kawaoi, 1974
N, N -O- Phenylenedimaleimide	Kato et al., 1975
m-Maleimidobenzoyl-N-hydroxy-succinimide ester	O'Sullivan et al., 1978b
2-Iminothiolane /4,4´-dithiodipyridine	King and Kochoumian, 1979
4-Hydroxy-3-nitro-methyl benzimidate hydrochloride	Muller and Pfleiderer, 1979
Noncovalent coupling	Yorde et al., 1976

N´-O- phenylenedimaleimide. Phenylenedimaleimide reacts selectively with sulphydryl groups and prior to coupling it is necessary to introduce these groups into the antibody. This has been accomplished by reduction of existing disulfide bridges (Kato et al., 1975) or by reaction with 4-methyl mercaptobutyrimidate (O'Sullivan et al., 1979b). Mercaptobutyrimidate selectively modifies amino residues and the extent of the reaction can be readily controlled. The antibodies were modified while bound to an immunoadsorbent. This has advantages in that binding sites of the antibody are protected during the reaction, excess reagent can be readily removed, and a purified antibody preparation is used. Following elution from the immunoadsorbent, the modified antibodies were reacted with the dimaleimide, unreacted maleimide was then removed by gel filtration, and the maleimide-substituted antibody reacted with β-galactosidase. Approximately 60% of the enzyme was conjugated to antibody with little or no loss of enzyme activity.

The fact that β-galactosidase contains sulfydryl residues while antibodies do not has formed the basis of a simple but effective technique for conjugating this enzyme to antibodies. Kitagawa and Aikawa (1976) used the reagent m-maleimidobenzoyl-N-hydroxysuccinimide (MBS) to couple a number of drugs and insulin to β-galactosidase. MBS contains both a hydroxysuccinimide ester and a maleimide residue; these residues react with amino and sulfydryl groups, respectively (Fig. 4). Both these reactions are selective and proceed under mild conditions. MBS has proved effective in conjugating β-galactosidase to antibody (O'Sullivan et al., 1978b). Maleimide residues were introduced into the antibody by reaction with MBS. Due to the lack of sulfydryl groups in antibody molecules, MBS cannot poly-

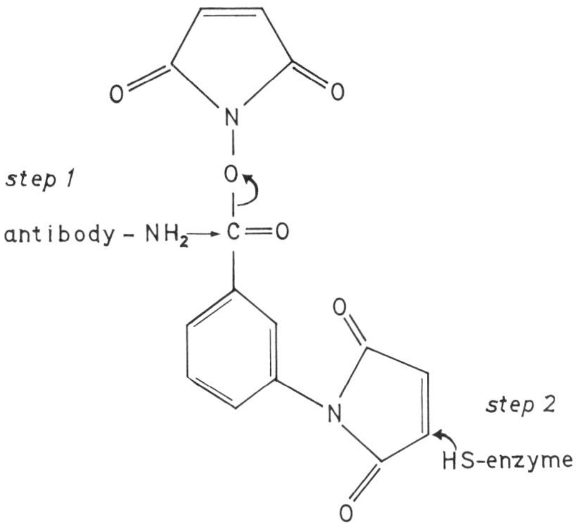

Enzyme Immunoassay

FIGURE 4 Preparation of enzyme-antibody conjugates using the cross-linking reagent MBS.

merize the antibody at this stage of the reaction. Following removal of unreacted MBS, the modified antibodies were coupled to β-galactosidase via sulfydryl groups present in the enzyme.

The following protocol was used to prepare galactosidase-antibody conjugates (O'Sullivan et al., 1979b). MBS (3.2 mg) dissolved in 15 µl of dioxan was mixed with 1.5 ml of 0.1 M phosphate buffer, pH 7.0, containing 1.5 mg of antibody and 0.05 M NaCl. The solution was held at 30°C for 1 hr, when unreacted MBS was removed by gel filtration on a Sephadex G-25 column (30 x 0.9 cm) equilibrated with 0.1 M phosphate buffer, pH 7.0, containing 0.01 M MgCl$_2$ and 0.05 M NaCl. Fractions eluting in the void volume were pooled (3 ml, 1.5 mg protein) and immediately mixed with 1.5 mg of β-galactosidase dissolved in 3 ml of 0.1 M phosphate buffer, pH 7.0, containing 0.01 M MgCl$_2$ and 0.05 M NaCl. This solution was maintained at 30°C for 1 hr, when the reaction was terminated by addition of sufficient 1 M β-mercaptoethanol to give a final concentration of 0.01 M β-mercaptoethanol. Nonconjugated antibody was removed from the label by gel filtration on a Sephadex G-200 column (90 x 2.0 cm) equilibrated with 0.01 M Tris-HCl buffer, pH 7.0, containing 0.01 M MgCl$_2$, 0.01 M β-mercaptoethanol, and 0.05 M NaCl (Fig. 5). The label eluted in the void volume of the column and was stored at 4°C in the above buffer which also contained 0.1% w/v BSA and 0.02% w/v NaN$_3$.

Prior to coupling the antibody was purified by affinity chromatography (O'Sullivan et al., 1979c). The enzyme (Boehringer-Mannheim) was

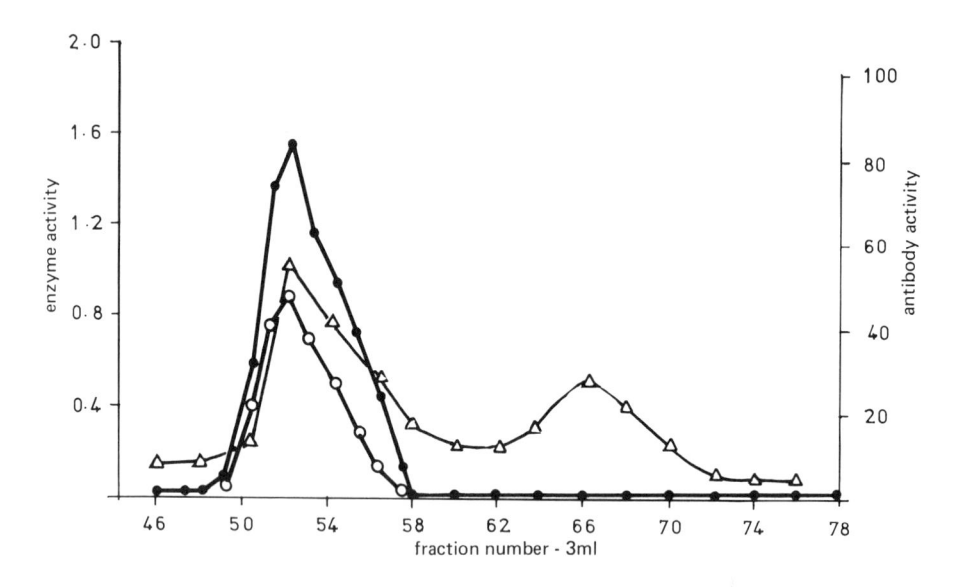

FIGURE 5 Purification of a galactosidase donkey antisheep IgG antibody conjugate by gel filtration. ●, Total enzyme activity (Δ absorbance/20 min); o, enzyme activity bound to a sheep IgG solid phase (Δ absorbance/20 min); Δ, antibody activity. Binding (counts/second) of ^{125}Iodine-labeled sheep IgG to a sheep IgG-solid phase in the presence of the donkey antisheep antibody.

obtained as a suspension (5 mg/ml), which was centrifuged at 1000 g for 15 min and the supernatant discarded. The enzyme contained an average of five sulfydryl groups per molecule. An average of three maleimide residues were introduced into each antibody molecule. The cross-linking reaction was highly effective, with 80% of the enzyme conjugated to immunoreactive antibody. Approximately 60% of the antibody was incorporated into the conjugate. Virtually no enzyme activity and only 10-15% of the antibody activity was lost during the reaction. Both the cross-link and the antibody were stable for at least 6 months. The activity of the enzyme decreased by 20% in this period, but this did not affect the performance of the label.

V. METHODOLOGY

The methodology of EIA, particularly of heterogeneous EIA, is similar in many respects to RIA. One restriction in EIA is the large size of the enzyme label. Separation techniques based on the difference in molecular size between the free and the antibody-bound label cannot be used in EIA.

Techniques based upon double antibody precipitation (Comoglio and Celada, 1976) or upon solid-phase systems (Van Weemen and Schuurs, 1975) have proved most popular. An ingenious separation system based upon magnetic polyacrylamide-agarose beads (Guesdon et al., 1978) has been described. Many of the separation techniques used in EIA have been reviewed (Schuurs and Van Weemen, 1977). To illustrate at least some of the many techniques employed in EIA, two assays will be described in more detail.

A. Competitive Enzyme Immunoassay

High-dosage methotrexate therapy is considered to be of value in the treatment of a number of cancers. The drug is normally rapidly excreted in the urine and high dosages are usually tolerated by the patient. However, if excretion of the drug is delayed, high plasma levels are maintained and the patient may suffer extremely serious toxic side effects. Fortunately, these patients can be rescued by prompt treatment with folinic acid. Thus a rapid diagnostic assay for methotrexate is of value in the clinical management of patients undergoing therapy with methotrexate.

A methotrexate EIA, as originally developed, incorporated a conventional second antibody separation technique (Marks et al., 1978). This assay was subsequently modified by using a preincubated first and second antibody complex to decrease the time necessary to complete the separation step. The complex was prepared by mixing methotrexate-specific antiserum (diluted 5000-fold) with an equal volume of donkey antisheep immunoglobulin antiserum (diluted 30-fold) for 24 hr at $4°C$. The complex was stable for at least 1 month when stored at $4°C$ in the presence of 0.02% w/v NaN_3. All reagents were diluted in 0.05 M sodium phosphate buffer, pH 7.4, containing 0.1 M NaCl and 0.1% w/v BSA.

To perform the assay, preincubated complex (700 µl), standards or samples (100 µl), and a β-galactosidase-methotrexate label (100 µl, containing 40 ng enzyme) were added to plastic tubes, mixed, and held at $30°C$ for 1 hr. Phosphate buffer (1.5 ml) was then added and the tubes centrifuged at 1000 g for 20 min at $4°C$. The supernatant was removed by aspiration, the precipitate was washed with 2 ml phosphate buffer, and then centrifuged and aspirated as above. Enzyme activity in the precipitate was measured by mixing the precipitate with 1 ml of 0.1 M phosphate buffer, pH 7.0, containing 2.3 mM O-nitrophenyl-β-D-galactopyranoside, 1 mM $MgCl_2$, and 0.01 M mercaptoethanol. The solution was held at $40°C$ for 1 hr, when the reaction was terminated with 1.5 ml of 0.2 M sodium carbonate. The absorbance at 420 nm of the solution was measured.

A typical standard curve obtained using the assay is illustrated in Fig. 6. Approximately 40% of the label was bound to antibody in the absence of standard. The assay was sensitive to around 100 pg methotrexate per tube. This is more than sufficient to measure clinically significant levels of methotrexate.

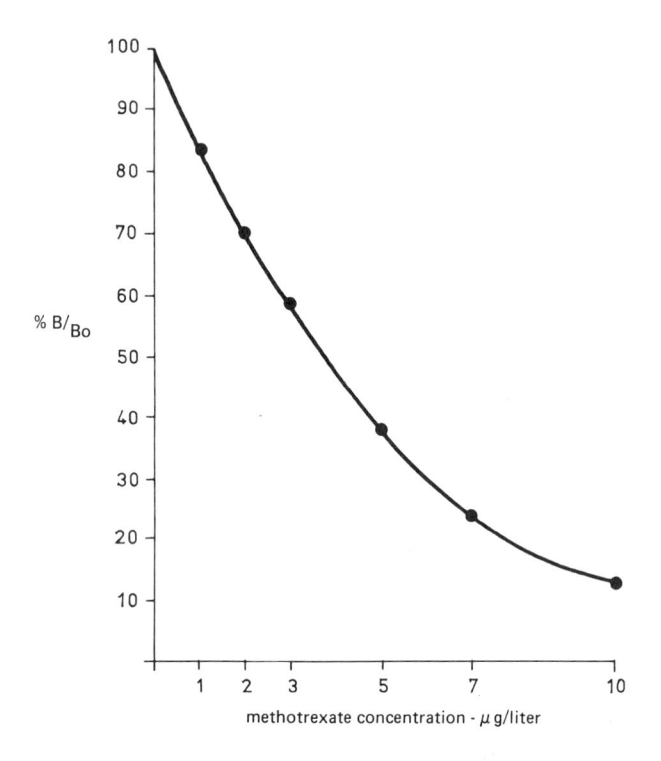

FIGURE 6 A typical methotrexate standard curve obtained using the methotrexate EIA. B, Label bound at stated methotrexate concentration; Bo, label bound at zero methotrexate concentration.

B. Enzyme-Linked Immunoadsorbent Assay

The enzyme-linked immunoadsorbent assay (ELISA) technique has proved very effective in detecting specific antibodies. The following protocol was used to detect antibodies to human growth hormone (HGH). An aliquot (100 µl) of a serial dilution of HGH antiserum was mixed with an HGH-substituted cellulose solid phase (100 µl, 10 µg cellulose). Unless otherwise stated all the reagents were diluted with 0.05 M sodium barbitone buffer, pH 8.0, containing 0.065 M NaCl and 0.5% w/v BSA. The reagents were held at room temperature for 1 hr, after which the solid phase was washed once with 2 ml of barbitone buffer containing 1M NaCl and 0.1% w/v BSA and once with barbitone buffer pH 8.0. Washing was performed by mixing the solid phase and buffer, centrifuging the tubes of 1000 g for 15 min, and then aspirating the supernatant from the solid phase. A β-galactosidase donkey antirabbit Fc conjugate (100 µl) was mixed with the aspirated solid phase and left overnight at room temperature. The solid phase was washed and aspirated as above.

Enzyme Immunoassay 53

Enzyme activity bound to the solid phase was measured by both spectrophotometric and fluorimetric techniques. The spectrophotometric method has been described in a previous section. The fluorimetric assay was performed by mixing the solid phase with 1 ml of 0.01 $\underline{M}$ phosphate buffer, pH 7.0, containing 0.1% w/v BSA and 1 x $10^{-4}\underline{M}$ 4-methylumbelliferyl-β-galactopyranoside. The reaction was terminated after 1 hr with 1.5 ml of 0.1 $\underline{M}$ glycine-NaOH buffer, pH 10.3. Fluorescence was measured in a Locarte MK4 fluorimeter. Approximately 200 ng and 5 ng of enzyme were used in the spectrophotometric and fluorimetric assays, respectively (O'Sullivan et al., 1979d). Antibodies were detectable at dilutions of 1:160,000 and 1:640,000 in the spectrophotometric and fluorimetric assays (Fig. 7). This corresponds to approximately 8 fmol and 2 fmol/tube of HGH antibody, respectively.

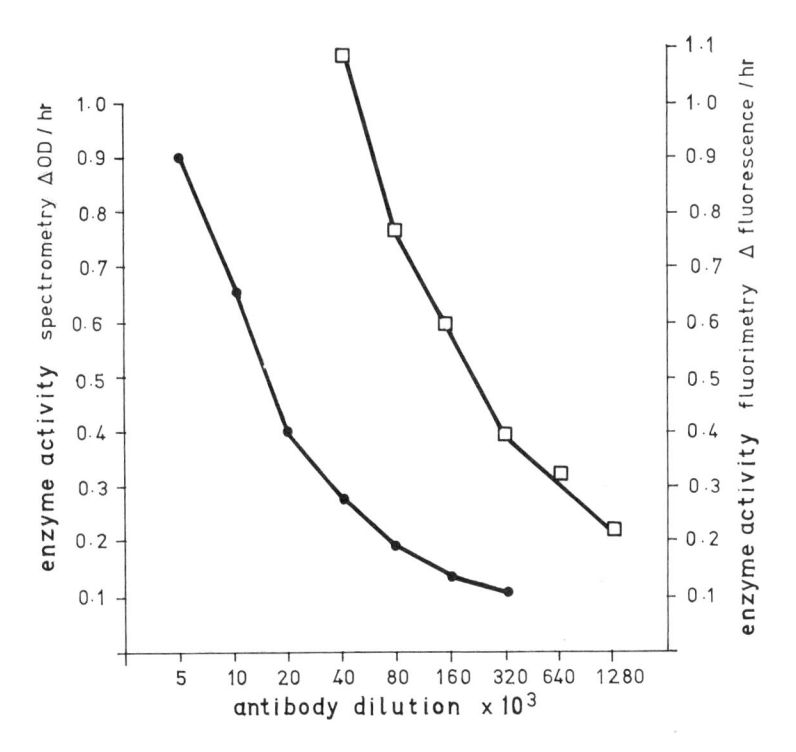

FIGURE 7 The detection of HGH-specific antibodies using a galactosidase-second-antibody conjugate. ●, Enzyme activity measured by spectrometry: total enzyme activity = Δ absorbance 3.6/hr, enzyme activity bound to solid phase in absence of antibody = Δ absorbance 0.07/hr; □, enzyme activity measured by fluorimetry: total enzyme activity = Δ fluorescence of 4.3 units/hr, enzyme activity bound to solid phase in absence of antibody = Δ fluorescence of 0.19 units/hr.

VI. APPLICATIONS

EIAs have been developed for a wide variety of compounds and extensive lists of applications have been published (Wisdom, 1976; Scharpe et al., 1976; Schuurs and Van Weemen, 1977; Voller et al., 1978; O'Sullivan et al., 1979a; Oellerich, 1980; Ngo and Lenhoff, 1981). The number of applications has expanded enormously in the last few years. A selection of assays which illustrates the range of applications of the technique is listed in Tables 4 and 5.

TABLE 4 Applications of EIA

Compound	Reference
Cortisol	Comoglio and Celada, 1976
Phenobarbital	Aoki et al., 1980
Testosterone	Hosoda et al., 1980
Progesterone	Joyce et al., 1977
Estriol	Korhonen et al., 1980
Triiodothyronine	O'Sullivan et al., 1978a
Methotrexate	Al-Bassam et al., 1979
Nortryptyline	Al-Bassam et al., 1978a
Thyroxine	Miyai et al., 1980
Digoxin	Brunk and Malmstadt, 1977
Aflatoxin B1	Pestka et al., 1980
β_2-microglobulin	Carlier et al., 1981
Ferritin	Page et al., 1980
Carcinoembryonic antigen	Maiolini et al., 1980
IgG	Arseneault, 1980
IgE	Guesdon et al., 1978
Insulin	Mattiasson and Nilsonn, 1977
Thyrotropin	Miyai et al., 1981
Human chorionic gonadotropin	Yorde et al., 1976
Fibronectin	Vuento et al., 1981
Double-stranded RNA	Field et al., 1980
Tomato ring spot virus	Converse, 1978
Adenovirus	Sarkkinen et al., 1980
Rubella virus	Vejtorp and Leerhoy, 1980
Foot-and-mouth virus	Crowther and Abu Elzein, 1979
<u>Mycoplasma hyperpneumoniae</u>	Nicolet and Paroz, 1980
Streptococcal antigen	Polin and Kennett, 1980
Clostridium botulinum toxin type A	Notermais et al., 1978
<u>Leishmania donovani</u> soluble antigen	Rassam and Al-Mudhaffer, 1980
Mouse mammary tumor virus	Dubucq et al., 1981

TABLE 5 Applications of EIA

Antibodies to	References
Cytomegalovirus	Ziegelmaier et al., 1981
Rubella virus	Ziegelmaier et al., 1981
Influenza virus	Jennings et al., 1981
Herpesvirus	Lindenschmidt, 1981
African swine fever virus	Hamdy et al., 1981
Salmonella virus	Beasley et al., 1981
Mycobacterium tuberculosis	Grange et al., 1980
Pneumoccocal capsular polysaccharide antigen	Koskela and Leinonen, 1981
Oral bacteria	Tolo et al., 1981
Plasmodium falciparum	Spencer et al., 1981
Schistosoma mansoni	Deedler and Kornelis, 1980
Trichinella spiralis	Taylor et al., 1980
Amebiasis	Agarwal et al., 1981
Chlamydia trachomatis	Cevenini et al., 1981
Platelets	Borzini et al., 1981
Histone	Aitkaci et al., 1981
Myelin	Calabrese et al., 1981
Monoclonal antibodies to human mononuclear cells	Douillard et al., 1980
Monoclonal antibodies to aflatoxin B1-modified DNA	Haugen et al., 1981
Monoclonal antibodies to cell surface antigens	Suter et al., 1980

VII. CHARACTERISTICS

The claimed advantages and possible disadvantages of enzyme labels are listed in Table 6. Unlike radioisotopes, the activity of enzymes can be altered by their environment. It is possible that factors present in biological fluids may affect the activity of the enzyme label, resulting in inaccurate assays. This is likely to be more of a problem in homogeneous assays, where the enzyme activity is measured in the presence of the sample. In addition, measurement of enzyme activity usually involves addition of substrate, a timed incubation period, addition of a reagent to terminate the reaction, and, finally, measurement of the absorbance of the solution. All these steps can potentially decrease the precision of the assay.

These effects will vary from enzyme to enzyme, so it is important to choose an enzyme with the properties listed in Table 1. The effect

TABLE 6 Evaluation of the Merits of Enzyme Labels

Possible disadvantages
  Factors present in biological fluids may affect enzyme activity causing inaccurate assays.
  Inherent imprecision of enzyme assays results in imprecise EIAs.
  The complexity of measuring enzyme activity leads to an increased work load.
  EIAs are insensitive.
  The preparation of enzyme labels is difficult.

Possible advantages
  Enzymes have a long shelf-life.
  No radiation hazard during label preparation.
  No problem of radioactive waste disposal.
  Equipment required to measure enzyme activity can be relatively inexpensive and is widely available.
  A separation step may not be required, as a result there is an increased potential for automating the assay and the assay can be extremely rapid.
  Ideal for qualitative screening assays.

of plasma on the activity of β-galactosidase has been examined (O'Sullivan et al., 1978a). Plasma samples (0.2 ml) from 29 hospital patients were mixed with β-galactosidase (200 ng) in 0.8 ml of 0.05 M barbitone buffer, pH 8.6, containing 0.1% w/v BSA, and left for 24 hr at 4°C. An aliquot (0.2 ml) from each sample was then assayed for enzyme activity and the activity compared with that of enzyme mixed with buffer alone. The results are presented in Table 7. None of these samples inhibited the activity of β-galactosidase. The coefficient of variation of the assay was approximately 2%, which is small compared with the total coefficient of variation in many immunoassays.

TABLE 7 The Effect of Plasma on β-Galactosidase Activity

	Enzyme activity (absorbance/hr)
Enzyme in buffer	0.770, 0.790[a]
Enzyme in plasma	0.785 ± 0.015[b]
Lowest value recorded	0.760, 0.770[a]
Highest value recorded	0.830, 0.830[a]

[a] Duplicate values
[b] Mean value of activity measured in the presence of plasma from 29 hospital patients

One objection often raised against enzyme labels is that enzyme activity is markedly affected by even small changes in temperature. In practice this is not a problem as the enzyme activity is usually measured in a batchwise fashion, that is, temperature variations will affect the activity of the label in the standards and samples to the same extent.

The precision of a methotrexate EIA (Al-Bassam et al., 1979) is comparable to that of many RIAs (Table 8). Methotrexate levels in plasma from 50 patients treated with methotrexate were evaluated by both EIA and RIA. There was a good agreement between results obtained using the two techniques (Fig. 8).

The specificity of an immunoassay is largely dictated by the nature of the antibody. This is clearly demonstrated by comparing the specificity of a methotrexate EIA and a RIA developed using the same antibody (Table 9). Similar results have been reported for a nortriptyline EIA and a nortryptyline RIA (Al-Bassam et al., 1978a).

The precision of many EIAs has been evaluated (Schuurs and Van Weemen, 1977) and it is apparent that the precision can be comparable to that of RIA. The performance of a triiodothyronine EIA has been evaluated in a study involving 11 laboratories (Braun et al., 1981). The conclusions were that (1) the specificity of the assay and its analytical range met diagnostic requirements; (2) the intra- and interassay precision was comparable to RIA values; (3) the recovery of added triiodothyronine was between 90 and 103%; (4) the results were comparable with those obtained using RIA; (5) there was no interference by hyperbilirubinemia or by high concentrations of bile salts, but interference can occur in very hemolytic and lipemic samples. In conclusion, the accuracy and precision of EIA can be affected by factors not encountered in RIA. In practice the effects would seem to be rather small, particularly if a suitable enzyme label is used. However, more data are required before a firm conclusion can be drawn on these points.

TABLE 8 Precision of a Methotrexate EIA

Quantity of methotrexate (µg/l)		Coefficient of variation (%)
Intraassay[a]	55	4.7
	506	3.0
	11,440	4.0
Interassay[b]	55	8.0
	510	7.7
	11,500	9.4

[a] Mean of 15 determinations in one assay
[b] Mean of determinations in five consecutive assays

TABLE 9 Cross-Reaction of Folate Analogues and Methotrexate Metabolites in EIA and RIA

Compound	Percentage cross-reaction in [a] EIA	RIA
Methotrexate	100	100
4-Amino-N^{10}-methylpteroic acid	40	31
Aminopterin	20	19
7-Hydroxymethotrexate	0.25	0.23
2,6-Diamino-6-methylpteridine	0.06	0.12
N^{10}-Methylfolic acid	0.01	0.03
Others[b]	< 0.0065	< 0.0065

[a] Cross-reaction expressed as

$$\frac{\text{Amount of methotrexate displacing 50\% of the label}}{\text{Amount of hapten displacing 50\% of the label}} \times 100$$

[b] Folic acid, dihydrofolic acid, tetrahydrofolic acid, folinic acid, p-aminobenzoic acid, and glutamic acid

The relative complexity of measuring the activity of enzyme labels is more of a problem. Techniques have been developed to simplify the methodology of EIA. These have included the use of microtiter plates (Voller et al., 1974), automation (Saunders et al., 1979), and dip sticks (Tyhach et al., 1981). The use of microtiter plates simplifies both the separation step and the measurement of the activity of the enzyme label. Both automated (Clem and Yolken, 1978) and inexpensive, portable, battery-operated spectrophotometers (Rook and Cameron, 1981) have been designed for direct measurement using microtiter plates. Such techniques have proved very popular in qualitative diagnostic assays (Voller et al., 1978). The preparation of antigen- and antibody-coated plastic supports has been reviewed (Parsons, 1981; Herrman, 1981). Care must be taken when using microtiter plates in quantitative assay as variability in the adsorptive properties of the surfaces of individual wells on the same plate has been reported (Kricka et al., 1980).

The performance of a two-site immunoenzymometric assay for IgE was evaluated at four centers (Bienvenu et al., 1982). The EIA was performed in microtiter plates. Enzyme activity was measured using a direct through-the-plate reading spectrometer. To improve reproducibility only the 60 inner wells on each plate were used. The evaluation demonstrated

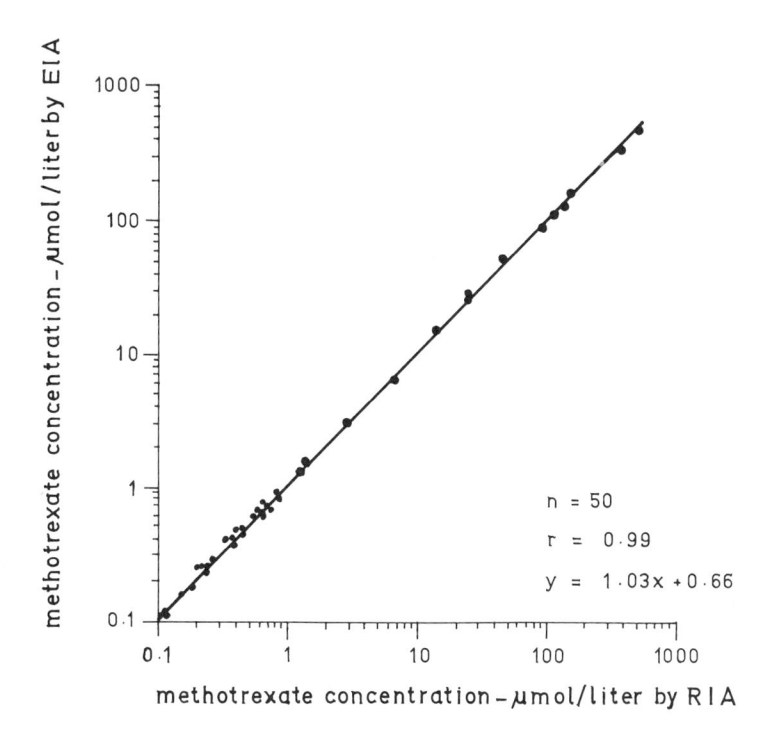

FIGURE 8 A comparison of plasma methotrexate values determined by EIA and RIA.

that IgE determination is perfectly feasible by EIA. Precision was satisfactory for IgE values between normal and pathological levels. The technique correlated well with two RIA methods.

It has been claimed that EIA lacks sensitivity when compared to RIA. One reason put forward to support this view is that the presence of the enzyme sterically hinders access of the antibody to the antigen. EIAs for nortryptyline, methotrexate, and triiodothyronine, which used enzyme-hapten labels, were somewhat less sensitive than the corresponding RIAs. However, the difference was small in all these assays. It has clearly been established that EIAs based upon enzyme-antibody conjugates can be extremely sensitive (Hamaguchi et al., 1976; O'Sullivan et al., 1979d). An EIA with a sensitivity of 1 attamole per tube was recently reported (Kato et al., 1981). It has also been claimed that the preparation of enzyme labels is difficult. In practice, the synthesis of enzyme-antibody conjugates is relatively straightforward. The complexity of preparing enzyme-hapten conjugates depends upon the nature of the hapten, it is not necessarily difficult in all instances.

VIII. CONCLUSIONS

Since the introduction of EIA in the early 1970s, interest in the technique has developed rapidly. Many substances can now be measured by the method and many more EIAs are likely to be developed in the near future. The key question is: In which areas does EIA offer advantages over RIA and other techniques? This question is not easy to answer; it may well be that EIA and RIA will prove to be complementary techniques. EIA would appear to be well-suited to the measurement of drugs using homogeneous assays, and for qualitative diagnostic assays using heterogeneous techniques. EIA may also prove to be of value in quantitative two-site immunoenzymometric assays for high-molecular-weight antigens, provided the precision and reliability of these assays prove to be satisfactory.

ACKNOWLEDGMENTS

The author thanks Mrs. Betty Muller for typing the manuscript. Many of the data used in this chapter were obtained at the Department of Biochemistry of the University of Surrey. The work was performed in collaboration with Dr. E. Gnemmi, Mr. A.D. Simmonds, Miss M. Simmons, Dr. G. Chieregatti, Dr. D.A. Morris, Dr. M.N. Al-Bassam, Professor V. Marks, and Professor J.W. Bridges.

REFERENCES

Agarwal, S. S., Sharma, P., Das, P., Ahmad, J., and Dutta, G. P. (1981). Micro-enzyme linked immunosorbent assay for serodiagnosis of amoebiasis. Indian J. Med. Res. 74: 219.

Aherne, G. W., Piall, E., and Marks, V. (1977). Development and application of a radioimmunoassay for methotrexate. Br. J. Cancer. 36: 608.

Aitkaci, A., Monier, J. C., and Mamelle, N. (1981). Enzyme-linked immunosorbent assay for anti-histone antibodies and their presence in systemic lupus erythematosus sera. J. Immunol. Methods. 44: 311.

Al-Bassam, M. N. (1979). The development of enzyme immunoassay for hormones and drugs. Ph.D. Thesis, University of Surrey pp. 229.

Al-Bassam, M. N., O'Sullivan, M. J., Gnemmi, E., Bridges, J. W., and Marks, V. (1978a). Double antibody enzyme immunoassay for nortriptyline. Clin. Chem. 24: 1590.

Al-Bassam, M. N., O'Sullivan, M. J., Gnemmi, E., Bridges, J. W., and Marks, V. (1978b). Nortriptyline enzyme-immunoassay. In Enzyme Labelled Immunoassay of Hormones and Drugs, S. B. Pal (ed), Walter de Gruyter, Berlin, pp. 375.

Al-Bassam, M. N., O'Sullivan, M. J., Bridges, J. W., and Marks. V. (1979). An improved methotrexate enzyme-immunoassay. Clin. Chem. 25: 1448.
Alder, F. L., and Chi-Tan, L. (1971). Detection of morphine by haemagglutination-inhibition. J. Immunol. 106: 1684.
Aoki, K., Terasawa, N., and Kuroiwa, Y. (1980). Enzyme immunoassay for phenobarbital. Chem. Pharm. Bull. Tokyo 28: 3291.
Arseneault, J. J. (1980). An enzyme immunoassay for the quantitation of IgG in serum and cerebrospinal fluid. Clin. Chim. Acta 107: 73.
Avrameas, S., and Ternynck, T. (1971). Peroxidase labelled antibody and Fab conjugate with enhanced intracellular penetration. Immunochemistry 8: 1175.
Beasley, W. J., Joseph, S. W., and Weiss, E. (1981). Improved serodiagnosis of Salmonella enteric fevers by an enzyme-linked immunosorbent assay. J. Clin. Microbiol. 13: 106.
Belanger, L., Sylvestre, C., and Dufour, D. (1973). Enzyme-linked immunoassay for alpha-fetoprotein by competitive and sandwich procedures. Clin. Chim. Acta 48: 15.
Bienvenu, J., Bienvenu, F., Michaud-Caudie, C., Quincy, C., Nicolas, J. P., Phang, P. P., Mandrand, B., and Trabuc-Jaubert, M. (1982). Enzyme-linked immunoadsorbent assay for serum IgE: Evaluation in four centres. Ann. Clin. Biochem. 19: 60.
Borrebaeck, C., Mattiasson, B., and Svensson, K. (1978). A rapid nonequilibrium enzyme immunoassay for determining serum gentamicin. In Enzyme Labelled Immunoassay of Hormones and Drugs, S. B. Pal (ed.), Walter de Gruyter, Berlin, p. 15.
Borzini, P., Tedesco, F., Greppi, N., Rebulla, P., Parravicini, A., and Sirchia, G. (1981). An immunoenzymatic assay for the detection and quantitation of platelet antibodies: The platelet β-galactosidase test. J. Immunol. Methods 44: 323.
Braun, V. S. L., Vogt, W., Borner, K., Delcourt, R., Ober, O., Ederveen, A. B., Haas, H., Kagedal, B., Kaltwasser, F., Luond, H., Oellerich, M., Haindl, H., Wagner, H., and Hengst, K. (1981). Bestimmung von Triiodothyronin in Serum mit einem heterogenen Enzymimunoasay: ergebnisse einer gemeinsamem erprobung. J. Clin. Chem. Clin. Biochem. 19: 911.
Brunk, S. D., and Malmstadt, H. V. (1977). Adaptation of the EMIT digoxin assay to a mini-disc centrifuged analyser. Clin. Chem. 23: 1054.
Burd, J. W., Carrico, R. J., Tetter, M. C., Buckler, R. T., Johnson, R. D., Boguslaski, R. C., and Christner, J. E. (1977a). Specific protein-binding reactions monitored by enzymatic hydrolysis of ligand-fluorescent dye conjugates. Anal. Biochem. 77: 56.

Burd, J. F., Wong, R. C., Feeney, J. E., Carrico, R. J., and Boguslaski, R. C. (1977b). Homogeneous reactant-labelled immunoassay for therapeutic drugs exemplified by gentamicin determinations in human serum. Clin. Chem. 23: 1402.

Cais, M., Dani, S., Eden, Y., Gandolfi, O., Horn, M., Isaaco, E. E., Josephy, Y., Sarr, Y., Slovin, E., and Snarsky, L. (1977). Metalloimmunoassay. Nature 270: 534.

Calabrese, V. P., Wallen, W., Castellano, G., Ward, L., Anderson, M. G., and DeVries, G. H. (1981). Enzyme-linked immunoadsorbent assay (ELISA) for antibodies to human myelin and Axolemma-enriched fractions. Neurosci. Letts. 21: 189.

Cambiaso, C. L., Leek, A. E., de Steenwinkel, F., Billen J., and Masson, P. L. (1977). Particle counting immunoassay (PACIA) 1. A general method for the determination of antibodies, antigens and haptens. J. Immunol. Methods 18: 33.

Carlier, Y., Colle, A., Tachon, P., Bout, D., and Capron, A. (1981). Quantification of Beta 2-Microglobulin by inhibition enzyme immunoassay. J. Immunol. Methods 40: 231.

Carrico, R. J., Christner, J. E., Boguslaski, R. C., and Yeung, K. K. (1976). A method for monitoring specific binding reactions with cofactor labelled ligands. Anal. Biochem. 72: 271.

Cevenini, R., Donati, M., Rumpianesi, F. (1981). Elementary bodies as single antigen in a micro-ELISA test for chlamydia trachomatis antibodies. Microbiologica 4: 347.

Clem, T. R., and Yolken, R. H. (1978). Practical colorimeter for direct measurement of microplates in EIA systems. J. Clin. Microbiol. 7: 55.

Comoglio, S., and Celada, F. (1976). An immunoenzymatic assay of cortisol using E. Coli β-galactosidase as label. J. Immunol. Methods 10: 161.

Converse, R. H. (1978). Detection of tomato ringspot virus in red raspberry by ELISA. Plant Dis. Rep. 26: 189.

Craven, G. R., Steers, E., and Anfinsen, C. B. (1965). Purification, composition and molecular weight of the β-galactosidase of E.Coli K12. J. Biol. Chem. 240: 2468.

Crowther, J. R., and Abu Elzein, E. M. E. (1979). Application of the enzyme-linked immunosorbent assay to the detection and identification of foot-and-mouth disease virus. J. Hyg. (Camb.) 83: 513.

Deedler, A. M., and Kornelis, D. (1980). A comparison of the IFA and the ELISA for the demonstration of antibodies against Schistosome gut-associated polysaccharide antigens in Schistosomiasis. Z. Parasitenkd. 64: 65.

Douillard, J. Y., Hoffman, T., and Herberman, R. B. (1980). Enzyme-linked immunosorbent assay for screening monoclonal antibody production: Use of intact cells as antigens. J. Immunol. Methods 29: 309.

Dubucq, M., Hendrick, J. C., Osterrieth, P. M., Francois, C., and Franchimont, P. (1981). Enzymoimmunoassay of the main core protein (p28) of mouse mammary tumour virus (MMTV). Eur. J. Cancer 17: 81.

Engvall, E., and Perlmann, P. (1971). Enzyme-linked immunoadsorbent assay (ELISA). Quantitative assay of immunoglobulin G. Immunochemistry 8: 871.

Engvall, E., and Perlmann, P. (1972). Quantitation of specific antibodies by enzyme-labelled anti-immunoglobulins in antigen-coated tubes. J. Immunol. 109: 129.

Erlanger, B. F. (1973). Principles and methods for the preparation of drug protein conjugates for immunological studies. Pharmacol. Rev. 25: 271.

Exley, D., and Abuknesha, R. (1978). A highly sensitive and specific enzyme-immunoassay method for oestradiol-17β. FEBS Letts. 91: 162.

Field, A. K., Davies, M. E., and Tytell, A. A. (1980). Determination of antibodies to double-stranded RNA by enzyme-linked immunoadsorbent assay (ELISA). Proc. Soc. Exp. Biol. Med. 164: 524.

Finley, P. R., Williams, R. J., and Lichti, D. A. (1980). Evaluation of a new homogeneous enzyme inhibitor immunoassay of serum thyroxine with use of a bichromatic analyzer. Clin. Chem. 26: 1723.

Gibbons, I., Skold, C., Rowley, G. L., and Ullman, E. F. (1980). Homogeneous enzyme immunoassay for proteins employing β-galactosidase. Anal. Biochem. 102: 167.

Gibbons, I., Hanlon, T. M., Skold, C. N., Russell, M. E., and Ullman, E. F. (1981). Enzyme-enhancement immunoassay: A homogeneous assay for polyvalent ligands and antibodies. Clin. Chem. 27: 1602.

Gnemmi, E., O'Sullivan, M. J., Chieregatti, G., Simmons, M., Simmonds, A., Bridges, J. W., and Marks, V. (1978). A sensitive immunoenzymometric assay (IEMR) to quantitate hormones and drugs. In Enzyme labelled Immunoassay of Hormones and Drugs, S. B. Pal (ed.) Walter de Gruyter, Berlin, p. 29.

Grange, J. M., Gibson, J., and Nassau, E. (1980). Enzyme-linked immunoadsorbent assay (ELISA): A study of antibodies to Myobacterium tuberculosis in the IgG, IgA and IgM classes in tuberculosis, sarcoidosis and Crohn's disease. Tubercle 61: 145.

Guesdon, J., Thierry, R., and Avrameas, S. (1978). Magnetic EIA for measuring human IgE. J. Allergy Clin. Immunol. 61: 23.

Haimovich, J., Hurwitz, E., Novik, N., and Sela, M. (1970). Use of protein-bacteriophage conjugates for detection and quantitation of proteins. Biochim. Biophys. Acta 207: 125.

Hamaguchi, Y., Kato, K., Fukui, H., Shirakawa, I., Okawa, S., Ishikawa, E., and Katunuma, N. (1976). Enzyme-linked sandwich immunoassay of macromolecular antigens using rabbit antibody-coupled glass rod as a solid phase. Eur. J. Biochem. 71: 459.

Hamdy, F. M., Colgrove, G. S., de Rodriquez, E. M., Snyder, M. L., and Stewart, W. C. (1981). Field evaluation of enzyme-linked immunoadsorbent assay for detection of antibodies to African swine fever virus. Am. J. Vet. Res. 42: 1441.

Haugen, A., Groopman, J. D., Hsu, I. C., Goodrich, G. R., Wogan, G. N., and Harris, C. C. (1981). Monoclonal antibody to aflatoxin β 1-modified DNA detected by enzyme immunoassay. Proc. Natl. Acad. Sci. USA 78: 4124.

Herrman, J. E. (1981). Quantitation of antibodies immobilized on plastic. In Methods in Enzymology, Vol., 73, Immunochemical Technique Part B, J. J. Langone and H. V. Vunakis (eds.), Academic Press, New York, p. 239.

Hosoda, H., Yoshida, H., Sakai, Y., Miyairi, S., and Nambara, T. (1980). Sensitivity and specificity in enzyme immunoassay of testosterone. Chem. Pharm. Bull. Tokyo 28: 3035.

Ishikawa, E. (1973). Enzyme immunoassay of insulin by fluorimetry of the insulin-glucoamylase complex. J. Biochem. (Tokyo) 73: 1319.

Jennings, R., Smith, T., and Potter, C. W. (1981). Use of the enzyme-linked immunoadsorbent assay (ELISA) for the estimation of serum antibodies in an influenza virus vaccine study. Med. Microbiol. Immunol. (Berlin) 169: 247.

Joshi, U., Raghaven, V., Zemse, G., Sheth, A., Borkar, P. S., and Ramachandran, S. (1978). Use of "penicillinase" in the estimation of human chorionic gonadotrophin and human placental lactogen by enzyme-linked immunoassay. In Enzyme Labelled Immunoassay of Hormones and Drugs, S. B. Pal (ed.), Walter de Gruyter, Berlin, p. 233.

Joyce, B. G., Read, G. F., and Fahmy, D. R. (1977). A specific enzyme immunoassay for progesterone in human plasma. Steroids 29: 761.

Kato, K., Hamaguchi, Y., Fukui, H., and Ishikawa, E. (1975). Coupling Fab' fragment of rabbit anti-human IgG antibody to β-D-galactosidase and a highly sensitive immunoassay of human IgG. FEBS Letts. 56: 370.

Kato, K., Suzuki, F., and Semba, R. (1981). Determination of brain enolase isozymes with an enzyme immunoassay at the level of single neurons. J. Neurochem. 37: 998.

Kennedy, J. H., Kricka, L. J., and Wilding, P. (1976). Protein-protein coupling and the applications of protein conjugates. Clin. Chim. 70: 1.

King. T. P., and Kochoumian, L. (1979). A comparison of different enzyme-antibody conjugates for enzyme-linked immunosorbent assay. J. Immunol. Methods 28: 201.

Kitagawa, T., and Aikawa, T. (1976). Enzyme coupled immunoassay of insulin using a novel coupling reagent. J. Biochem. (Tokyo) 79: 233.

Kleine, T. O. (1978). Lower the cost of EMIT for carbamazepine by its adaptation to a mechanized microtitre system. Clin. Chim. Acta 82: 193.

Korhonen, M. K., Juntunen, K. O., and Stenman, U. H. (1980). Enzyme immunoassay of estriol in pregnancy urine. Clin. Chem. 26: 1829.

Koskela, M., and Leinonen, M. (1981). Comparison of ELISA and RIA for measurement of pneumococcal antibodies before and after vaccination with 14-valent pneumococcal capsular polysaccharide vaccine. J. Clin. Pathol. 34: 93.

Kricka, L. J., Carter, T. J. N., Burt, S. M., Kennedy, J. H., Holder, R. L., Holliday, M. I., Telford, M. E., and Wisdom, G. B. (1980). Variability in the adsorption properties of microtitre plates used as solid supports in enzyme immunoassay. Clin. Chem. 26: 741.

Lauer, R. C., and Erlanger, B. F. (1974). An enzyme-immunoassay of antibody specific for adenosine using β-galactosidase. Immunochemistry 11: 533.

Leute, R., Ullman, E. F., and Goldstein, A. (1972). Spin immunoassay of opiates and narcotics in urine and saliva. JAMA 221: 1231.

Lindenschmidt, E. G. (1981). Experiences with the ELISA at the serodiagnostics of infections with viruses of the Herpesgroup. Immun. Infekt. 9: 140.

Maiolini, B., Ferrua, B., and Masseyeff, R. (1975). Enzyme immunoassay of human alphafoetoprotein. J. Immunol. Methods 6: 355.

Maiolini, R., Ferrua, B., Quaranta, J. F., Pinoteau, A., Euller, L., Ziegler, G. and Masseyeff, R. (1978). A sandwich method for EIA: Quantification of rheumatoid factor. J. Immunol. Methods 20: 25.

Maiolini, R., Bagrel, A., Chavance, C., Krebs, B., Herbeth, B., and Massayeff, R. (1980). Study of an enzyme immunoassay kit for carcinoembryonic antigen. Clin. Chem. 26: 1718.

Marks, V., O'Sullivan, M. J., Al-Bassam, M. N. and Bridges, J. W. (1978). A double antibody enzyme-immunoassay for methotrexate. In Enzyme Labelled Immunoassay for Hormones and Drugs, S. B. Pal (ed.), Walter de Gruyter, Berlin, p. 419.

Marks, V., Mould, G. P., O'Sullivan, M. J., and J. D. Teale. (1980). Monitoring of drug disposition by immunoassay. In Progress in Drug Metabolism, Vol. 5, J. W. Bridges and L. F. Chasseaud (eds.), John Wiley New York, p. 255.

Mattiasson, B., and Nilsonn, H. (1977). An enzyme immunoelectrode assay of human serum albumin and insulin. FEBS Letts. 78: 251.

Mattiasson, B., Svensson, K., Borrebaeck, C., Jonsson, S., and Knonvall, G. (1978). Non-equilibrium enzyme immunoassay of gentamicin. Clin. Chem. 24: 1770.

Miyai, K., Ishibashi, K., and Kawashima, M. (1980). Enzyme immunoassay of thyroxine in serum and dried blood samples on filter paper. Endocrinol. Jpn. 27: 375.

Miyai, K., Ishibashi, K., and Kawashima, M. (1981). Two-site immunoenzymometric assay for thyrotropin in dried blood samples on filter paper. Clin. Chem. 27: 1421.

Muller, J., and Pfleiderer, G. (1979). A new method of conjugating proteins for enzyme immunoassay. J. Appl. Biochem. 1: 301.

Nakane, P. K., and Kawaoi, A. (1974). Peroxidase-labelled antibody. A new method of conjugation. J. Histochem. Cytochem. 22: 1084.

Nakane, P. K., Sir Ram J., and Pierce, G. B. (1966). Enzyme-labelled antibodies for light and electron microscope localization of antigens. J. Histochem. Cytochem. 14: 789.

Ngo, T. T., and Lenhoff, H. M. (1980). Enzyme modulators as tools for the development of homogeneous enzyme immunoassays. FEBS Letts. 116: 285.

Ngo, T. T., and Lenhoff, H. M. (1981). Recent advances in homogeneous and separation-free enzyme immunoassays. Appl. Biochem. Biotechnol. 6: 53.

Ngowi, F. (1979). Development of an assay for 6-methyl prednisolone. M.Sc. Thesis, University of Surrey.

Nicolet, J., and Paroz, P. (1980). Tween 20 soluble proteins of Mycoplasma hypopneumoniae as antigen for an enzyme linked immunoadsorbent assay. Res. Vet. Sci. 29: 305.

Notermais, S., Dufrenne, J., and Schothorst, M. (1978). Enzyme-linked immunoadsorbent assay for the detection of Clostridium botulinum toxin type A. Jpn. J. Med. Sci. Biol. 31: 81.

Oellerich, M. (1980). Enzyme immunoassays in clinical chemistry: Present status and trends. J. Clin. Chem. Clin. Biochem. 18: 197.

O'Sullivan, M. J., and Marks, V. (1981). Methods for the preparation of enzyme-antibody conjugates for use in enzyme immunoassay. In Methods in Enzymology, vol. 73, Immunochemical Techniques, Part B, J. J. Langone and H. V. Vunakis (eds.), Academic Press, New York, p. 147.

O'Sullivan, M. J., Gnemmi, E., Morris, D., Al-Bassam, M. N., Simmons, M., Bridges, J. W., and Marks, V. (1978a). An enzyme-immunoassay for triiodothyronine. In Enzyme Labelled Immunoassay of Hormones and Drugs, S. B. Pal (ed.), Walter de Gruyter, Berlin, p. 301.

O'Sullivan, M. J., Gnemmi, E., Morris, D., Chieregatti, G., Simmons, M., Simmonds, A. D., Bridges, J. W., and Marks, V. (1978b). A simple method for the preparation of enzyme-antibody conjugates. FEBS Letts. 95: 310.

O'Sullivan, M. J., Bridges, J. W., and Marks, V. (1979a). Enzyme immunoassay: A Review. Annal. Clin. Biochem. 16: 221.

O'Sullivan, M. J., Gnemmi, E., Morris, D., Chieregatti, G., Simmonds, A. D., Simmons, M., Bridges, J. W., and Marks, V. (1979b). Comparison of two methods of preparing enzyme-antibody conjugates. Application of these conjugates for enzyme-immunoassay. Anal. Biochem. 100: 100.

O'Sullivan, M. J., Gnemmi, E., Chieregatti, G., Morris, D., Simmonds, A. D., Simmons, M., Bridges, J. W., and Marks, V. (1979c). The influence of antigenic properties on the conditions required to elute antibodies from immunoadsorbents. J. Immunol. Methods 30: 127.

O'Sullivan, M. J., Gnemmi, E., Simmonds, A. D., Chieregatti, G., Heyderman, E., Bridges, J. W. and Marks, V. (1979d). A comparison of the ability of β-galactosidase and horse-radish peroxidase enzyme-antibody conjugate to detect specific antibodies. J. Immunol. Methods 31: 247.

Page, M., Theriault, L., and Nilsoon, M. (1980). Solid phase ELISA for serum ferritin. Scand. J. Clin. Lab. Invest. 40: 641.

Parsons, G. H. (1981). Antibody-coated plastic tubes in radioimmunoassay. In Methods in Enzymology, vol. 73, Immunochemical Techniques, Part B, J. J. Langone and H. V. Vunakis (eds.), Academic Press, New York, p. 224.

Pestka, J. J., Gaur, P. K., and Chu, F. S. (1980). Quantitation of Aflatoxin B1 and Aflatoxin B1 antibodies by an enzyme-linked immunosorbent microassay. Appl. Environ. Microbiol. 40: 1027.

Polin, R. A., and Kennett, R. (1980). Use of monoclonal antibodies in an enzyme-linked inhibition assay for rapid detection of streptococcal antigen. J. Pediatr. 97: 540.

Rassam, M. B., and Al-Mudhaffer, S. A. (1980). The micro-ELISA sandwich technique for the quantitation of Leishmania Donovani soluble antigen. Ann. Trop. Med. Parasitol. 74: 591.

Riad-Fahmy, D., Read, G. F., Joyce, B. G., and Walker, R. F. (1981). Steroid immunoassays in endocrinology. In Immunoassays for the 80s, A. Voller, A. Bartlett, and D. Bidwell (eds.), MIT Press Lancaster, p. 205.

Rook, G. A. W., and Cameron, C. H. (1981). An inexpensive, portable, battery-operated photometer for the reading of ELISA tests in microtitration plates. J. Immunol. Methods 40: 109.

Rowley, G. L., Rubenstein, K. E., Huisjen, J., and Ullman, E. F. (1975). Mechanisms by which antibodies inhibit hapten-malate dehydrogenase conjugates. J. Biol. Chem. 250: 3759.

Rubenstein, K. E., Schneider, R. S., and Ullman, E. F. (1972). Homogeneous enzyme immunoassay. A new immunological technique. Biochem. Biophys. Res. Commun. 47: 846.

Sarkkinen, H. K., Tuokko, H., and Halonen, P. E. (1980). Comparison of enzyme-immunoassay and radioimmunoassay for detection of human rotaviruses and adenoviruses from stool specimens. J. Virol. Methods 1: 331.

Saunders, G. C., Campbell, S., Sanders, W. M., and Martinez, A. (1979). Automation and semiautomation of enzyme immunoassay instrumentation. In Immunoassays in the Clinical Laboratory, R. M. Nakamura, W. R. Dito and E. S. Tucker (eds.), Alan R. Liss, New York, p. 119.

Schall, R. F., Fraser, A. S., Hansen, H. W., Kern, C. W., and Tenso, H. J. (1978). A sensitive manual EIA for thyroxine. Clin. Chem. 24: 1801.

Scharpe, S. L., Cooreman, W. M., Blomme, W. J., and Laekeman, G. M. (1976). Quantitative enzyme immunoassay. Clin. Chem. 22: 733.

Schroeder, H. R., Vogelhut, P. O., Carrico, R. J., Boguslaski, R. C., and Buckler, R. T. (1976). Competitive protein binding assay for biotin monitored by chemiluminescence. Anal. Chem. 48: 1933-1937.

Schuurs, A. H. W. M., and Van Weemen, B. K. (1977). Enzyme-immunoassay. Clin. Chim. Acta 81: 1-7.

Spencer, H. C., Collins, W. E., Warren, M., Jeffery, G. M., Mason, J., Huong, A. Y., Stanfill, P. S., and Skinner, J. C. (1981). The enzyme-linked immunoadsorbent assay (ELISA) for malaria. III. Antidoby response in documented plasmodium falciparum infections. Am. J. Trop. Med. Hyg. 30: 747.

Suter, L., Bruggen, J., and Sorg, C. (1980). Use of enzyme-linked immunoadsorbent assay for screening of hybridoma antibodies against cell surface antigens. J. Immunol. Methods 39: 407.

Taylor, S. M., Kenny, J., Mallon, T., and Davidson, W. B. (1980). The micro-ELISA for antibodies to Trichinella spiralis: Elimination of false positive reactions by antigen fractionation and technical improvements. Z. Veterinarmed B. 27: 764.

Tolo, K., Schenck, K., and Brandtzaeg, P. (1981). Enzyme-linked immunoadsorbent assay for human IgG, IgA, and IgM antibodies to antigens from anaerobic cultures of seven oral bacteria. J. Immunol. Methods 45: 27.

Tyhach, R. J., Rupchock, P. A., Pendergrass, J. H., Skjold, A. C., Smith, P. J., Johnson, R. D., Albarella, J. P., and Profitt, J. A. (1981). Adaptation of prosthetic-group-label homogeneous immunoassay to reagent-strip format. Clin. Chem. 27: 1499.

Ullman, E. F., Blakemore, J., Leute, R. K., Eimstad, W., and Jaklitsch, A. (1975). Homogeneous enzyme immunoassay for thyroxine. Clin. Chem. 21: 1011.

Ullman, E. F., Schwarzberg, M., and Rubenstein, K. E. (1976). Fluorescent transfer immunoassay. A general method for the determination of antigens. J. Biol. Chem. 251: 4172.

Van der Waart, M., and Schuurs, A. H. W. M. (1976). Towards the development of a radioenzyme-immunoassay (REIA). Z. Anal. Chem. 279: 142.

Van Weemen, B. K., and Schuurs, A. (1971). Immunoassay using antigen-enzyme conjugates. FEBS Lett. 15: 232.

Van Weemen, B. K., and Schuurs, A. H. W. M. (1975). The influence of heterologous combinations of antiserum and enzyme-labelled estrogen on the characteristics of estrogen EIAs. Immunochemistry 12: 667.

Velan, B., and Halmann, M. (1978). Chemiluminescence immunoassay, a new sensitive method for determination of antigens. Immunochemistry 15: 331.

Vejtorp, M., and Leerhoy, J. (1980). Comparison of the sensitivity of ELISA and the haemagglutination-inhibition test for routine diagnosis of rubella. Acta Pathol. Microbiol. Scand. B. 88: 349.

Voller, A., Bidwell, D. E., Huldt, G., and Engvall, E. (1974). A microplate method of enzyme linked immunosorbent assay and its application to malaria. Bull. WHO 51: 209.

Voller, A., Bartlett, A., and Bidwell, D. E. (1978). Enzyme immunoassay with special reference to ELISA technique. J. Clin. Pathol. 31: 507.

Vuento, M., Salonen, E., Pasanen, M., Stenman, U-H. (1981). Competitive enzyme immunoassay for human plasma fibronectin. J. Immunol. Methods 40: 101.

Wei, R., and Reibe, S. (1977). Preparation of a phospholipase C-antihuman IgG conjugate and inhibition of its enzyme activity by human IgG. Clin. Chem. 23: 1386.

Wisdom, G. B. (1976). Enzyme immunoassay. Clin. Chem. 22: 1243.

Wong, R. C., Burd, J. F., Carrico, R. J., Buckler, R. T., Thoma, J., and Boguslaski, R. C. (1979). Substrate-labeled fluorescent immunoassay for phenytoin in human serum. Clin. Chem. 25: 686.

Yalow, R. S., and Berson, S. A. (1959). Assay of plasma insulin in human subjects by immunological methods. Nature 184: 1684.

Yorde, D. E., Sasse, E. A., Wang, T. Y., Hussa, R. O., and Garancis, J. C. (1976). Competitive enzyme linked immunoassay with use of soluble enzyme/antibody immune complexes for labelling. 1. Measurement of human choriogonadotropin. Clin. Chem. 22: 1372.

Ziegelmaier, R., Behrens, F., and Enders, G. (1981). Class-specific determination of antibodies against cytomegalo (CMV) and rubella virus by ELISA. J. Biol. Stand. 9: 23.

4
High Sensitivity, Pulsed-Light Time-Resolved Fluoroimmunoassay

SALIFU DAKUBU, ROGER EKINS, THOMAS JACKSON,
AND NICHOLAS J. MARSHALL
Middlesex Hospital Medical School, London, England
London England

I. INTRODUCTION

Immunological reactions have been applied to the qualitative and semiquantitative assay of antigens and antibodies for many years. Radioimmunoassay (RIA) and related "saturation assay" procedures introduced more than two decades ago (Ekins, 1960; Yalow and Berson, 1960) provided means for the specific, quantitative, and highly sensitive measurement of hormones in body fluids and were relatively easily and economically adapted to the assay of large numbers of samples. They provided a major improvement on the specificity and sensitivity of the hormone assays then available. Later, radioisotopically labeled antibody or "immunoradiometric" (IRMA) techniques were developed (Wide et al., 1967; Miles and Hales, 1968a), primarily in an attempt to improve upon the sensitivity of immunoassays based upon the "saturation assay" principle. However, the theoretical justification for the claim that methods of this kind would prove more sensitive than conventional RIAs was to remain debatable for some years (Woodhead et al., 1974). Meanwhile, although the extreme sensitivity of both RIA and IRMA techniques clearly constituted one of the primary reasons for their widespread adoption, some of the other advantages of these methods (e.g., their specificity, economy, convenience, etc.) led to their subsequent application to measurements of many substances where their high potential sensitivity is neither needed nor fully exploited.

Recent efforts have been increasingly directed towards replacement of the radioactive markers used in these techniques by nonradioisotopic labels. Purely commercial incentives have often prompted manufacturers to move in this direction. In addition various legal, social, and logistic objections have been raised to the use of radioactive labels in an immunoassay context, of which the most commonly invoked are the potential health and environmental hazards associated with the use of radioactivity and the limited shelf-life of the labeled reagents themselves. Another pursuasive argument for the adoption of nonradioisotopic labels has centered on the opportunity they provide for the development of "nonseparation" assay techniques which obviate the problems (particularly irksome in the context of assay automation) arising in the separation of the radiolabeled products of the reactions relied on in conventional RIA. There nevertheless exist fundamental advantages in a reliance on radioactive nuclides as reagent labels. These have served to sustain their continued use in the immunoassay field for two decades and therefore cannot be lightly disregarded. For example, gamma-emitting isotopes are totally unaffected by chemical and physical variations in the microenvironment; also sample measuring equipment is inexpensive, reliable and now widely available. The most commonly used isotope, ^{125}I, may easily and conveniently be coupled to a protein or peptide without significant change of the latter's molecular and immunological properties, and the label can be detected with extreme sensitivity since its specific activity is high and the background radiation levels against which it is measured are normally extremely low.

In summary, although the development of an alternative, nonradioisotopic labeling technique relying, for example, on a fluorescent, electron spin, or enzyme label to quantify the immunological reactions involved in immunoassay procedures is often, in principle at least, relatively straightforward, the sensitivity with which such labels can be detected, their "ruggedness," and their ease of use seldom match those of the commonly used radioisotopes. For these reasons, the "alternative" nonradioisotopic immunoassay methods have been widely applied only to those analytes (such as drugs and certain hormones) which exist in biological fluids at relatively high concentrations or in situations where qualitative rather than quantitative results are sufficient. Thus, although the nonisotopic immunoassays may occasionally offer very real practical and logistic benefits, their relative lack of sensitivity has meant that they have not provided a serious challenge to the radioisotope techniques in those areas requiring high assay sensitivity and in which the latter made their original impact. In this chapter we shall therefore address ourselves primarily to the quest for nonradioisotopic assays of a sensitivity as great as, or greater than, that achievable with the radioisotopically based methods.

A. Methodological Constraints on Immunoassay Sensitivity

It is particularly important, in the context of the present discussion, to identify the factors that critically limit assay sensitivity and to attempt to distinguish those that are a fundamental characteristic of an individual methodology per se from those specifically related to the particular form of label employed. In any informed discussion of the relative merits of alternative labels it is, in our view, obligatory to understand the basic principles underlying the various immunoassay methodologies to which they are applied, and the differing forms of assay design which stem from the

Fluoroimmunoassay 73

application, in practice, of these underlying analytical principles.

Despite recent efforts to rationalize the nomenclature which has proliferated to describe the wide spectrum of immunoassay and closely related assay techniques developed in the past 20 years, it remains unsatisfactory. Nevertheless it is now generally recognized that all such methods reflect one of two different analytical approaches:

1. Techniques variously described as <u>competitive</u>, <u>inhibition</u>, <u>saturation</u>, or <u>limited reagent</u> assays, which generally rely on observation of the distribution of the (labeled) antigen following its reaction with an antibody (see Fig. 1) (notably "conventional" RIA (Fig. 2))

2. Techniques, commonly termed <u>immunometric</u> assays (see Chapter 7), which rely on the observation of the distribution of (labeled) antibody following its reaction with the analyte (Fig. 3) (e.g., IRMA, Fig. 4A)

It should be emphasized that immunoassay methodologies may be further differentiated on the basis of a number of other special features. For example, immunoassays may be categorized as <u>equilibrium</u> or <u>disequilibrium</u> (or as <u>conjoint</u> or <u>disjoint</u>), reflecting the order of addition to incubation mixtures of the individual reagents used. They may also be distinguished on the basis of the particular fraction of the labeled reagent

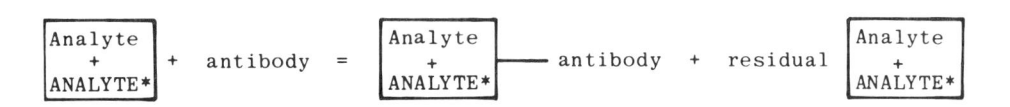

| Analyte + ANALYTE* | + antibody = | Analyte + ANALYTE* | —— antibody + residual | Analyte + ANALYTE* |

Response variable : Final distribution of labeled |ANALYTE*| between

, complex and residuum

e.g. Radioimmunoassay (RIA)

FIGURE 1 Basic principle of "labeled analyte" techniques. The response variable in assays relying on this approach essentially includes any convenient representation of the final distribution of the <u>labeled analyte tracer</u> present in the system (e.g., "% of labeled analyte bound"; "free/bound"; "bound/free" etc.) following its reaction with the specific reagent employed (e.g., antibody). Note that the labeled analyte tracer is not required to be chemically identical to the unlabeled material, but must react with identical reagent (binding) sites with comparable affinity. Such techniques are variously designated as <u>saturation assay</u>, <u>competitive protein binding assay</u>, <u>limited</u> reagent assay, etc. since the optimal concentration of antibody required to maximize sensitivity tends towards zero.

1

Ab Analyte

● labeled analyte

2 **Separate 'free' and 'bound'**

'Bound'   'Free'

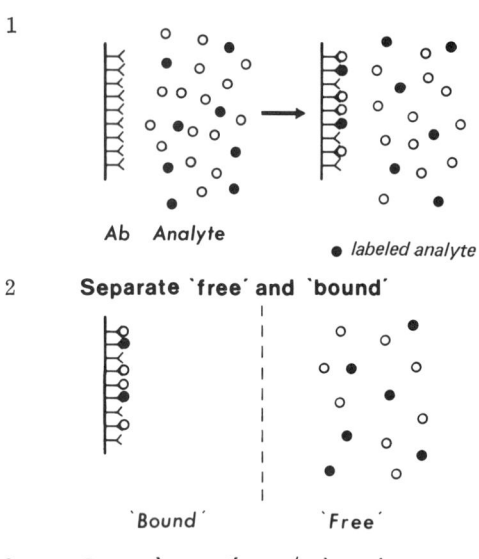

3 **Count 'bound' and/or 'free'**

RIA / Saturation analysis ·

$[Ab]_{Opt} \rightarrow 0$

FIGURE 2 Typical radioimmunoassay protocol. This technique, which exemplifies the "labeled analyte" approach portrayed in Fig. 1, relies on antibody as the specific reagent and radiolabeled analyte tracer.

(i.e., bound or free) which is ultimately observed (though this extremely important distinction has generally attracted less emphasis than it warrants, e.g., Figs. 4A and 5). These and other elements in the particular assay "strategy" adopted by the analyst may be variously combined to yield a wide variety of overall assay designs, each displaying particular sensitivity and specificity characteristics. Moreover, each such design requires individual analysis if its maximal sensitivity potential is to be realized and the optimal combination of reagents which should be employed distinguished.

It is, however, clearly inappropriate to attempt, in this chapter, a detailed appraisal of all the possible variations in assay design currently employed, albeit (as suggested above) an understanding of the basic attributes of the most commonly adopted immunoassay strategies is a necessary prerequisite to any meaningful discussion of the relative merits of various labels. We shall therefore confine ourselves to a consideration of the two fundamental approaches represented in Figs. 1 and 3 as exemplified in Figs. 2 and 4B.

Fluoroimmunoassay

```
Analyte +  [antibody + ANTIBODY*]  =  Analyte — [antibody + ANTIBODY*]  +  residual [antibody + ANTIBODY*]
```

Response variable : Final distribution of labeled ANTIBODY* between

complex and residuum

e.g. Immunoradiometric assay (IRMA)

FIGURE 3 Basic principle of "labeled reagent" techniques. The response variable in this form of assay includes a convenient representation of the final distribution of the labeled reagent following its reaction with the analyte (e.g., % [of labeled reagent] bound; "amount bound", "% free etc."). Note that the design of assays falling into this category is especially dependent on the particular way in which the final distribution of the labeled reagent is observed, that is, which fraction of the labeled reagent is ultimately measured. Assays in which the analyte-bound fraction is estimated require an optimal concentration of antibody tending towards infinity (see Fig. 4); conversely when the residual "free" or "unbound" fraction is estimated, the optimal antibody concentration tends to zero and the system may be classified as "competitive." (See Fig. 5.)

Theoretical analysis of the equilibrium "RIA-like" assays (Fig. 2), and the two-site "IRMA-like" approach (Fig. 4B) shows important differences in the optimal reagent concentrations which yield maximal assay sensitivity in the two cases, and, as a direct consequence, differing constraints on the sensitivity achievable in each of these forms of assay. In particular, if we restrict our consideration to those assay "strategies" in which the bound fraction of the labeled reagent is observed, and the use of an "ideal" labeled reagent is assumed (i.e., one possessing an "infinite" specific activity characterized by zero "nonspecific binding"), entirely opposing conclusions emerge regarding the optimal antibody concentration required in the two forms of assay. In the "RIA-like" assays, maximal sensitivity is attained as the concentration of antibody approaches zero. Assays falling into this category may therefore be termed limited reagent or competitive assays. In contrast, in a labeled antibody or "IRMA-like" method relying on final measurement of the bound antibody, maximal sensitivity is achieved as the concentration of antibody approaches "infinity." Assays of this class may be termed excess reagent or noncompetitive assays (Ekins, 1976). This fundamental distinction in assay design is the key factor which differentiates the overall performance characteristics of the two forms of technique. For example, as a concomitant of the differences in optimal concentrations of antibody, it is demonstrable that the incubation time required to reach maximal sensitivity becomes extended towards infinity in "competitive" assays and converges towards zero in those assays of the "noncompetitive, immuno-

1

Analyte
☐⊣ labeled antibody

2 Add immunoadsorbent

Supernatant

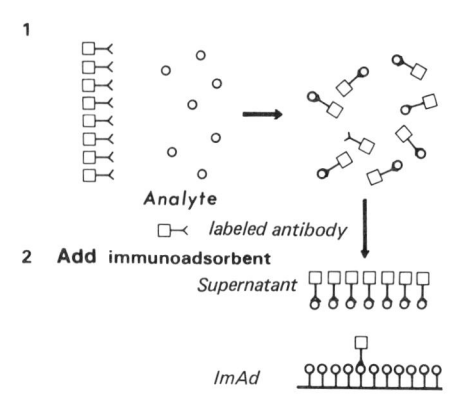

ImAd

3 Separate

4 Count Ab-analyte complex

IRMA

$[Ab]_{Opt} \to \infty$

FIGURE 4A Original immunoradiometric assay (IRMA) as described by Miles and Hales (1968). The radioactivity residing in the analyte-antibody complex (i.e., the activity in the supernatant) is estimated.

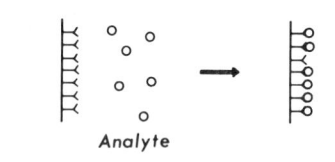

1

Analyte

2 Add labeled Ab

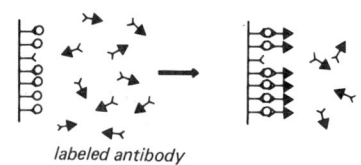

labeled antibody

3 Separate

4 Count adsorbed Ab

Two-site IRMA

$[Ab]_{Opt} \to \infty$

FIGURE 4B Typical "two-site" or "sandwich" immunoradiometric assay (IRMA) protocol. Assuming the labeled antibody finally bound to analyte constitutes the measured assay response, this assay design (like that shown in (A)) requires the use of a high concentration of labeled antibody (tending to infinity) and is "noncompetitive."

Fluoroimmunoassay

1 [diagram: labeled antibody + ImAd + Analyte → complex]

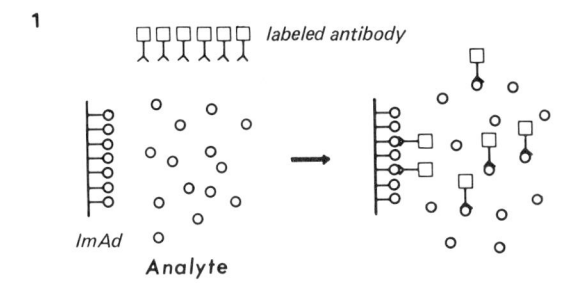

2 **Separate**

3 **Count ImAd**

Conjoint IRMA / **Saturation analysis**

$$[Ab]_{Opt} \to 0$$

FIGURE 5 "Conjoint" IRMA technique in which residual "free" labeled antibody is sequestered onto an immunosorbent and <u>the radioactivity residing in this fraction is estimated</u>. Such an approach requires the use of a <u>low</u> concentration of labeled antibody (tending towards zero) and may be termed "competitive" (or labeled antibody "saturation analysis"). The theoretical sensitivity of such a system is identical to that of a conventional RIA.

metric" design as defined above. The specificity characteristics of the antibody used with respect to cross-reacting antigens are also profoundly dependent on the relative concentrations of antibody and analyte in the assay system. It is readily demonstrable, for example, that the "excess reagent" labeled antibody techniques are intrinsically less specific than the "competitive" labeled analyte methods (although this particular disadvantage may be overcome by the use of the "sandwich" or "two-site" assay designs exemplified in Fig. 4B).

The contrast in the optimal amounts of antibody employed is of particular significance in relation to the sensitivities potentially obtainable in each form of assay. Although the question of the relative sensitivities of "limited reagent" and "excess reagent" methods can only satisfactorily be addressed in strict mathematical terms (see, for example, Ekins et al.,

1968; Rodbard and Weiss, 1973; Schuurman and de Ligny, 1979; Jackson et al., 1983), some perception of the sensitivity characteristics of these contrasting analytical approaches can, it is hoped, be conveyed in this presentation without recourse to detailed mathematical analysis.

Clearly, in any immunoassay procedure the avidity with which the antibody binds to analyte (i.e., the equilibrium constant of the reaction between them), the specific activity of the label, the extent of misclassification of free labeled reagent (whether antigen or antibody) as bound (and vice versa), and the precision with which the individual reagent manipulations required in the assay are performed all constitute important factors influencing the overall sensitivity of the method, regardless of which particular assay design is adopted. However the relative importance of these factors differs significantly as a result of the contrasting amounts of antibody which are optimal in each case. In a typical "RIA-like" assay procedure, the convergence of the optimal antibody concentration towards zero implies that the equilibrium constant of the reaction between analyte and antibody is the dominant factor in determining the extent of reaction between the two, and hence in constraining the ultimate sensitivity obtainable in this form of assay design. The care with which the various manipulations required in procedures of this kind are carried out (e.g., pipetting of antibody and of labeled analyte, separation of the products of the reaction, etc.) also constitutes an equally important determinant of assay sensitivity. Thus it is readily demonstrable (Ekins et al., 1968; Ekins and Newman, 1970) that the ultimate sensitivity attainable in "competitive" or "RIA-like" techniques is given by the quotient ϵ/K, where ϵ is the relative experimental error in the measurement of the response variable (i.e., fraction of labeled antigen bound to antibody), and K is the equilibrium constant of the analyte/antibody reaction. Since, in practice, it is virtually impossible to reduce the errors associated with the various manipulations involved in such techniques to a level below approximately 1% (i.e., $\epsilon = 0.01$), and since the equilibrium constants of antigen-antibody reactions seldom, if ever, surpass 10^{12} liters/$\underline{M}$, the maximal sensitivity achievable with "RIA-like" immunoassays is of the order of $10^{-14}\underline{M}$. It must be emphasized that this sensitivity limit <u>cannot be overcome regardless of the nature of the marker used to label the analyte in this type of procedure</u>, nor, indeed, have assays reflecting this analytical approach ever, in practice, displayed higher sensitivities.

In contrast, in a noncompetitive "IRMA-like" assay, the affinity of the antibody is of lesser importance since, in principle at least, the effects of low antibody affinity can be offset by an increase in the concentration of labeled antibody employed in the system. However, any increase in the amount of antibody used will usually be accompanied by a concomitant increase in the absolute amount of "free" antibody misclassified as antigen "bound" (i.e., the "nonspecific binding" of labeled antibody), an effect which inevitably serves to reduce the ability of the assay system to distinguish small amounts of antigen. Thus, assuming a predefined total incubation time, there will exist (as in RIA) an optimal amount of antibody resulting in maximal sensitivity, the value of which will primarily depend on the fractional nonspecific binding of antibody and the equilibrium constant of the antibody employed (Jackson et al., 1983). The upshot of these considerations is that, though a high antibody affinity is clearly advantageous insofar as it leads to a reduction in the optimal concentration of antibody that must be used (hence in the level of nonspecifically bound antibody, implying in turn a corresponding increase in assay sensitivity), the affinity of the antibody is generally of lesser importance in IRMA-like assays than in the "competitive" or "RIA-like" techniques. A further particular and important implication of the use of higher concentrations of antibody in "IRMA-like" assays is that the precision of antibody pipetting is generally of minor consequence. Taken together, these two considerations imply, as is discussed more fully below, that the noncompetitive labeled antibody techniques are potentially able to attain higher sensitivities than are yielded by the competitive or saturation assay methods exemplified by RIA.

It should perhaps be emphasized at this point that the reasons for the superiority of the labeled antibody approach are somewhat different from those originally advanced by Miles and Hales (1968b) in claiming a sensitivity advantage for IRMA techniques with respect to RIA. Rather, they stem entirely, as indicated above, from the differences in optimal antibody concentrations (usually) characterizing the two approaches. For example it is readily demonstrable (Rodbard and Weiss, 1973) that the sensitivity attainable with a labeled antibody technique is identical to that of a labeled analyte technique, such as RIA, if the "free" rather than the bound fraction of labeled antibody is measured (see Fig. 5). Under these conditions, the optimal amount of labeled antibody yielding maximum sensitivity approaches zero as in a conventional RIA. This emphasizes the assertion that the superior potential sensitivity of the different analytical strategies that we have considered stems entirely from concomitant differences in optimal assay design, not, as is commonly supposed, as a direct consequence of the labeling of antigen or antibody per se.

These considerations illustrate the manner in which, although the same set of factors fundamentally govern assay sensitivity in the two basic forms of assay design, the interplay between these factors is profoundly influenced by the divergence towards zero and infinity respectively in the optimal antibody concentrations required to maximize assay sensitivity in each case. Thus, in practice, the equilibrium constant of the antibody/analyte reaction, and the error in the manipulations involved primarily govern assay sensitivity in the limited reagent, "RIA-like" assays. In the excess-reagent, labeled antibody assays, on the other hand, the extent of nonspecific binding of labeled antibody constitutes one of the most important determinants of assay sensitivity since this parameter fundamentally governs the "noise" level against which small signals must be determined and hence the overall assay detection limit. However a second and critically important determinant of assay sensitivity in this form of assay design is the sensitivity of the signal detection system per se (i.e., the ability of the measuring system to detect small amounts of labeled antibody). This inevitably leads us to a consideration of the relative merits of different labels and the constraints these intrinsically impose upon assay sensitivity.

B. Label-Related Constraints on Immunoassay Sensitivity

As indicated above, in addition to the fundamental constraints imposed by the particular assay strategy adopted, the sensitivity of an immunoassay may be affected by factors directly related to the labeling system used. In the preceding discussion of the potential sensitivity of "limited reagent" and "excess reagent" methodologies, the simplifying assumption was implicitly made that the "detectability" of the labeled reagent is infinite and that errors in the measurement of the label itself are therefore not important factors determining assay sensitivity and precision. In practice, of course, this ideal can never be achieved. Clearly the "detectability" of the labeled reagent, defined in essence by the "specific activity" of the label (i.e., the number of observable events/unit time/unit mass) and by the instrumental background and error (i.e., the "noise" in the detection system employed) - also imposes limits on assay sensitivity. For example, the specific activity of ^{125}I, the most commonly used radiolabel, is such that the minimum number of labeled molecules that can be observed with reasonable precision (using the usual type of laboratory scintillation counter and a realistic counting time) is approximately 10^7/ml. However, since the sensitivities of RIA techniques are generally limited to molecular concentration levels of this order of magnitude for the fundamental reasons discussed above (1 ml of a 10^{-14} M solution contains 6×10^6 molecules), the specific activity of an ^{125}I-labeled antigen does not constitute a major practical limitation on assay sensitivity in RIAs relying on antibodies with affinities of $10^{12} M^{-1}$ or less. Thus, by a fortunate technical coincidence, the maximal potential sensitivity of the "RIA-like" techniques is reached (or is virtually reached) using ^{125}I-labeled antigens, and the substitution of other labels, even were they to display greatly increased specific activity, could not and would not, in these circumstances, effect any major improvement in assay sensitivity. In practice, however, most nonradioisotopic labels cannot be measured at such low concentrations as ^{125}I, implying that immunoassays based upon their use are inevitably less sensitive than the corresponding radioisotopically based techniques.

In contrast, the _potential_ sensitivities of excess-reagent-labeled antibody assays, (assuming extremely low levels of "nonspecific binding" of antibody) lie some orders of magnitude below 10^7 molecules/ml. For this reason the "detectability" of ^{125}I labels can impose an important practical constraint upon the sensitivity of an assay relying on ^{125}I-labeled antibody.

This point is illustrated and emphasized in Fig. 6. In this figure, both the "potential" sensitivities and the highest sensitivities that can reasonably be expected using ^{125}I-labeled reagents in both types of assay are shown. Clearly, the use of an "ultra" high specific activity, nonradioisotopic label in a "competitive" immunoassay design can, at best, provide only a marginal improvement in sensitivity in comparison with the corresponding ^{125}I-labeled-analyte technique when realistic analyte-antibody affinities are assumed. In contrast the opportunity for sensitivity enhancement in assays of immunometric design is very great. The use of a label of very high specific activitiy combined with a methodology resulting in extremely low levels of nonspecific binding of labeled antibody thus provides the basis for the development of assays surpassing the sensitivities currently attainable both by RIA and IRMA techniques by several orders of magnitude.

Another point of importance revealed in this figure is that, assuming a low level of nonspecific binding, greater sensitivity can generally be achieved (using an antibody of a defined equilibrium constant) by adopting a labeled antibody design rather than one relying on labeled analyte. However the sensitivity advantage of a noncompetitive IRMA technique with respect to the corresponding RIA based upon the same antibody reduces with increasing affinity of the antibody used, so that, with antibodies displaying very high affinities (e.g., of the order of 10^{12} $\underline{M}^{-1}$ and above), there is relatively little to be gained by the use of IRMA methodology. With antibodies possessing equilibrium constants in the range of 10^9-10^{11} $\underline{M}^{-1}$, IRMA methods can be anticipated to yield sensitivities some 10 to 100-fold greater than the equivalent RIA. This prediction has found broad confirmation in a number of experimental studies (Hunter, 1982). Clearly, considerations of this kind have important implications with regard to the optimal use of monoclonal antibodies, whose affinities generally fall within this range. Indeed, inspection of Fig. 6 reveals that a noncompetitive, radio-labeled monoclonal antibody assay design may be expected to yield an assay sensitivity comparable to that of a conventional RIA based on an antibody some two orders of magnitude greater in affinity.

C. Alternative Labels for "Ultra"-Sensitive
   Immunometric Assays

Several different nonradioisotopic labeling systems which have been proposed for the development of highly sensitive immunometric assays deserve comment. The use of enzymes as labels in such assays has long been considered an attractive alternative since each enzyme molecule can catalyze the reaction of many substrate molecules. However, the detection limit of the substrate or products of an enzyme reaction using spectrophotometry is generally relatively high so that despite enzyme-amplification effects, immunoenzymometric assays employing this approach have not usually

Labeled antigen ("RIA-like")

Labeled antibody ("IRMA-like")

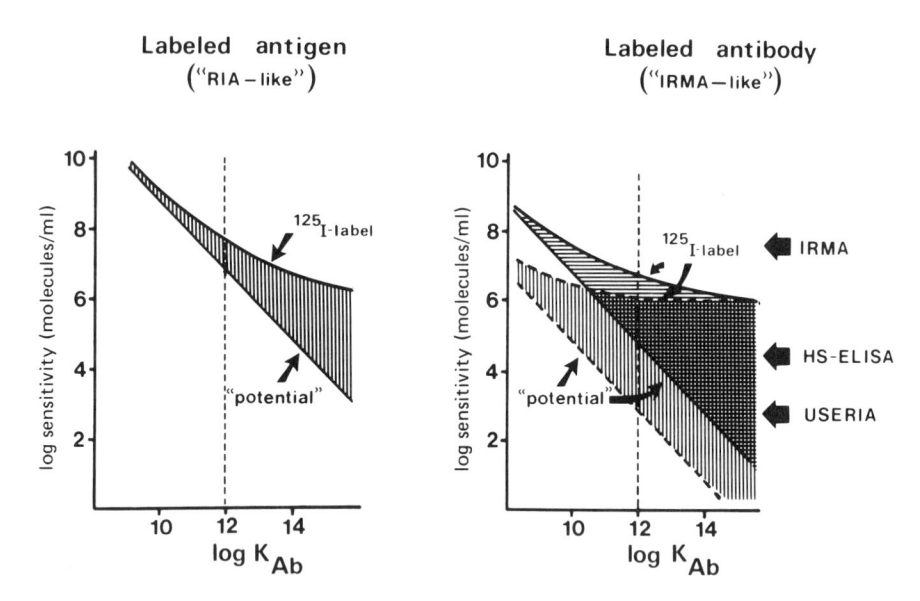

FIGURE 6 Potential and "actual" sensitivities attainable in competitive ("RIA-like") and noncompetitive, labeled antibody ("IRMA-like") assay techniques, assuming in the case of the "actual" sensitivities, the use of mono-^{125}I-labeled antigen or antibody. Calculations have been based on the assumption of 1% manipulation errors ($\epsilon = 0.01$) in the case of the "RIA-like" assays, and of levels of nonspecific binding of the order of 1% (upper curves) and 0.01% (lower curves) in the case of "IRMA-like" assays. Note that, using antibodies displaying affinities less than $10^{12}\underline{M}^{-1}$, the "sensitivity gap" (shaded area) between potential and practically attainable sensitivities in RIA techniques is minor. A much wider interval exists in the case of IRMA-like techniques, particularly when n.s.b. levels are reduced to very low levels. Note that a 10 to 100-fold increase in sensitivity can be expected when using the same antibody in a noncompetitive IRMA method as compared with the equivalent competitive RIA technique, assuming low levels of n.s.b. (Note, however, that the consumption of antibody in the noncompetitive techniques will be far greater.) The arrows indicate the sensitivity levels generally attained with IRMA techniques per se, and the claimed sensitivity yielded by IRMA-like methods relying on enzyme-labeled antibodies and fluorogenic (HS-ELISA) or radioactive (USERIA) substrates. (From Harris et al., 1979; Shalev et al., 1980.)

exhibited higher sensitivity than immunoradiometric methods. Ultrasensitive immunoenzymometric assays have nevertheless been reported which rely on fluorogenic (Shalev et al., 1980) or radioactive (Harris et al., 1979) substrates which yield products that can be measured at very low concentrations. The claimed sensitivities of these techniques has reached about 10^4 and 10^3

molecules/ml, respectively, in general agreement with the theoretical predictions depicted in Fig. 6. The more obvious disadvantages of labels of this type, are that an additional, sometimes lengthy, incubation under well-controlled conditions (to ensure constancy of the enzyme "amplification factor") is required for the enzyme activity measurement; in addition applications are likely to be limited to liquid-phase detection systems (see discussion in Section III).

Another highly promising technique involves the use of chemiluminescent labels. Chemiluminescent assay techniques are more extensively reviewed in Chapter 5. Chemiluminescent phenomena may be usefully compared with radioactive decay. The observable event in the radioactive decay of an isotope (i.e., the emission of a β-particle or γ-ray) occurs continuously at a well-defined rate which is unaffected by physical or chemical conditions. A radioactive counter may detect about 50-80% of these disintegrations, but the fraction of the isotope which decays in an acceptable counting time in the case of the nuclides commonly employed for immunoassay purposes is necessarily relatively small. In the case of ^{125}I, for example, only about 0.001% of the radioactive atoms present will disintegrate in a 1-min period. For tritium, the proportion is some two orders of magnitude less than this. In contrast, the quantum efficiency of a chemiluminescent label (i.e., the fraction of the labeled molecules which emits a detectable photon) can be very high, even approaching 100%; moreover the observable event in chemiluminescence, the emission of the photon, can be chemically controlled so that virtually all the photon emission events take place within a very brief interval. Thus, although the efficiency of detection of these events in currently available instruments may be low, (e.g., 1-10%), the sensitivity of measurement of chemiluminescent labels may nevertheless be considerably better than ^{125}I, with clear implications for the overall sensitivity of labeled antibody methods. Indeed, currently developed immunochemiluminometric assays have been shown to compare extremely favorably with the sensitivity of conventional IRMA techniques (Woodhead et al., 1981).

The main factors presently inhibiting the further development of chemiluminescent methods appear to be instrumental: the chemiluminescent reaction is routinely initiated while the sample is in the counting position (adding a degree of complexity to the instrument design); moreover, the efficiency of conventional detectors is still somewhat low and their background signal rather high. Nevertheless advances in instrument design and the possibility of "recycling" the chemiluminescent label, (thus amplifying the number of photons emitted) make this one of the more promising nonradioisotopic labels currently under development.

At the molecular level, fluorescent methods are also closely analogous to the radioisotopic and chemiluminescent measurement techniques insofar as each form of label yields some form of detectable radiation in response to the input of energy from an external source. In the case of "conventional" fluorescence, photons of light are directed at the fluorescent marker, the incident photons are absorbed, and longer wavelength photons are emitted by the fluorophore. Thus, while the energy released in radioactive decay derives from prior exposure of atoms to neutron irradiation, and the energy of the photons emitted by a chemiluminescent label is released as a result of a chemical reaction, the energy for fluorescence photons derives from concurrent, in situ, irradiation by a light source. Typically, the fluorescent label returns to its ground state after photon emission and the excitation/emission cycle can again take place. "Recycling" of the fluorescent label can thus lead to the emission of many photons from each labeled molecule within a short space of time, implying an extremely high effective specific activity.

Unfortunately, in contrast to the events involved in radioisotopic decay, the efficiency of the conversion of incident photons by a fluorescent substance is susceptible to influence by many factors. Irreversible chemical changes in the fluorophore may be caused by the incident light (bleaching), and energy may be dissipated by alternative routes (quenching) without emission of photons. Moreover the natural fluorescence of many biological substances, sample containers, etc., leading to high and variable blank signals often constitutes a major impediment to the attainment of high sensitivity. For these and other reasons, the fluorescence immunoassay methods have not generally displayed sensitivities comparable to those of the radioisotopic techniques, and have consequently made relatively little impact on the immunoassay field until recently.

In the following section, we discuss one way in which some of the practical problems associated with conventional fluorescence techniques are being overcome, yielding a new methodology which is already showing promise of providing immunoassay techniques with sensitivities significantly superior to those obtained with the current generation of radioisotopically based methods.

II. TIME-RESOLVED FLUORESCENCE: GENERAL PRINCIPLES

As discussed above, in immunoassays relying on fluorescent labels, the lower limit of detection of the fluorophore per se is limited by the magnitude of the background fluorescence emitted by serum constituents, incubation tubes, components of the instrument itself, etc., in addition to stray light deriving from the exciting source. These effects severely limit, in turn, the sensitivity of immunoassays utilizing conventional fluorophores (Wieder, 1978; Soini and Hemmila, 1979). Clearly the development of techniques that distinguish the specific fluorescence of the fluorophore from background fluorescence deriving from extraneous sources is of critical importance in the evolution of fluoroimmunoassays of ultrahigh sensitivity.

The method commonly adopted to identify the specific fluorescence yielded by a fluorophore relies on isolating, by means of filters etc., light of a wavelength characterizing the fluorescent emission of interest. This approach, combined with the restriction, by similar means, of the wavelength range of the exciting light, will clearly reduce the magnitude of background signals; it is nevertheless usually insufficiently discriminatory to eliminate them completely. Meanwhile, a complementary approach of great promise relies on distinguishing the fluorescent signals of interest on the basis of their temporal decay characteristics following intermittent exposure of the fluorescent material to short pulses of incident light.

When a fluorophore is excited by pulsed radiation, fluorescence is emitted following each pulse with an intensity that decreases exponentially with time in a characteristic manner (Ware, 1972). Moreover, as shown in Fig. 7, an electronically "gated" detection system may be used to accumulate photons emitted over any selected time interval immediately following extinction of the incident light source. Such a system may, in principle, be employed to identify the fluorescent signals emitted by fluorophores characterized by different decay times in a manner closely analogous to the approach that has often been employed to resolve mixtures of radioisotopes on the basis of their differing half-lives (Wieder, 1978; Soini and Hemmila, 1979). The simplest situation clearly exists when a sample contains two fluorophores, one of which displays a fluorescent decay time significantly longer than the other. In these circumstances it is possible to measure the signal originating from the fluorophore displaying the longer decay time by permitting the more rapidly decaying fluorescence to die away to an insigni-

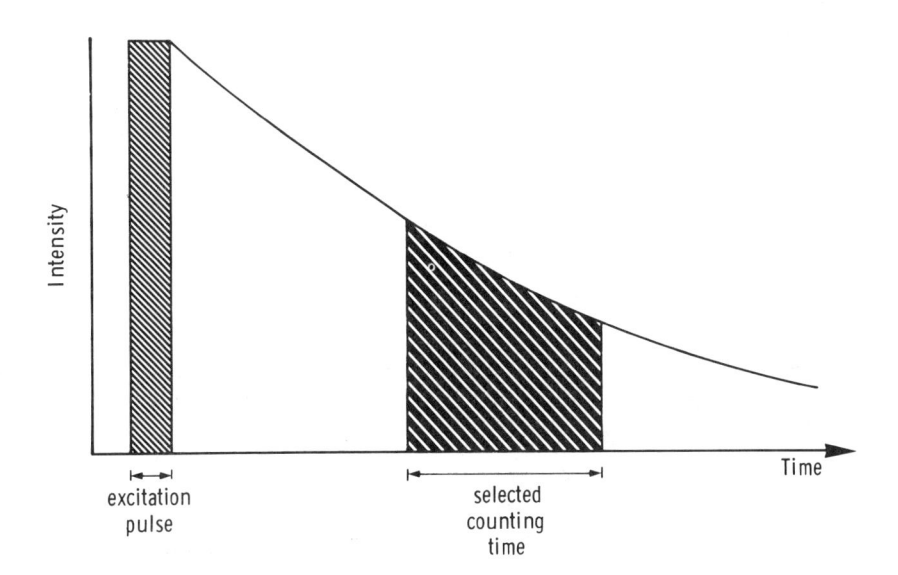

FIGURE 7 Selected counting of fluorescence photons following a pulse of excitation energy.

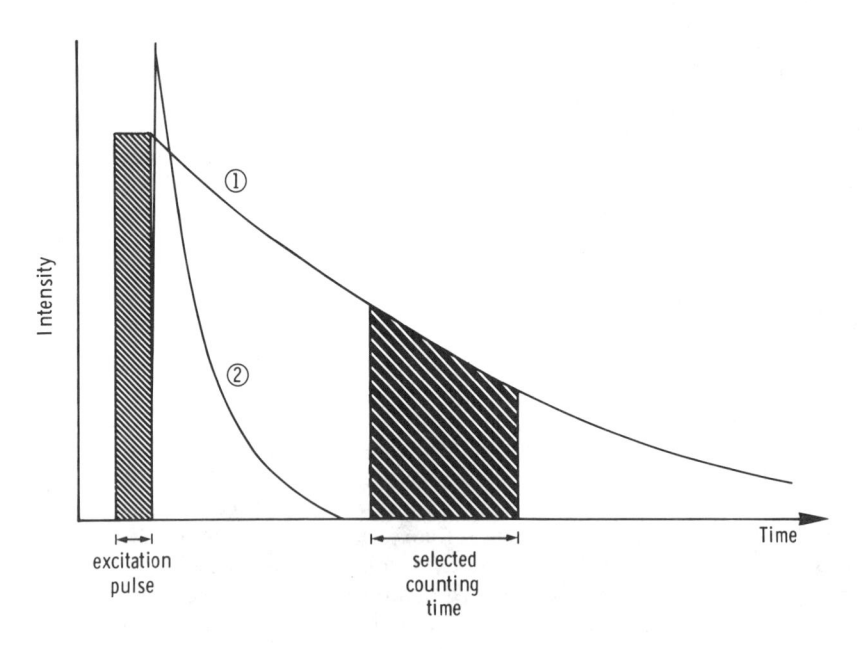

FIGURE 8 Discrimination of fluorescence signal from fluorescence of shorter lifetime by selection of photon counting time.

ficant level before commencing photon measurement (see Fig. 8). This simple principle of "time resolution" can be exploited to reduce or eliminate background fluorescence if this is characterized by a much shorter lifetime than that of the fluorophore of interest as shown in Fig. 8.

The fluorescence associated with serum proteins and many other common organic substances is characterized by lifetimes of the order of 10 nsec (the fluorescence lifetime being defined as the time required for the fluorescence emission to decay to 1/e of its initial intensity following excitation). The lifetimes of some of the fluorophores commonly employed in immunochemistry are listed in Table 1. Clearly the simple time resolution method illustrated in Fig. 8 cannot readily be exploited to distinguish between background fluorescence and that characterizing many of the fluorophores, such as FITC or DNS-Cl, listed in Table 1. Indeed, even in the case of NPM (which possesses a lifetime of 100 nsec) a significant contribution might be expected from background sources if a substantial proportion of the specific fluorescence deriving from the fluorophore per se were to be encompassed within the selected measurement interval of the detection system. In contrast, the fluorescent lanthanide chelates listed in the table form a relatively unique group of fluorophores because of their much extended fluorescence decay times, which are some 3-4 orders of magnitude longer

TABLE 1 Fluorescence Lifetimes of Some Common Fluorophores

	Fluorescence decay time	
Nonspecific background	10	ns
Fluorescein isothiocyanate (FITC)	4.5	ns
Dansyl chloride (DNS-Cl)	14	ns
N-3-pyrene maleimide (NPM)	100	ns
Rare Earth Chelates	1	μs-ms

than those characterizing most background fluorescence. The prolonged decay times of these compounds essentially stem from the delays incurred by the internal transfer processes whereby light energy absorbed by the organic moiety is conveyed to the chelated rare earth atoms which constitute the ultimate source of the emitted fluorescent photons. Meanwhile, a second useful characteristic of the lanthanide chelates is their large Stokes shift — that is, the difference in the wave lengths of the fluorescent and (optimal) exciting radiations. For example, europium emits fluorescence in a narrow band of wavelengths at around 613 nm; maximal excitation of the chelate occurs using incident light of a wavelength of 340, implying a shift of approximately 270 nm. The combination of a large Stokes shift and an extended fluorescence decay time provides the basis for the construction of relatively simple and inexpensive pulsed light, time-resolving fluorometers capable of yielding exceptionally high signal-to-noise ratios which are the essential prerequisite of high-sensitivity fluorometric measurements. In the following sections, we review some of the technical problems associated with the development of the fluoroimmunoassays based upon this principle.

A. Basic Immunoassay Strategy

Earlier in this chapter, we have emphasized some of the fundamental advantages of the noncompetitive labeled antibody immunoassay techniques over assays based on the competitive or saturation assay principle.

Although fluorescent markers in general, and the chelated rare earth fluorophores in particular, may be employed as the basis for both labeled analyte and labeled antibody-based immunoassay techniques, the need for short-incubation, high-sensitivity assays has inevitably focused our own developmental activity onto methods relying on the noncompetitive labeled antibody analytical approach. For these reasons, the following discussion centers primarily on the use of the lanthanide chelates as antibody-labeling reagents. One interesting and important feature of these fluorophores in this context is that their excitation band between 300 and

350 nm overlaps the emission wavelengths of the principal fluorescent amino acids in antibody protein (in particular tyrosine, emission 303 nm). This implies that when antibody protein is labeled with rare earth metal chelate, the potential exists for direct energy transfer from the protein to the chelate (Forster, 1959). One possible consequence of such energy transfer is that the effective absorption coefficient of the protein-chelate complex may be significantly greater than that of the complex alone, resulting in an increase in the sensitivity of detection of the metal ion by several orders of magnitude. Such an effect has been observed by Leung and Meares (1977).

B. Bifunctional Labeling Reagents

The successful exploitation of the lanthanide chelates in fluoroimmuno-assay is governed by two considerations. The first of these is the choice of a bifunctional reagent which, while strongly chelating the metal ion, also attaches to the molecule to be labeled (i.e., antibody) without impairment of its essential immunochemical properties. The second consideration relates to the transformation, if necessary, of the metal chelate into a highly fluorescent enolate with an appropriate fluorescence lifetime. The characteristics of this process depend on the nature of the bifunctional group initially selected to effect the labeling.

If the bifunctional reagent is a good chromophore and a good energy donor for the metal ion, as is, for example, a β-diketone (see next section), then the final labeled complex will be highly fluorescent when initially prepared. On the other hand, the selected bifunctional reagent may be of such a nature that the labeled product is intrinsically poorly fluorescent. In this case, the fluorescence measurement may be made in a solution containing chelating chromophores which are good donors (such as β-diketones) using conditions designed to ensure the transfer of the metal ions from their original state of chelation to the new highly fluorescent form.

Bifunctional chelating reagents which endow the labeled molecular species with high intrinsic fluorescence will yield products that clearly behave, and may be used, as conventional fluorophores in every respect quite aside from the particular advantages deriving from their prolonged fluorescence decay and large Stokes shift. We shall refer to these as bifunctional reagents of the first class. Different considerations apply in the case of those chelating reagents which require measurement to be made following their exposure to a "cocktail" containing a donor-chelating chromophore such as a β-diketone. These we shall designate as bifunctional reagents of the second class. In this circumstance a balance must be struck between, on one hand, the requirement to attain maximum chelation of the metal ion with β-diketone and, on the other, the need to minimize loss of the excitation energy consequent upon the presence in solution of a large excess of the chelating material. Bifunctional reagents of the first class may also profit from excitation energy transfer because fluorescence measurements are made with the chelate still attached to the protein molecule as discussed above. Reagents of the second class do not possess this advantage.

The considerations that guide the choice of materials for labeling and measurement are discussed in the following sections.

C. Lanthanide Chelate Fluorescence

Certain rare earth metal salts (most particularly those of europium and terbium), while fluorescing only weakly when in aqueous solution, yield intense fluorescence when complexed with certain organic ligands. This phenomenon derives, as indicated above, from an internal transfer process in which the energy of the incident light photons absorbed by the organic moiety is transferred to the ligated rare earth atom which is, in turn, raised to an excited state. The return of the metal ion to its ground state results in fluorescent emission characteristic of the metal itself and independent of the nature of the organic chelate. The principal fluorescent emission wavelength for europium is 613 nm; terbium yields two emission lines at 490 and 545 nm. Other characteristics of the fluorescence process—quantum efficiency, maximum excitation, fluorescence decay time, etc. — are strongly dependent on the nature of the organic chelating moiety and the solvents used, the presence of synergistic components in the system, etc.

The generally accepted view of the energy transfer processes involved is depicted in Fig. 9. Initially, orbitals of the ligand absorb incident energy, with consequent excitation from the singlet ground state to vibrational levels of the first singlet excited state. Energy is then transferred by intersystem crossing from singlet to the triplet excited state of the ligand, and subsequently transferred from the excited triplet state of the ligand to the resonance levels of the metal ion eg. Eu^{3+}. The ligand is the donor and the Eu^{3+} ion the acceptor, the energy difference between the excited triplet state of the ligand and the resonance level of the metal ion being such as to permit intramolecular energy transfer (Halverson et al., 1964). Finally the excited Eu^{3+} ion emits a photon. In the case of Eu^{3+} the transition giving rise to fluorescence is from the ^5D excited state to the various sublevels of ^7F ground state. Clearly the frequency of the emitted light is characteristic of the metal ion itself, but both its intensity and lifetime are greatly influenced by the nature of the chelating material.

It has been found that β-diketone structures (Fig. 10) give particularly favorable complexes, due primarily to a combined rigidity and good Π electron delocation (Halverson et al., 1964; Vallarino et al., 1979). However, substitution of different groups at the R_1 and R_2 positions of the basic structure markedly influences the fluorescence characteristics of the final complex. Inclusion of good electron donors, and the use of bulky substituents can lead to enhanced fluorescence yields. Thus the β-diketone,

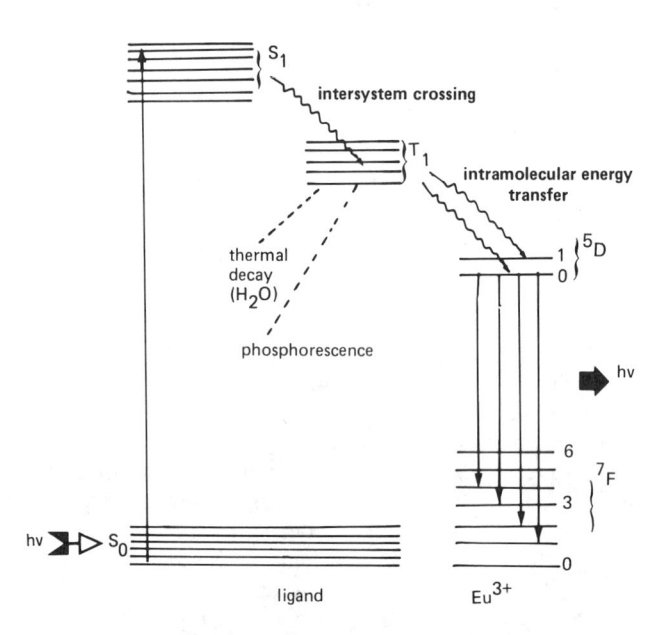

FIGURE 9 Schematic diagram of the energy transfer processes of the chelate leading to Eu^{3+} metal ion fluorescence.

FIGURE 10 The basic structure of a β-diketone, for example, in trifluoroacetylacetone, $R_1 = CH_3$, $R_2 = CF_3$..

thenoyltrifluoroacetone (Fig. 11) is widely used in the fluorimetric determination of Eu and Tb.

In Fig. 9 a potential problem of "leakage" of energy in the system is also portrayed. Energy from the excited triplet state of the ligand may be lost either by phosphorescence or by thermal excitation of water molecules, with consequent quenching of the final signal. This is an obvious problem if, as in a conventional immunoassay, the fluorescently labeled molecule exists in an aqueous environment, due to the ready inclusion of the oxygen of water molecules into the coordination shell of Eu^{3+}. In these circumstances, the hydrogen-bonded solvent thus becomes an effective energy sink. This problem is minimized by the inclusion of synergistic agents in the fluorophore complex (Halverson et al., 1964). The synergistic agent typically includes the structure $R_3P = O$ or $R_3S = O$, where R is a bulky saturated hydrocarbon group which, together with the bulky radicals of the β-diketone, completes an energy insulating sheath for the final complex. The latter typically comprises three ligands with a β-diketone structure, with the octacoordinated shell of Eu^{3+} completed by two molecules of the synergistic agent, each of which supplies an oxygen atom for coordination. Thus a typical complex of the EuL_3S_2 type is $Eu(TFAC)_3(TOPO)_2$ where TFAC is trifluoroacetylacetone and TOPO is trioctylphosphine oxide, the latter being the synergistic agent (Fig. 12).

Utilizing such fluorophores, and a fluorometer with a pulsed xenon light source operating in a time-resolution mode, it is possible to assay Eu^{3+} ion with a detection limit below $10^{-13}\underline{M}$ (Fig. 13).

D. Labeling Antibodies for Time-Resolving Fluoroimmunoassay

We shall now consider the attachment to antibody proteins of the two classes of bifunctional reagents referred to above. We first examine reagents of the first class: those that render the labeled protein highly fluorescent without subsequent modification. One compound used successfully for this purpose has been orthophenanthroline isothiocyanate. Vallarino et al. (1979) have prepared and used the complex shown in Fig. 14. The fluorophore is the octacoordinated Eu^{3+} complexed with three β-diketone ligands and one 1,10 orthophenanthroline isothiocyanate, utilizing two coordinating

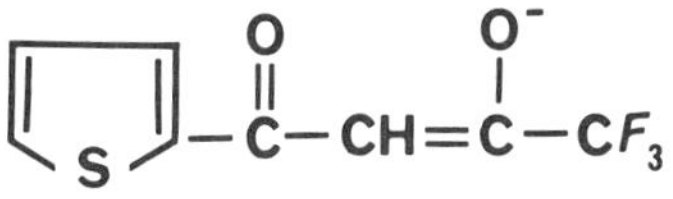

FIGURE 11 The β-diketone thenoyltrifluoroacetone (TTA).

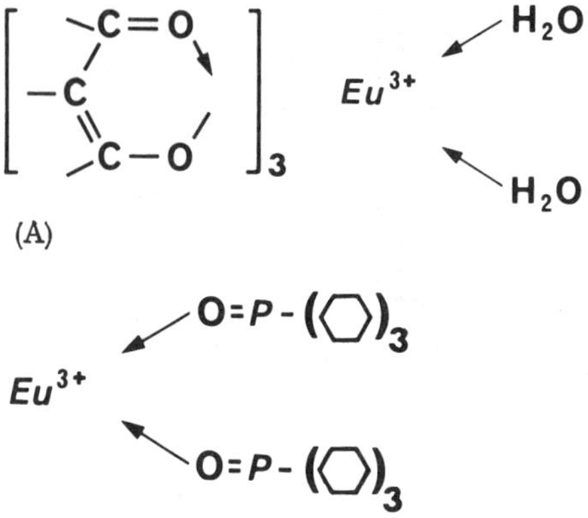

(A) [structure: -C(=O)-C=C(-O-)-]₃ Eu³⁺ with H₂O, H₂O

(B) Eu³⁺ with O=P-(⌬)₃, O=P-(⌬)₃

FIGURE 12 The chelation of Eu^{3+} with a β-diketone illustrating the role of the synergistic agent Trioctylphosphine oxide (TOPO) in eliminating water to provide a stable fluorophore. (A) Water molecules completing the octa-coordination of the Eu^{3+} ion. (B) Explanation of how the water molecules are replaced by TOPO thus insulating the Eu^{3+} complex from solvent effects.

nitrogens to complete the chelation. The isothiocyanate substituent in the orthophenanthroline structure allows coupling of the fluorophore to proteins. Vallarino et al. (1979) have suggested that an additional advantage of this system is that orthophenanthroline is a chromophore with characteristics favorable for energy transfer to the central metal ion. This bifunctional agent then acts to couple the fluorophore to the protein and also as a synergistic agent enhancing the already high fluorescence intensity due to the tris β-diketone complex.

Wieder (1978) has also described an interesting bifunctional reagent of the first class (Fig. 15) which comprises a modified TTA molecule. The aminomethyl group substituted into the structure facilitates linkage to proteins. This forms the potential basis of a generalized methodology utilizing β-diketones. Wieder (1978) reports that the reagent, when chelating Eu^{3+} and measured in organic solvents at room temperature, can display a quantum efficiency of about 0.5 and a decay lifetime of about 0.3 msec.

However, neither of these reagents appear to have been employed for labeling antibodies for practical use in immunoassay; indeed both authors have suggested that special precautions would have to be taken to solubilize the fluorophore complex in aqueous systems while protecting the fluorophore from the quenching effects of water molecules.

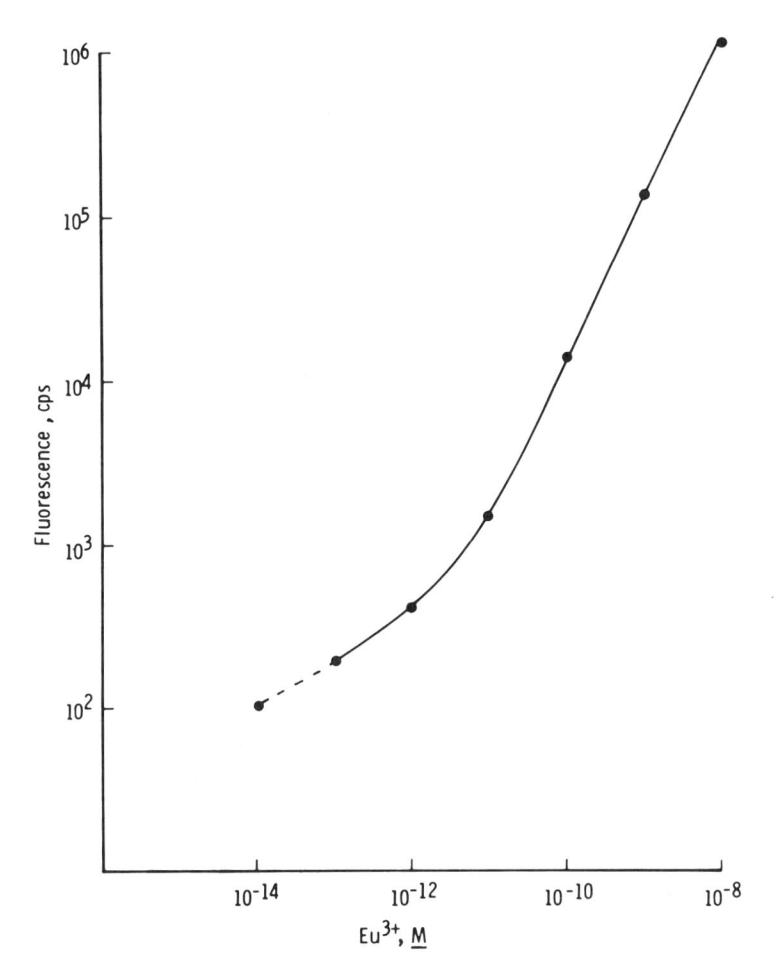

FIGURE 13 Fluorescence measurement from serial dilutions of Eu^{3+}. $EuCl_3$ was diluted in "enhancement" solution, containing the β-diketone α-naphthoyltrifluoroacetone, NTA (20 $\mu\underline{M}$ NTA, 100 $\mu\underline{M}$ TOPO, 0.1% Triton X-100, in 0.05 $\underline{M}$ acetate buffer, pH 4.0). Mean value of triplicate determinations are shown with the background subtracted (background = 1.7 x 10^3 cps). The coefficient of variation ranges from 0.16% at 10^{-8} $\underline{M}$ to 2.7% at 10^{-14} $\underline{M}$. The measurements were made with a prototype fluorometer employing a xenon flash lamp (Courtesy of Wallac Oy, P.O. Box 10, SF20101, Turku 10, Finland.)

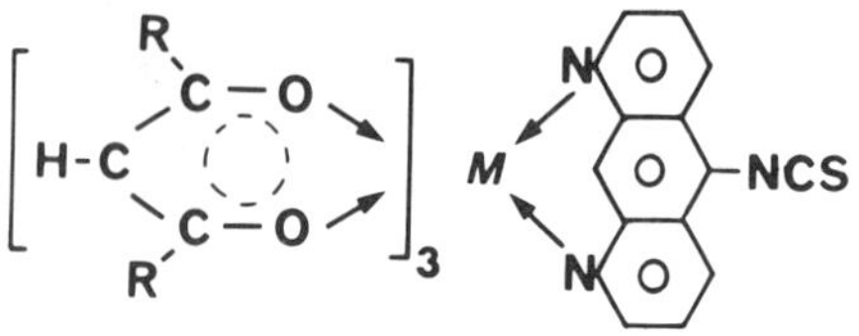

FIGURE 14 The bifunctional reagent 1,10 orthophenanthroline isothiocyanate completing the chelation of a rare earth metal ion with a β-diketone.

A good example of a bifunctional reagent of the second class has been described by Sundberg et al. (1974) and Leung and Meares (1977) who employ an analogue of ethylenediaminetetracetic acid (EDTA) (Fig. 16). The EDTA forms a hexadentate ligand which chelates the central metal ion while the radical R may be activated to permit its attachment to protein, for example, by the diazo reaction in the case where R represents NH_2. When a protein has been labeled with this reagent it is necessary, as indicated earlier, to assay the metal ion in a solution containing a β-diketone. In this circumstance the fluorescence yield is enhanced to the same order as that of a β-diketone complex.

Using the reagent 1-(p-benzenediazonium) EDTA-Eu^{3+} we have labeled a variety of specific antibodies and developed several prototype fluoroimmunoassays (Marshall et al., 1981). For example, donkey anti-rabbit IgG has been labeled with Eu^{3+} chelate and used as the second antibody in a sandwich-type assay for rabbit IgG. The experimental procedure is outlined in Fig. 17: a typical dose-response curve representing results based on this approach is shown in Fig. 18.

Clearly, labeled immunoglobulins may also be used in indirect sandwich assays. Such an approach has been applied to the detection and measurement of hepatitis B surface antigen (HBs antigen) as shown in Fig. 19. A direct sandwich assay for HBs antigen in which antibodies to HBs antigen have been labeled is illustrated in Figure 20. These three assays demonstrate the generality of the method based on the EDTA analog.

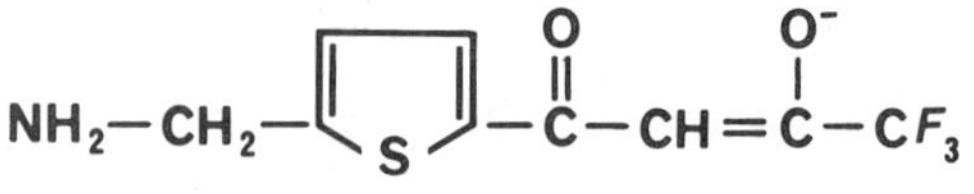

FIGURE 15 A modified thenoyltrifluoroacetone (TTA) molecule as a bifunctional reagent for labeling proteins with fluorescent lanthanide metal ions.

Fluoroimmunoassay 95

$$^{-}OOCCH_2\diagdown\diagup CH_2COO^{-}$$
$$N-CH-CH_2-N$$
$$^{-}OOCCH_2\diagup\diagdown CH_2COO^{-}$$
$$|$$
$$\bigcirc$$
$$|$$
$$R$$

FIGURE 16   EDTA analogues as bifunctional reagents for labeling proteins with rare earth ions.

III. PRESENT APPLICATIONS AND FUTURE PROSPECTS

Since the initial preparation of this chapter, several other practical examples of the application of the concepts and methodology discussed in the earlier sections have been described (Meurman et al., 1982; Pattersson et al., 1983; Siitari et al., 1983). Each is based on the use of "class 2" chelates as defined above: elicitation of the fluorescence signal relies on exposure of the labeled antibody to a development "cocktail" resulting in rapid transfer of europium from solid-phased antibody into solution prior to its fluorescence measurement. This technique, the exact methodological details of which are shortly to be published (Hemmila et al. 1984), was originally developed and exploited in our own laboratory largely because of the improved precision which it yielded using the relatively simple prototype pulsed-light time-resolving fluorometer (manufactured by LKB/Wallac Instruments Ltd.) initially at our disposal. Though not ideal, this approach obviates the problems of optical geometry and instrument design which are encountered

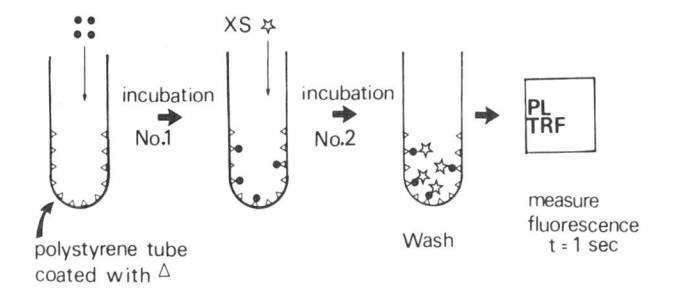

FIGURE 17   Sequence of steps in time-resolved fluoroimmunoassay of rabbit IgG.   △ = anti-rabbit IgG; ☆ = Eu^{3+} labeled anti-rabbit IgG; ● = Rabbit IgG.

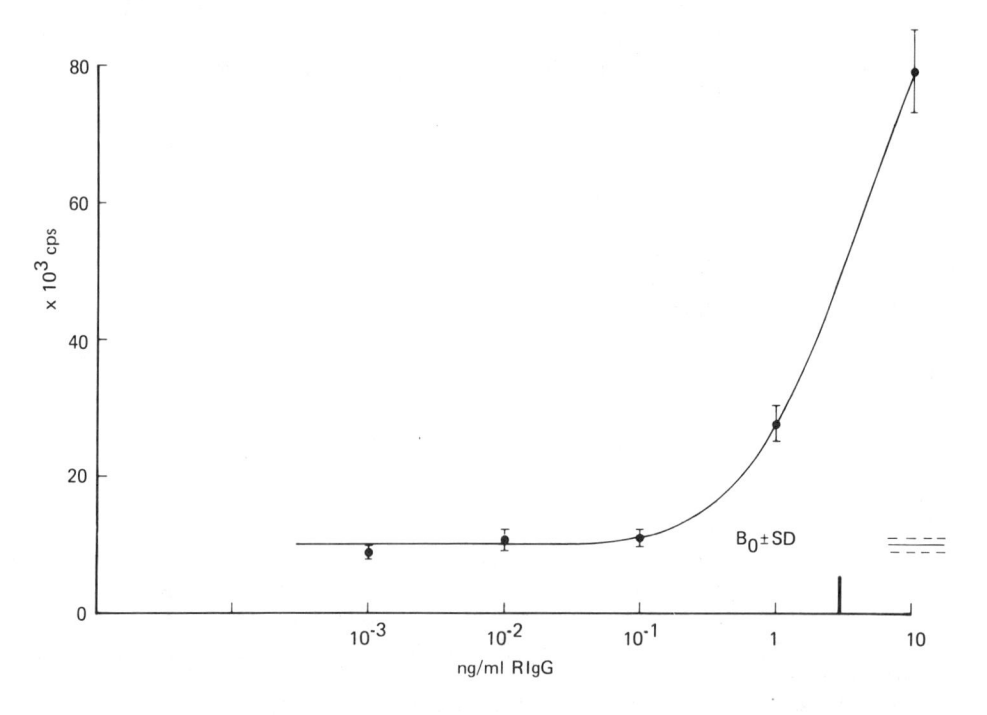

Figure 18 Typical calibration curve in the assay of rabbit IgG performed as illustrated in Fig. 17. The means of triplicate determinations with their standard deviation are shown. B_0 is the fluorescence signal corresponding to zero concentration of rabbit IgG.

in accurate measurements of fluorescence on solid surfaces; it has subsequently permitted the development of a number of satisfactory immunoassays closely comparable in performance to the corresponding radioisotopically based method.

It is evident, however, that future developments in this field must include the perfection of techniques whereby precise fluorescence measurements can be carried out directly on solid surfaces. This will necessitate intensive research on fluorescent materials with appropriate emission characteristics and possessing the chemical properties required to permit their sensitive measurement in aqueous media while remaining coupled to antibody. Such research will also inevitably entail the development of improved (albeit inexpensive) instrumentation providing high-intensity pulsed illumination of the fluorescent target material while minimizing instrument-generated background "noise." With regard to the latter, the technical problems which are likely to arise include the long-lived fluorescence and phosphorescence encountered in optical components per se, and the relatively

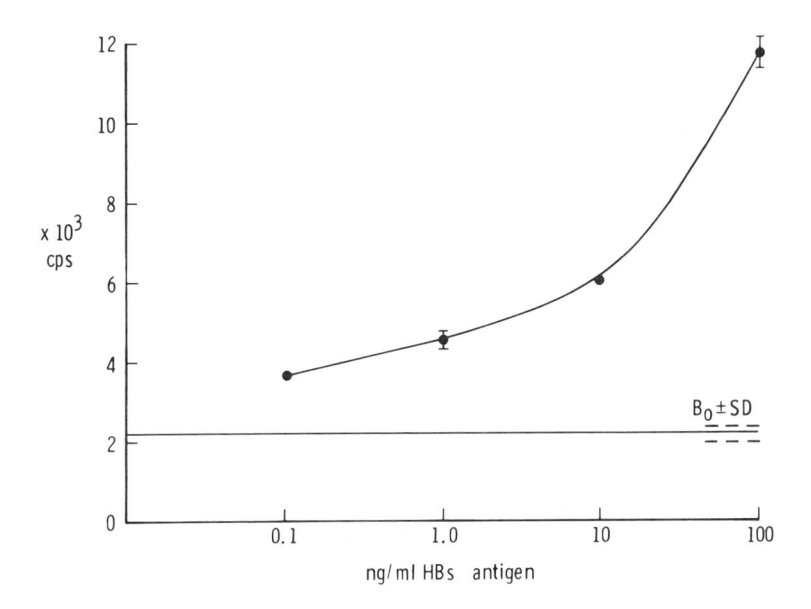

FIGURE 19 Typical calibration curve in an HBs antigen assay obtained by the indirect sandwich assay method using Eu^{3+}-labeled donkey antirabbit IgG. The tube coating antibody was horse anti-HBs antigen; the second antibody is rabbit anti-HBs antigen. Means of duplicate determinations are shown together with standard deviations.

slow rate of extinction of the pulsed-xenon lamps on which the present generation of pulsed fluorometers is largely based. The anticipated development of inexpensive pulsed nitrogen lasers possessing very short cut-off times can be expected to result in a substantial improvement on the assay sensitivities attainable with present instrumentation.

If we assume that these technical hurdles can be surmounted, one of the major potential advantages deriving from fluorescent immunoassay techniques is the scope they offer for multiple analyte detection and measurement. This is likely to become increasingly important in such contexts as, for example, the immunodiagnosis of infectious diseases.

The development of a scanning, time-resolved fluorometer or, alternatively, of a system of light guides simultaneously viewing selected areas of a plastic surface coated with an array of antibodies of differing specificity, will clearly permit the rapid measurement of multiple analytes in the same sample. The possibility of such development is one of the many attractive features of the techniques described in this chapter, and distinguishes the fluorescence immunoassay methods from many of the alternative techniques, whether based on radioisotopic or nonradioisotopic labels.

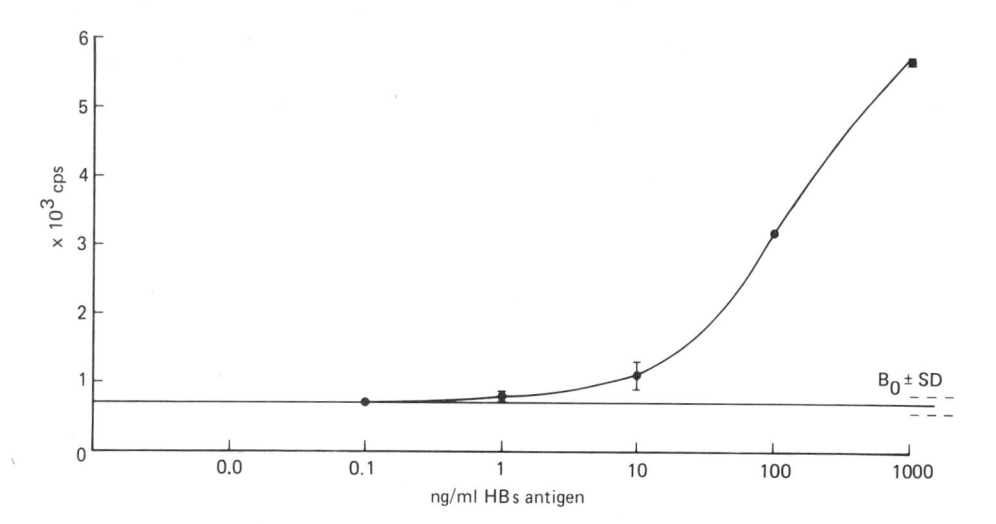

FIGURE 20 Typical calibration curve for HBs antigen assay by the direct sandwich method as shown in Fig. 17. Tubes were coated with horse anti-HBs antigen; the labeled reagent was Eu^{3+} goat anti-HBs antigen. Means of duplicate determinations are shown with standard deviations.

In summary, we have endeavored, in this chapter, to indicate the fundamental reasons why future developments in the immunoassay field are likely to be based on the use of labeled antibodies rather than on labeled analytes as has become customary in the past two decades. We have also attempted to indicate that the full potential of the labeled antibody approach can only be realized using labels possessing effective specific activities far higher than those characterizing commonly used radioisotopes. Finally we have discussed a form of nonradioisotopic label which we believe to be a strong contender among the alternative labels capable of fulfilling this particular requirement. Although these techniques are still in a relatively early stage of development, considerable resources are currently being directed toward the perfection of the technologies involved; we therefore have little doubt that the pulsed-light, time-resolving immunoassays will be shown to compare extremely favorably with current radioisotopically based techniques, and perhaps render the latter largely obsolete within the foreseeable future.

ACKNOWLEDGMENTS

The concepts and developments discussed in this chapter represent the outcome of discussions between Dr. Erkki Soini and Professor Roger Ekins which took place in the mid-1970s. Since these initial discussions, LKB/Wallac have developed and placed at our disposal a prototype time-resolving fluorometer and a variety of europium chelates without which the experimental studies referred to in this chapter could not have been performed. We are happy to record many fruitful discussions and interchanges with members of the LKB/Wallac team since the inception of this collaborative project, including, in particular Dr. Erkki Soini, Dr. Timo Lovgren, and Ilkka Hemmila.

REFERENCES

Ekins, R. P. (1960). The estimation of thyroxine in human plasma by an electrophoretic technique. Clin. Chim. Acta 5: 453.

Ekins, R. P., and Newman, B. (1970). Theoretical aspects of saturation analysis. Acta Endocrinol. Suppl. 147: 11.

Ekins, R. P., Newman, B., and O'Riordan, J. L. H. (1968). Theoretical aspects of "saturation" and radioimmunoassay. In Radioisotopes in Medicine: In Vitro Studies, R. L. Hayes, F. A. Goswitz, and B.E.P. Murphy (eds.), Oak Ridge Symposia, USAEC, Oak Ridge, Tennessee, p. 59.

Ekins, R. P. (1976). General principles of hormone assay. In Hormone Assays and Their Clinical Application, J. A. Loraine and E. T. Bell (eds.), Churchill Livingstone, Edinburgh, p. 1.

Forster, Th. (1959). Transfer mechanism of electronic excitation, Disc. Faraday Soc. 27: 7.

Harris, C. C., Yolken R. H., Krokan, H., and Hsu, I. C. (1979). Ultrasensitive enzymatic radioimmunoassay: Application to detection of cholera toxin and rotavirus. Proc. Natl. Acad. Sci. USA 76: 5336.

Halverson, F., Brinen, J. S., and Leto, J. R. (1964). Photoluminescence of lanthanide complexes, II. Enhancement by an insulating sheath. J. Chem. Phys. 41: 157.

Hemmila, I., Dakubu, S., Mukkala, V-M., Siitari, H., and Lovgren, T. (1984). Europium as a label in time-resolved immunofluorometric assays. Anal. Biochem. (In press.)

Hunter, W. M. (1982). Recent advances in radioimmunoassay and related procedures. In Radioimmunoassay and Related Procedures in Medicine, IAEA, Vienna, p. 3.

Jackson, T. M., Marshall, N. J., and Ekins, R. P. (1983). Optimisation of immunoradiometric (labeled antibody) assays. In Immunoassays for Clinical Chemistry, W. M. Hunter and J. E. T. Corrie (eds.), Churchill Livingstone, Edinburgh, p. 557.

Leung, C. S-H., and Meares, C. F. (1977). Attachment of fluorescent metal chelates to macromolecules using "bifunctional" chelating agents. Biochem. Biophys. Res. Commun. 75: 149.

Marshall, N. J., Dakubu, S., Jackson, T., and Ekins, R. P. (1981). Pulsed-light, time-resolved fluoroimmunoassay. In Monoclonal Antibodies and Developments in Immunoassay, A. Albertini and R. Ekins (eds.), Elsevier/North Holland Biomedical Press, Amsterdam, p. 101.

Meurman, O. H., Hemmila, I. A., Lovgren, T. N. E., and Halonen, P. E. (1982). Time-resolved fluoroimmunoassay: A new test for Rubella antibodies. J. Clin. Microbiol. 16: 920.

Miles, L. E. M. and Hales, C. N. (1968a). Labeled antibodies and immunological assay systems. Nature 219: 186.

Miles, L. E. M. and Hales, C. N. (1968b). An immunoradiometric assay of insulin. In Protein and Polypeptide Hormones, Part 1, M. Margoulies (ed.), Excerpta Medica Foundation, Amsterdam, p. 61.

Pattersson, K., Siitari, H., Hemmila, I., Soini, E., Lovgren, T., Hanninen, V., and Tanner, P. (1983). Time-resolved fluoroimmunoassay of human choriogonadotropin. Clin. Chem. 29: 60.

Rodbard, D., and Weiss, G. H. (1973). Mathematical theory of immunometric (labeled antibody) assay. Anal. Biochem. 52: 10.

Schuurman, H. J., and de Ligny, C. L. (1979). Physical models of radioimmunoassay applied to the calculation of detection limit. Anal. Chem. 51(1): 2.

Shalev, A., Greenberg, G. H., and McAlpine, P. J. (1980). Detection of attograms of antigen by a high sensitivity enzyme-linked immunosorbent assay (HS-ELISA) using a fluorogenic substrate. J. Immunol. Methods 38: 125.

Siitari, H., Hemmila, I., Soini, E., Lovgren, T., and Koistinem, V. (1983). Detection of hepatitis B surface antigen using time-resolved fluoroimmunoassay. Nature 301: 258.

Soini, E., and Hemmila, I. (1979). Fluoroimmunoassays: Present status and key problems. Clin. Chem. 25: 353.

Sundberg, M. W., Meares, C. F., Goodwin, D. A., and Diamanti, C. I. (1974). Selective binding of metal ions to macromolecules using bifunctional analogues of EDTA. J. Med. Chem. 17: 1304.

Vallarino, L. M., Watson, B. D., Hindman, D. H. K., Jagodic, V., and Leif, R. C. (1979). Quantum dyes: A new tool for cytology automation. In The Automation of Cancer Cytology and Cell Image Analyses, N. J. Pressman and G. L. Wied (eds.), Tutorials of Cytology, Chicago, p. 31.

Ware, W. (1972). Techniques for fluorescence lifetime measurement and time-resolved emission spectroscopy. In Fluorescence Techniques in Cell Biology, A. A. Thaer and M. Seruetz (eds.), Springer-Verlag, Berlin.

Wide, L., Bennick, H., and Johansson, S. G. O. (1967). Diagnosis of allergy by an in vitro test for allergen antibodies. Lancet 2: 1105.

Wieder, I. (1978). Background rejection in fluorescence immunoassay. In Immunofluorescence and Related Staining Techniques, Proceedings of the 6th International Conference, W. Knapp (ed.), Elsevier Biomedical Press, Amsterdam, p. 67.

Woodhead, J. S., Addison, G. M., and Hales, C. N. (1974). The immunoradiometric assay and related techniques. Br. Med. Bull. 30: 44.

Woodhead, J. S., Simpson, J. S. A., Weeks, I., Patel, A., Campbell, A. K., Hart, R., Richardson, A., and McCapra, F. (1981). Chemiluminescent labeled antibody techniques. In Monoclonal Antibodies and Developments in Immunoassay, A. Albertini and R. P. Ekins (eds.), Elsevier/North Holland, Amsterdam, p. 135.

Yalow, R. S., and Berson, S. A. (1960). Immunoassay of endogenous plasma insulin in man. J. Clin. Invest. 39: 1157.

5

Immunoassays Using Chemiluminescent Labels

IAN WEEKS, ANTHONY K. CAMPBELL AND J. STUART WOODHEAD
Welsh National School of Medicine, Cardiff, Wales
FRANK McCAPRA
School of Chemistry and Molecular Sciences,
University of Sussex, Falmer, England

I. INTRODUCTION

The impact of the radioimmunoassay (RIA) technique of Yalow and Berson (1960) and of the competitive protein-binding assays described by Ekins (1960) has been far-reaching during the past two decades, enabling the sensitive and specific quantitation of biologically important molecules. In the years preceding the introduction of these techniques the determination of such species in the peripheral circulation or in the extracellular fluids, if at all possible, was performed using time-consuming and often unreliable bioassays.

  A variation of the radioimmunoassay methodology was subsequently introduced by Miles and Hales (1968) in which labeled antibodies rather than labeled antigens are used. The immunoradiometric assay (IRMA) utilizes excess reagents, in contrast to the limiting reagent condition of radioimmunoassay, and is hence kinetically and thermodynamically more favorable for immune complex formation than the latter condition. Such assays have advantages over RIA in terms of speed, sensitivity, working range, and, in some instances, selectivity. Further, the preparation and storage of labeled antibodies is more advantageous than that of labeled antigens because of their stability and within-class similarity. In spite of these properties the development of immunoradiometric assays for use in routine clinical diagnosis has been relatively slow, a fact which is undoubtedly due to the requirement for large quantities of antibody. However, recent advances in monoclonal antibody technology have stimulated interest in labeled antibody methodology with all its associated advantages.

  Although high-performance immunoradiometric assays for polypeptides are available, disadvantages arise from the use of ^{125}I as a label, since it is a high-energy γ-emitter with only a 60-day half-life. For this reason, it causes radiolytic damage to molecules into which it is incorporated and is also the subject of increasing legislative control in many countries. Tritium suffers less from these constraints but has a much lower sensitivity of detection and relies on the use of liquid scintillation counting for quantitation. The problems associated with the use of radionuclide labels have stimulated much interest in the development of immunoassays which do not utilize such probes. In general nonisotopic label immunoassays (Voller et al., 1981) lack the sensitivity required for the precise quantitation of circulating polypeptide hormones and several types are not suited for routine clinical application.

Fluorescent (Smith et al., 1981) and enzyme probes (Wisdom, 1976) have found wide application as labels for immunoassay and are used for the determination of the concentrations of a number of biologically important molecules present at relatively high concentrations in the peripheral circulation (such as drugs and steroid hormones). At present there is much interest in the use of time-resolved fluorescence measurements using lanthanide series chelates in the hope that assays of greater sensitivity may be derived from this methodology than is possible with conventional fluorescence systems.

In contrast to other nonisotopic probes, chemiluminescent molecules can be detected easily at very low levels ($<10^{-18}$ mol) and have demonstrated their potential use as probes in immunoassay in the past 3 years.

II. CHEMILUMINESCENCE

A. General Considerations

Chemiluminescence is the phenomenon observed when the vibronically excited product of an exoergic chemical reaction reverts to its ground state with photonic emission. This emission is distinguished from fluorescence and phosphorescence since the excited states are populated directly as a result of the chemical reaction rather than by the prior photophysical absorption of light (Fig. 1). A large number of molecules are capable of undergoing chemiluminescent reactions (Fig. 2), all the reactions studied to date being of an oxidative nature. Chemiluminescence is exemplified by "light-sticks" which, when initiated, glow for several hours. These utilize bis-phenyl oxalates which emit light in the presence of hydrogen peroxide in a suitable solvent system. The emission, however, is not directly visible and is only observed following energy transfer to a fluorescent molecule such as fluorescein or rhodamine which emits in the visible region of the spectrum.

The best known chemiluminescent reactions of biochemical interest (excluding the bioluminescent reactions of luciferin-luciferase systems) are those of luminol and its derivatives. In aqueous media, light emission is stimulated by alkaline hydrogen peroxide and a catalyst. Figure 3 illustrates

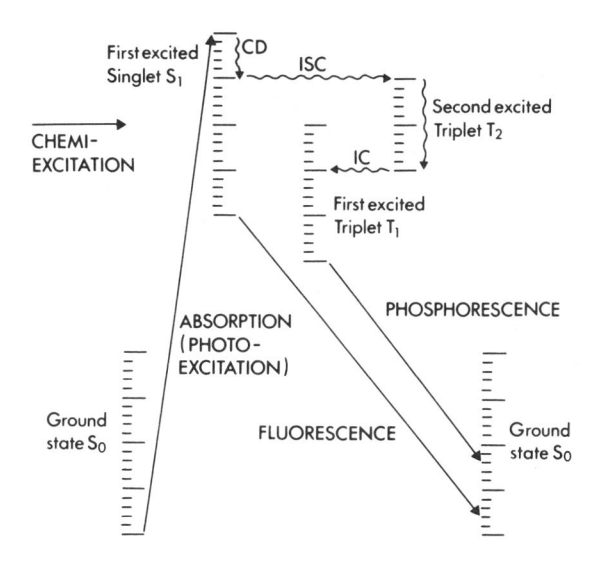

FIGURE 1 Typical Jablonski diagram showing the photochemical and photophysical pathways of chemiluminescence, fluorescence, and phosphorescence. ⟶, Radiative transitions; ⤳, nonradiative transitions; ISC, intersystem crossing; IC, internal conversion; CD, collisional deactivation.

the oxidative excitation reactions of luminol and an aryl acridinium ester. A wide range of substances are capable of catalyzing the luminol reaction (Campbell and Simpson, 1979), from simple species such as transition metal cations to macromolecules such as horseradish peroxidase. This catalytic requirement has led to the development of assay systems in which the catalyst is coupled to the relevant immunogen and quantified by luminol titration (Olsson et al., 1979). However, serum factors such as heme compounds can greatly affect such assays because of their catalytic effects on luminol chemiluminescence; most immunochemical applications involve the coupling of luminol itself to a component of the immune reaction. The most efficient catalyst has been found to be microperoxidase (Schroeder and Yeager, 1978) which is now used universally for luminol label immunoassays where high sensitivities of detection are required. A number of immunological assays have been developed for steroids using derivatives of luminol as the label. The aromatic amino group by which luminol can be derivatized is functionally important with regard to the emission characteristics of the molecule (Schroeder and Yeager, 1978). Much of the early work involved with the coupling of luminol to immunogens produced a drastic reduction in quantum yield (Simpson et al., 1979), that is, the number of photons emitted per mole of luminol.

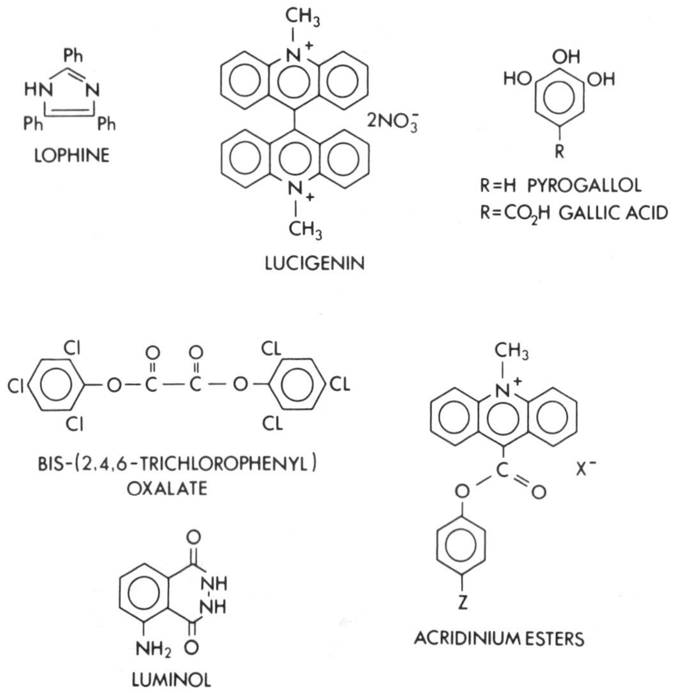

FIGURE 2 Some well-known chemiluminescent molecules.

Recently derivatives of isoluminol have gained popularity since they are less affected by structural alteration than is luminol itself (Table 1).

Of the other chemiluminescent compounds so far studied, only acridinium esters (McCapra et al., 1977) have attracted attention as probes in immunoassays (Simpson et al., 1981; Weeks et al., 1983c). This class of molecules offers several advantages over luminol resulting from the mechanism of the chemiluminescent reaction, which is postulated to be a concerted-multiple bond cleavage. This occurs in the presence of dilute alkaline hydrogen peroxide to yield the vibronically excited molecule 10-methylacridone which emits light on reverting to its ground state (Fig. 3). During the reaction the emissive species dissociates from the rest of the molecule. Thus the emission characteristics are relatively independent of structural changes to the phenolic moiety which can be coupled to the immunoreactant or can constitute part of its structure without affecting the quantum yield of the derivative.

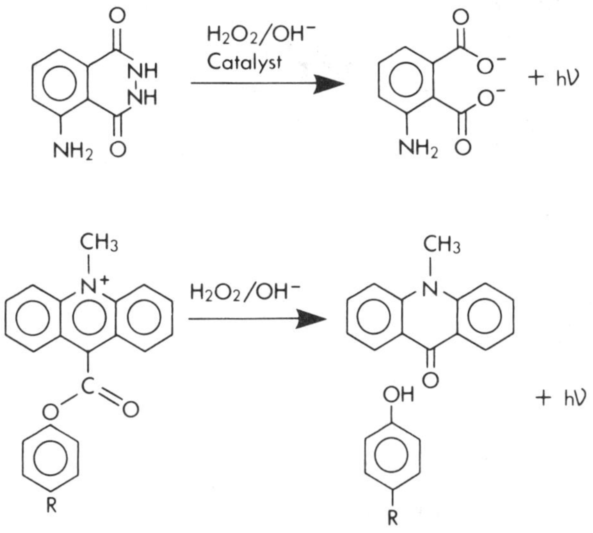

FIGURE 3 Chemiluminescent reactions of luminol and acridinium esters.

B. Quantitation

Measurement of bio- and chemiluminescence was originally carried out by the use of scintillation counters with their photomultiplier tubes taken out of coincidence. Recently several commercial luminometers have become available though none has been specifically designed for chemiluminescence immunoassay. For this reason much of the work has been done on in house equipment utilizing plastic tubes or microtiter plates. Generally commercial equipment relies on analog current measurement rather than digital photon counting. However, digital photon counters are available, including the Biolumat 950 (Berthold, Wildbad, West Germany) on which our most recent work has been carried out.

Chemiluminescence has different measurement requirements from bioluminescence which is often much longer lived. The former is commonly quantified in terms of the peak height of the chemiluminescent reaction profile or the integral of the profile over a given time period. Integration is less subject to errors due to mixing artefacts or slight changes in the reaction rate and provides a reproducible means of luminescence measurement. Optimization of initiating reagent concentrations can be used to obtain the total counts from the system in a very short time (< 1 sec in the case of certain acridinium esters) since the rate of the reaction is highly dependent on these parameters. Thus quantitation of luminescence emission from an immunoassay system can be considerably more rapid than gamma counting.

TABLE 1 Detection Limits of Some Isoluminol Derivatives Using H_2O_2 Hematin

Molecule	Detection limit (pM)
Isoluminol	30
N,N-Diethyl isoluminol	1
N-aminobutyl-N-ethyl isoluminol (ABEI)	2
N-aminobutyl isoluminol	20
N-(amino-2-hydroxypropyl) isoluminol	20
N-(amino-2-hydroxypropyl)-N-ethyl isoluminol	5

Source: Schroeder et al., 1978.

Suitable equipment must be available for immunoassays to capitalize on this speed of detection. Recently, a microprocessor-controlled automated luminometer has been introduced by Berthold (Biolumat 950). This system can handle up to 400 sample tubes carried in a continuous flexible belt, with automated injection and data reduction facilities (Fig. 4).

III. IMMUNOASSAYS USING CHEMILUMINESCENT PROBES

A. Labeled Haptens

Considerable work has been published on chemiluminescence immunoassays for haptens, particularly steroids. The pioneering work of Schroeder et al. (1978) involved competitive binding immunoassays for thyroxine and biotin using luminol-based chemiluminescent derivatives. Such assays also demonstrated the potential use of chemiluminescence immunoassays as homogeneous systems, that is, systems which require no separation, in addition to the more familiar heterogeneous systems. It has been demonstrated that luminol luminescence can be enhanced on the binding of the labeled hapten to antibody, thus the greater the concentration of analyte, the lower the chemiluminescence intensity. This technique has been used recently for the measurement of plasma steroids and their urinary metabolites using antibody-enhanced chemiluminescence based on isoluminol derivatives of progesterone (Kohen et al., 1979), estriol-16α-glucuronide (Kohen et al., 1980a), and cortisol (Kohen et al., 1980b). Such homogeneous assays are prone to interference from other molecules and are relatively insensitive because of the high chemiluminescence background and the high concentrations of labeled molecules necessary.

Heterogeneous assays have found greater applicability than homogeneous assays and several steroid assays have been reported, for example, plasma progesterone (Kohen et al., 1981), estradiol (Kim et al., 1982), testosterone (Kim et al., 1983) and urinary estrone (Weerasekera et al.,

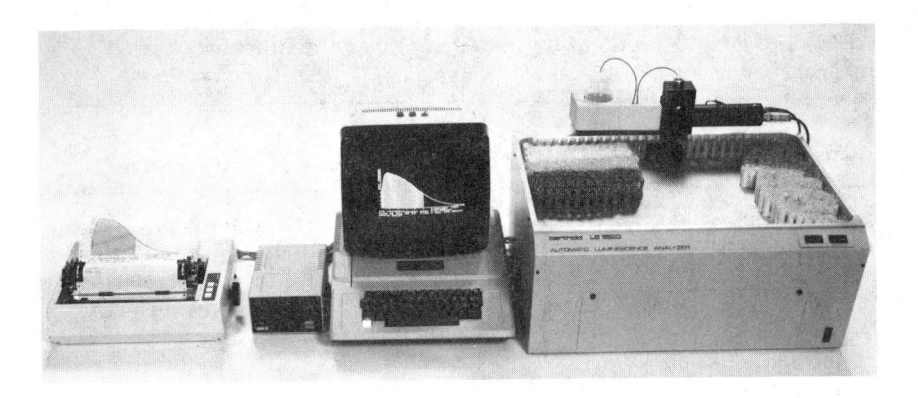

FIGURE 4 Biolumat 950 automated luminometer.

1982), estriol (Barnard et al., 1981), and pregnanediol (Eshhar et al., 1981) glucuronides. The use of solid-phase antibody methodology in this context greatly reduces the problem of nonspecific interference and has enabled clinically useful immunoassays to be established (Table 2). Earlier chemiluminescence immunoassays for steroids involved bound and free antigen separation by dextran-coated charcoal, such as those reported for cortisol (Pazzagli et al., 1981a) and progesterone (Pazzagli et al., 1981b) in a manner parallel to conventional steroid RIA.

B. Labeled Polypeptides

Little work has been reported so far concerning chemiluminescence immunoassays for proteins. There are a number of reasons for this. Firstly, proteins in general cannot be subjected to the rigorous conditions required for many chemical reactions without undergoing loss of immunological activity or complete denaturation. Small molecules are generally amenable to such conditions and thus have greater chemical versatility. Secondly, and more importantly, interference effects are often encountered with serum. Small-molecule assays usually involve an extraction procedure which effectively "cleans up" the environment of the species of interest thus minimizing interference. Interference can arise from a number of sources. Luminol, in particular, is subject to serum effects since its chemiluminescent reaction is catalyzed by many inorganic and organic species (Campbell and Simpson, 1979), such as heme proteins. More importantly, the gross absorption spectrum of serum overlaps with the emission spectrum of many chemiluminescent compounds thus decreasing the emission to an extent governed by the constituents of the serum.

For these reasons, the use of solid-phase techniques has proved invaluable, since the solid-phase chemiluminescent complexes produced in

TABLE 2 Assay Detection Limits of Some Chemiluminescent Steroid Immunoassays Using ABEI-Steroid Conjugates

Steroid	Detection limit of assay (pg)	Reference
Progesterone	4.0	Kohen et al., 1981
Estradiol	1.5	Kim et al., 1982
Testosterone	1.0	Kim et al., 1983
Estrone-3-glucuronide	1.5	Weerasekera et al., 1982
Estriol-16α-glucuronide	17.6	Barnard et al., 1981
Pregnanediol-3α-glucuronide	30.0	Eshhar et al., 1981

the immune reaction are washed free of the interfering factors before being quantified luminometrically.

In addition to the problems of interference, the synthesis of chemiluminescent protein derivatives is far from straightforward. Early work made apparent the balance between the characteristics of chemiluminescence emission and molecular structure in that modification of luminol, for coupling to protein, resulted in a dramatic decrease in quantum yield (Simpson et al., 1979). More subtle protein effects have recently been observed by Schroeder et al. (1981) in their development of an immunochemiluminometric assay of hepatitis B surface antigen ($HB_s Ag$). Here a derivative of aminohexyl ethyl isoluminol (AHEI) was used which possesses a long unconjugated, aliphatic side chain. While reaction at the site remote from the isoluminol nucleus would not be expected to bring about a change in chemiluminescent activity, a substantial decrease in emission was observed. This was attributed to quenching by the microenvironment of the isoluminol moiety. This dependence on the microenvironment, manifest either as enhancement or quenching, forms the basis of the homogeneous assays mentioned earlier. It is important to appreciate the extent of such interactions as they may have important effects on the results of assays where the bound fraction is quantified luminometrically.

Simpson et al. (1979) showed the potential of chemiluminescent polypeptide immunoassays using rabbit IgG and sheep (antirabbit IgG) antibodies in model systems. Although labeled proteins of low specific activity were produced, they served to show that the technique could be applied to clinically relevant immunoassays. Here, diazoluminol-labeled sheep (antirabbit IgG) antibodies were used as an indirect label in a coated tube two-site immunoassay for human α-fetoprotein. Much of the early work of Simpson et al. (1979) has been extended to encompass the investigation of the use of the long side-chain derivatives of isoluminol as labels in model

IgG systems using antibody-coated tubes (Cheng et al., 1982). The two-site ICMA for HBsAg reported by Schroeder et al. (1981) provides an indication of the clinical applicability of isoluminol-based derivatives in labeled antibody systems. The assay is quoted as achieving a sensitivity of 2 ng using antibody-coated microtiter plates and a specially constructed luminometer. Here the labeled antibodies were prepared using an N-succinimidyl derivative of AHEI. Such esters are commonly used in protein chemistry since they react with primary and secondary aliphatic amines to yield stable amide bonds (Anderson et al., 1964). The polypeptide immunoassays described thus far have all involved the use of coated tubes and wells or luminometric quantitation of the soluble, free-label, fraction. For labeled antibody assays of high sensitivity as are required for the quantitation of many peptide hormones, such systems are less suitable because of the requirement for high-capacity antibody solid-phase on which labeled immune complexes can be reliably quantified and also the relative loss of reaction speed. Problems can occur in this area of chemiluminescence assays because of the potential interference of certain solid-phase matrices on luminol emission (J.S.A. Simpson, 1980, personal communication).

The use of acridinium esters as chemiluminescent probes has been reported (Simpson et al., 1981), but only recently has a systematic study been undertaken (Weeks et al., 1983a; 1983c). Acridinium esters are analogous to lucigenin but can be more effectively derivatized via a phenyl moiety or arranged so that a phenyl ester is produced on coupling. In contrast to luminol or its variants, acridinium esters have no requirement for a catalyst and undergo a chemiluminescent reaction in the presence of dilute alkaline hydrogen peroxide (Fig. 3). This property offers certain advantages, in particular the absence of the rigorous oxidizing conditions required for luminol chemiluminescence yields a lower chemical background and hence a higher signal-to-noise ratio. Further, less serum interference occurs because of the absence of catalytic effects and is generally limited to light absorption by the serum.

Inspection of the chemiluminescent reaction of acridinium esters makes further advantages apparent. The reaction is postulated to occur via a concerted multiple-bond cleavage mechanism to yield vibronically excited 10-methylacridone which undergoes photonic emission at approximately 430 nm on reverting to its ground state. This mechanism involves dissociation of the excited species from the rest of the molecule which renders the nature of the emission largely independent of any structural modification of R and of any microenvironmental effects.

N-succinimidyl derivatives of acridinium salts have been used to label monoclonal antibodies to human α_1-fetoprotein (Weeks et al., 1983b; 1983c). A two-site immunochemiluminometric assay has been developed with this label which is suitable for prenatal screening for fetal neural tube defects by maternal serum α-fetoprotein (AFP) measurement. The assay utilizes polyclonal sheep (anti-AFP) antibodies covalently coupled to a diazonium derivative of finely divided cellulose and has a sensitivity of 100 pg (~1.4 fmol) with a working range that easily encompasses the normal and elevated ranges encountered between 14 and 18 weeks; gestation (Fig. 5). The performance is comparable to that of an analogous ^{125}I system (Weeks et al., 1981) but has the advantage that the labeled antibodies are more stable (in excess of several months) and can be quantified in less than 10 sec.

An interassay precision profile (Fig. 6) shows the assay to have an effective working range of ~9-300 ng/ml at a 10% CV acceptance limit and ~5-1000 ng/ml at a 15% CV acceptance limit.

Currently this methodology is being used to develop assays for ferritin, thyrotropin (TSH), and human chorionic gonadotropin (hCG). The high sensitivity attainable with these techniques is exemplified by a two-site immunochemiluminometric assay for TSH using monoclonal antibodies (Fig.

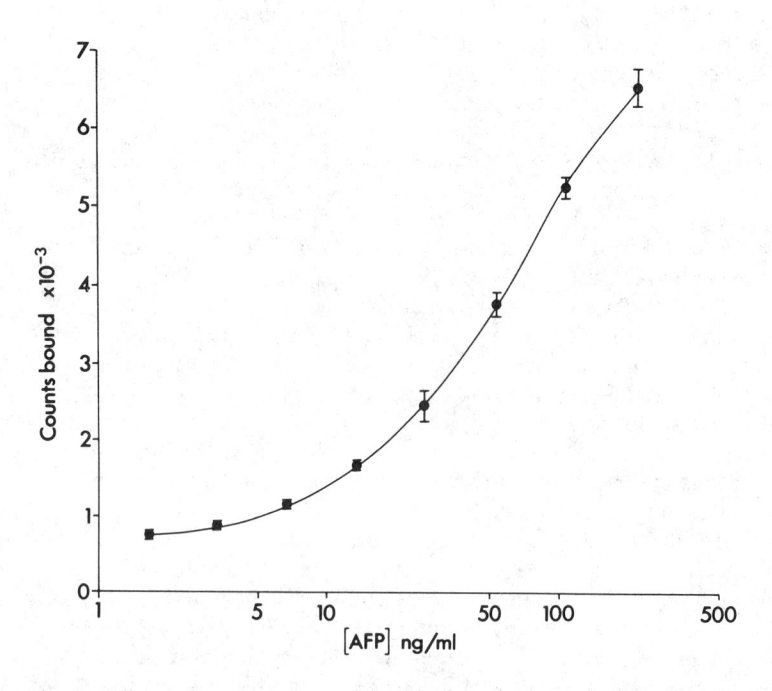

FIGURE 5 Dose-response curve for an AFP two-site ICMA (uncertainties represented by the standard deviations of triplicate measurements).

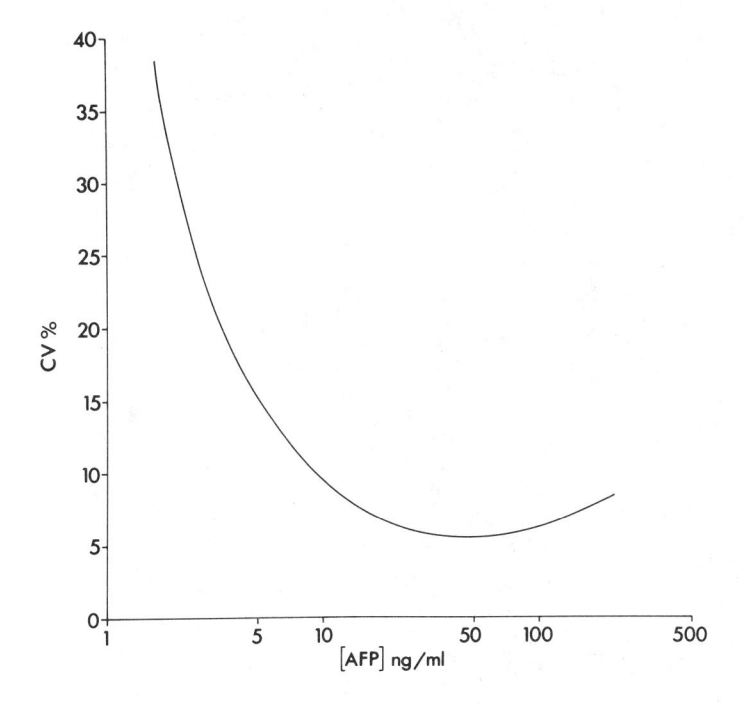

FIGURE 6   Interassay CV precision profile for AFP two-site ICMA.

7). Precision profiles show the sensitivity of this assay to be 0.01 mU/liter and it is currently being used to study TSH levels in thyrotoxic subjects which is not possible using conventional immunoassay techniques.

IV. CONCLUSION

Chemiluminescent probes are capable of yielding immunoassays of similar or even superior performance to those which normally utilize ^{3}H or ^{125}I. Labeled antibody methods utilizing stable, high-specific-activity chemiluminescent monoclonal antibodies which can be rapidly quantified, provide the basis of the rapid, highly sensitive, and specific assays of the future. The introduction of the appropriate automated equipment for high throughout assays utilizing chemiluminescent probes will assist in establishing the place of these techniques in every clinical chemistry laboratory. This combination of assays and equipment will yield systems better characterized and controlled than present immunoassays.

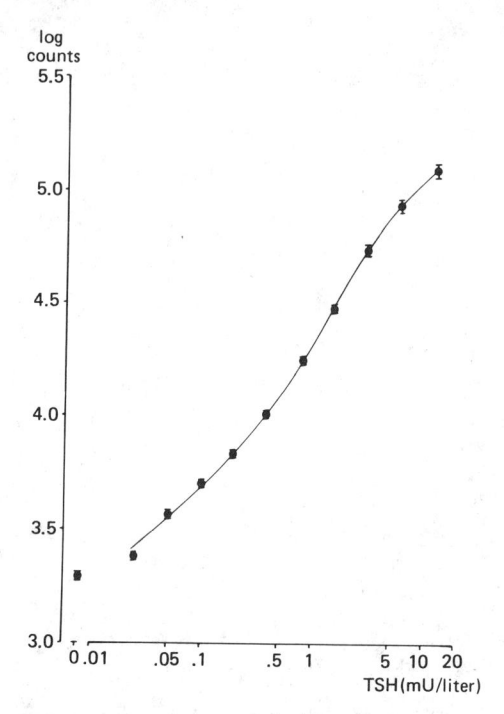

FIGURE 7 Dose-response curve for a TSH two-site ICMA (error limits are ± SD of triplicate measurements).

REFERENCES

Anderson, G. W., Zimmerman, J. E., and Callahan, F. (1964). The use of esters of N-hydroxysuccinimide in peptide synthesis. J. Am. Chem. Soc. 86: 1839.

Barnard, G. J., Collins, W. P., Kohen, F., and Lindner, H. R. (1981). The measurement of urinary estriol-16α-glucuronide by a solid phase chemiluminescence immunoassay. J. Steroid Biochem. 14: 941.

Campbell, A. K. and Simpson, J. S. A. (1979). Chemi- and bio-luminescence as an analytical tool in biology. Techn. Metabol. Res. B213: 1.

Cheng, P. J., Hemmila, I., and Lövgren, T. (1982). Development of solid phase immunoassay using chemiluminescent IgG conjugates. J. Immunol. Methods 48: 159.

Ekins, R. P. (1960). The estimation of thyroxine in human plasma by an electrophoretic technique. Clin. Chim. Acta 5: 453.

Eshhar, Z., Kim, J. B., Barnard, G., Collins, W. P., Glad, S., Lindner, H. R., and Kohen, F. (1981). Use of monoclonal antibody to pregnanediol-3α-glucuronide for the development of a solid phase chemiluminescence immunoassay. Steroids 38: 89.

Kim, J. B., Barnard, G. J., Collins, W. P., Kohen F., and Lindner, H.R. (1983). Solid-phase chemiluminescence immunoassay of plasma testosterone. J. Steroid Biochem. 18: 625.

Kim, J. B., Barnard, G. J., Collins, W. P., Kohen, F., Lindner, H. R., and Eshhar, Z. (1982). Measurement of plasma estradiol-17 β by solid-phase chemiluminescence. Clin. Chem. 28: 1120.

Kohen, F., Pazzagli, M., Kim, J. B., Lindner, H. R., and Boguslaski, R. C. (1979). An assay procedure for plasma progesterone based on antibody-enhanced chemiluminescence. FEBS. Letts. 104: 201.

Kohen, F., Kim, J. B., Barnard, G., and Lindner, H. R. (1980a). An assay for urinary estriol-16 -glucuronide based on antibody-enhanced chemiluminescence. Steroids 36: 405.

Kohen, F., Pazzagli, M., Kim, J. B., and Lindner, H. R. (1980b). An immunoassay for plasma cortisol based on chemiluminescence. Steroids 36: 421.

Kohen, F., Kim, J. B., Lindner, H. R., and Collins, W. P. (1981). Development of a solid-phase chemiluminescence immunoassay for plasma progesterone. Steroids 38: 73.

McCapra, F., Tutt, D. E., and Topping, R. M. (1977). Assay method utilizing chemiluminescence. British Patent No. 1,461,877.

Miles, L. E. M., and Hales, C. N. (1968). Labelled antibodies and immunological assay system. Nature 219: 186.

Olsson, T., Brunius, G., Carlsson, H. E., and Thore, A. (1979). Luminescent immunoassay (LIA): A solid-phase immunoassay monitored by chemiluminescence. J. Immunol. Methods 25: 127.

Pazzagli, M., Kim, J. B., Messeri, G., Kohen, F., Boletti, G. F., Tommasi, A., Salerno, R., and Serio, M. (1981a). Luminescent immunoassay (LIA) of cortisol 2 – Development and validation of the immunoassay monitored by chemiluminescence. J. Steroid Biochem. 14: 1181.

Pazzagli, M., Kim, J. B., Messeri, G., Martinazzo, G., Kohen, F., Franceschetti, F., Tommasi, A., Salerno, R., and Serio, M. (1981b). Luminescent immunoassay (LIA) for progesterone. Clin. Chim. Acta 115: 287.

Schroeder, H. R., and Yeager, F. M. (1978). Chemiluminescent yields and detection limits of some isoluminol derivatives in various oxidation systems. Anal. Chem. 50: 1114.

Schroeder, H. R., Boguslaski, R. C., Carrico, R. J., and Buckler, R. T. (1978). Monitoring specific protein binding reactions with chemiluminescence. In Methods in Enzymology 57, M. De Luca (eds.), Academic Press, New York, p. 424.

Schroeder, H. R., Hines, C. M., Osborn, D. D., Moore, R. P., Hurtle, R. L., Wogoman, F. F., Rogers, R. W., and Vogelhut, P. O. (1981). Immunochemiluminometric assay for hepatitis β surface antigen. Clin. Chem. 27: 1378.

Simpson, J. S. A., Campbell, A. K., Ryall, M. E. T., and Woodhead, J. S. (1979). A stable chemiluminescent labeled antibody for immunological assays. Nature 279: 646.

Simpson, J. S. A., Campbell, A. K., Woodhead, J. S., Richardson, A., Hart, R., and McCapra, F. (1981). Chemiluminescent labels in immunoassay. In Proceedings of the 2nd Int. Symp. on Bioluminescence and Chemiluminescence, M. De Luca and W. D. McElroy (eds.), Academic Press, New York, p. 673.

Smith, D. S., Al-Hakiem, H. H., and Landon, J. (1981). A review of fluoroimmunoassay and immunofluorometric assay. Ann. Clin. Biochem. 18: 253.

Voller, A., Bartlett, A., and Bidwell, D. E. (eds.), (1981). Immunoassays for the 80's. MTP Press Ltd., Lancaster, England.

Weeks, I., Kemp, H. A., and Woodhead, J. S. (1981). Two-site assays of human α_1-fetoprotein using ^{125}I-labeled monoclonal antibodies. Biosci. Rep. 1: 785.

Weeks, I., McCapra, F., Campbell, A. K., and Woodhead, J. S. (1983). Immunoassays using chemiluminescence labeled antibodies. In Immunoassays for Clinical Chemistry, W. M. Hunter and J. E. T. Corrie (eds.), Churchill Livingstone, Edinburgh, 525.

Weeks, I., Beheshti, I., McCapra, F., Campbell, A. K., and Woodhead, J. S. (1983a). Acridinium esters as high specific chemiluminescent labels in immunoassay. Clin. Chem. 29: 1474.

Weeks, I., Campbell, A. K., and Woodhead, J. S. (1983b). Two-site immunochemiluminometric assay for human α_1-fetoprotein. Clin. Chem. 29: 1480.

Weerasekera, D. A., Kim, J. B., Barnard, G. J., Collins, W. P., Kohen, F., and Lindner, H. R. (1982). Monitoring ovarian function by a solid-phase chemiluminescence immunoassay. Acta Endocrinol. 101: 254.

Wisdom, G. B. (1976). Enzyme-immunoassay. Clin. Chem. 22: 1243.

Yalow, R. S., and Berson, S. A. (1960). Immunoassay of endogenous plasma insulin in man. J. Clin. Invest. 39: 1157.

6
Nephelometric Methods

JOHN T. WHICHER
Bristol Royal Infirmary, Bristol, England

DAVID E. PERRY
Westminster Medical School, London, England

I. INTRODUCTION

In the last 10 years there has been an enormous increase in the use of immunochemical techniques for the measurement of proteins in biological fluids. This has led to a diversification of methods and a greater emphasis on automation. For proteins present at concentrations above 1 mg/liter, gel-based techniques such as radial immunodiffusion and Laurell rocket electrophoresis are being replaced by fluid phase systems such as nephelometry and turbidimetry. These techniques, because many can be automated, are faster, and may be cheaper and more precise and for these reasons are increasing in popularity.

The measurement of antibody-antigen complexes by their ability to scatter light — nephelometry — was first described by Libby in 1938 in a paper entitled "The photonreflectometer: an instrument for the measurement of turbid systems" (1938a) with a second paper on its application for determining the potency of antipneumococcal serum (1938b). Though these works were widely quoted as a turbidimetric assay, the instrument was in fact a nephelometer and considerable use was made of it during the next 10 years for investigating antisera to bacterial products. It was not until 1947 that Chow used it for quantitating a human protein, albumin, using an antiserum raised in the rabbit. This was followed by Gitlin and Edelhoch in 1951 who described a similar assay using an equine antiserum. The main problem at the time was the difficulty in purifying and characterizing plasma proteins, with the result that few specific antisera were available. However, Schultz and Schwick (1959), working in the Behring Institute, did show as early as 1959 that many plasma proteins could be quantitated immunochemically by measuring the decrease in light transmitted by antibody-antigen complexes: turbidimetry. During the next 10 years turbidimetry was widely used for quantitating albumin but despite the availability of improved antisera the technique was not further developed and was hampered by the lack of adequately sensitive photometers. Ritchie and colleagues, in 1969, brought nephelometry to the forefront again with an automated system which soon became available as the Technicon Automated Immunoprecipitation System. This instrument revolutionized protein assay in the routine clinical laboratory as it allowed considerably improved speed and precision. It also provided the stimulus to Hellsing and Laurent (1964) and others to investigate polymer enhancement of the antibody-antigen reaction which greatly reduced the reaction time of the assay. The Technicon continuous-flow concept proved very suitable for immunochemical measurement of proteins as it was automated and reaction was precisely controlled by the length of the mixing coils.

Knowledge of the kinetics of the antibody-antigen reaction and the factors affecting it advanced considerably during the next few years, in particular as a result of the work of Savory and co-workers (1974) and Killingsworth and Savory (1973). This understanding led to the development of a number of commercial discrete nephelometric systems employing various refinements such as laser light sources and microprocessor-controlled kinetic measurement. During the latter part of the 1970s nephelometry enjoyed its heyday while, with the advent of the 1980s, it was clear that with the knowledge gained using nephelometry turbidimetry could be equally successfully employed for specific protein measurement. The increased availability of sensitive automated photometers such as discrete enzyme analyzers and centrifugal analyzers speeded this development. Such methods have the advantage of utilizing equipment widely available in the laboratory and also useful for other purposes. However, despite these threats, nephelometry is likely to become more important in the future with the exciting development of nephelometric inhibition assays for haptens such as drugs and steroids, providing an important alternative to methods such as enzyme multiplied immunoassay technique (EMIT).

II. THE PHYSICS OF LIGHT SCATTERING

The early work of Tyndall in 1854 and Rayleigh in 1871 demonstrated that particles from diatomic size upwards scatter light when suspended in liquids or gases. This is because the electromagnetic light radiation causes electron clouds within the molecule to oscillate in synchrony with the frequency or wavelength of the incident radiation. The degree to which this will occur is known as the polarizability of the molecule (α). The oscillating electrons then radiate light, predominantly of the same wavelength, from the particle in all directions. For a very small particle the reradiated light waves are in phase and reinforce each other resulting in a symmetrical, though not spherical, envelope of scattered light (Fig. 1).

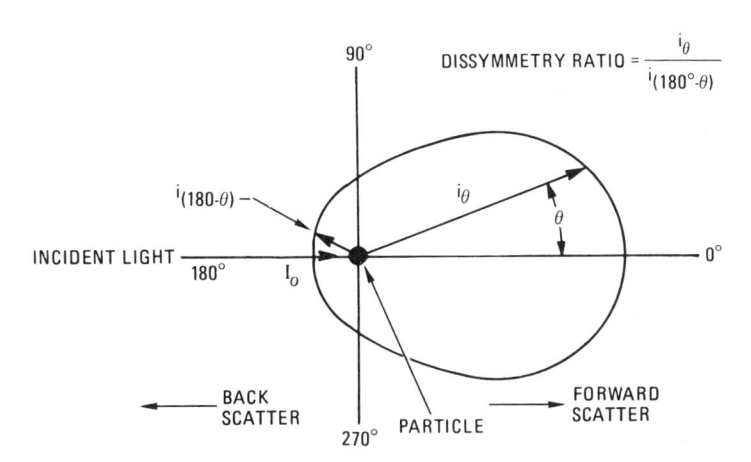

FIGURE 1 The angular distribution of light scattered by a particle of a diameter of less than or similar size to the wavelength of the incident light (Rayleigh-Debye Scattering). (From Kusnetz and Mansberg, 1978.)

For larger macromolecular particles interference between reradiated light waves from various parts of the particle results in a series of maxima and minima (Fig. 2). As the particle increases in size more light is scattered at a forward angle and this asymmetrical reradiation of light (dissymmetry ratio) may be used for measuring particle dimension. This is true for particles in which the molecular arrangement results in a random distribution of scattering centres. If, as in a crystal, they are close together and regularly spaced then the regular phase relationships may cause destructive interference of scattered light which results in an appearance of great clarity, as in a diamond.

The mathematics of light scattering were studied in detail by Rayleigh (1881, 1910) Mie (1908), and Debye (1915). Rayleigh showed that for small particles of greatest dimension less than 1/20 of the wavelength of the illuminating light, the intensity 1_θ of scattered light at a distance r from a particle of polarizability is

$$1 = 1_\theta \frac{8\pi^4 \alpha^2}{\lambda^4 r} (1 + \cos^2\theta)$$

where 1_0 is the intensity of the incident light, λ is the wavelength of illuminating light, and θ is the angle at which scatter is measured from the transmitted light path (Fig. 1) (Tanford, 1961; Van Holde, 1971).

Mie, in 1908, further extended the mathematical concepts of light scattering to include particles of all sizes. Debye studied particles of a size around that of the wavelength of the incident light and confirmed that scattering was dependent on both size and shape. He further developed

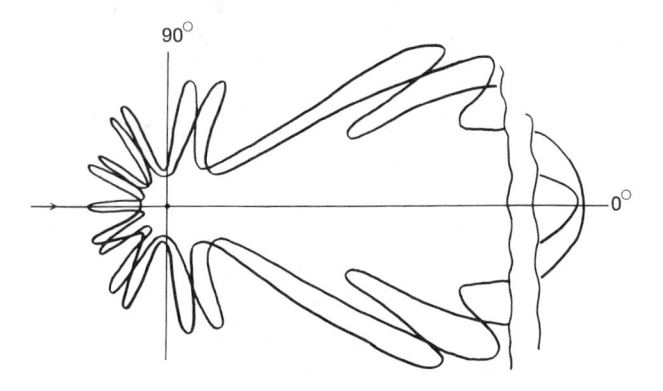

FIGURE 2 The envelope of light scattered from a large particle of size greater than the wavelength of the incident light. Maxima and minima result from destructive interference between reradiated light waves from different molecular constituents within the particle.

Rayleigh's theories to account for particle size based upon the radius of gyration. It is clear from the theories developed by these workers that the angle of maximum scatter is dependent upon particle shape becoming more asymmetrical and distributed in a forward direction with increasing molecular size. As the intensity of the light scatter is proportional to $1/\lambda^4$, short-wavelength light is much more effectively scattered than long, thus accounting for the blueness of the sky and the redness of the transmitted light of a sunset.

When light is scattered by a suspension of particles in solution the overall intensity of the light scattered is the sum of the intensities of the light scattered from all the individual particles provided that the suspension is sufficiently dilute to avoid interference between particles. Such dilute suspensions of particles result in a very small proportion of the incident light being scattered.

III. OPTICAL REQUIREMENTS FOR NEPHELOMETRY

The optical requirements of nephelometric systems for measuring the concentration of particles such as antibody-antigen complexes suspended in solution are derived from our knowledge of light-scattering theory.

The small proportion of light scattered by a sufficiently dilute suspension of particles results in the need for high-intensity light sources and sensitive photodetectors. This has encouraged the use of lasers which provide a very high intensity light source concentrated over a small area (0.5 - 0.8 mm diameter). They are cheap and light and do not require cooling as other light sources of a similar intensity usually do. Commercial nephelometers all use helium-neon lasers producing light of a wavelength 632.8 nm.

Stray light presents a problem in nephelometry as light reflected from internal surfaces such as cuvettes may enter the photodetector increasing the signal-to-noise ratio and thus decreasing the sensitivity of the instrument (Kusnetz and Mansberg, 1978). This is minimized again by the use of lasers providing a highly collimated beam of light without the requirements for complex optics. Other approaches have been used to minimize interference such as the use of two photodetectors placed at 90° to each other and to the light source in order to compensate for asymmetrical light scatter which may arise from the cuvette and particles of dust. Electrical filtration of large-amplitude scatter of the sort which may arise from dust particles has also been employed in some instruments. Such problems are minimized by the filtration of reagents and the use of clean dust-free cuvettes.

During the early phases of the antibody-antigen reaction the median dysymmetry ratio will be low, as IgG alone has a d/λ of 0.07. However, as large complexes are formed aggregates generally exceed d/λ of 0.1 with a high dysymmetry ratio (Fig. 1) (Kusnetz and Mansberg, 1978). Scatter is thus very much greater at forward angles approaching 0° and a number of nephelometers use a forward angle of between 5° and 40°. Optical systems detecting light at far forward angles are greatly aided by the collimated light of a laser. It is also important to appreciate that endogenous light-scattering particles in biological fluids such as proteins and lipids have a d/λ of less than 0.1 and show little angular dependence. Thus β-lipoprotein has a d/λ of 0.06 and albumin of 0.02. Forward angle-scattering measurement thus decreases the interference due to some endogenous light-scattering materials though not of course that due to endogenous immune complexes.

The optimum wavelength for scatter by antigen-antibody complexes is in the region of 460 nm (Anderson and Sternberg, 1978) with decreasing intensity at longer wavelengths. It is thus unfortunate that the cheap commercially available helium-neon laser produces light of a considerably longer wavelength with consequent loss in sensitivity. The helium-cadmium laser will be more suitable with a wavelength of 441.6 nm but at the moment it is expensive and not widely available.

IV. FORMATION OF LIGHT-SCATTERING COMPLEXES BY ANTIGEN-ANTIBODY REACTION

A. Precipitin Curve

Antibody combines with antigen to form complexes, the size of which varies with the ratio of the reactants. An antiserum contains a spectrum of antibody molecules which bind with varying degrees of tightness (affinity) to a variety of determinants on the antigen molecule. The overall binding constant for all the antibody molecules in an antiserum is known as avidity. The interaction of antibody with antigen follows the laws of mass action with antibody avidity determining the overall forward rate of reaction:

$$Ab + Ag \rightleftharpoons AbAg$$

The bonds formed between antibody and antigen are noncovalent and depend upon a "closeness of fit" and result from such forces as Van der Waals forces, hydrogen bonding, and hydrophobic interaction. There is thus continuous association and dissociation occurring with rearrangement of binding sites (Steward, 1977). Macromolecular antigens and their antisera have a multiplicity of different interacting determinants and in the presence of an excess of antibody result in a cross-linking of antibody molecules with antigen and the formation of large lattice-like complexes. If an increasing amount of antigen is added to an excess of antibody an increasing number of relatively constant-sized complexes are formed which are stable in solution, (Buffone et al., 1975) though in time they will aggregate to form visible precipitates. In the presence of an excess of antigen, complexes will decrease in size as they will be solubilized by antigen competing for available antibody-binding sites until the antibody molecules are largely present as binary complexes with two antigen molecules. Such small complexes scatter less light than the larger complexes of antibody excess. The result of this interaction in nephelometry is a nephelometric precipitin curve very similar to the Heidelberger type of precipitin curve derived from measuring precipitated antibody-antigen complexes (Heidelberger and Kendall, 1935) (Fig. 3). It is therefore important for a nephelometric assay that the antibody-antigen ratio is always on the side of antibody excess if a quantitative relationship between antigen concentration and light-scattering signal is to be maintained. If this is not the case two widely divergent antigen concetrations may give rise to the same nephelometric response.

The kinetics of the formation of antibody-antigen complexes after combination of antibody with antigen are shown in Fig. 4 (Steward, 1977). The peak rate of formation of light-scattering complexes usually occurs at between 20 sec and 1 min after mixing, though reaction will continue for some minutes. It is important to realize that high-avidity antibodies will react rapidly while low-avidity species may continue to form complexes for several hours after the initiation of the reaction. Light-scattering complexes tend to aggregate with time and thus scatter less light. They also eventually precipitate out of solution.

The peak rate of complex formation is dependent on antigen concentration and shows a relationship similar to that of the precipitin curve itself (Anderson and Sternberg, 1978). The peak rate of reaction is dependent on antibody avidity but usually occurs within the first 60 sec after mixing antibody and antigen and is thus a useful parameter for the rapid measurement of antigen concentration. It has, however, been clearly

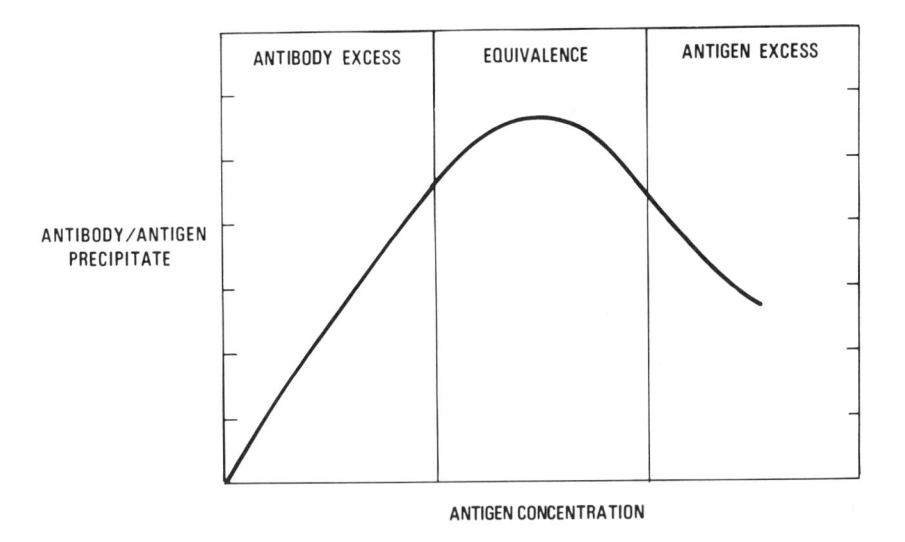

FIGURE 3 The quantitative precipitin curve generated by adding increasing amounts of antigen to a solution of antibody and measuring antibody-antigen complexes by nephelometry. (From Whicher and Blow, 1980.)

shown that the high-affinity antibodies within the antiserum react more rapidly (Cambiaso et al., 1974a). This means that the peak rate may depend on a subpopulation of antibody molecules thus resulting in differences in specificity between methods quantitating antigen using peak rate (kinetic) and the amount of precipitate formed at equilibrium (endpoint) (Ritchie, 1972; Anthony et al., 1980). It is also clear that the interval before peak rate occurs lengthens with decreasing antigen concentration. This is important because kinetic assays must be designed to measure over a time interval which includes peak rate for the antigen concentrations likely to be encountered. This is most successfully achieved by continuous electronic monitoring of the scatter signal (Anderson and Sternberg, 1978).

B. Factors Modifying the Antibody-Antigen Reaction

The antibody-antigen reaction can be thought of as comprising three distinct stages: (1) primary interaction with the formation of binary complexes; (2) cross-linking of binary complexes to form macromolecular lattice-like complexes; (3) aggregation of these complexes to form a precipitate. Both the initial interaction and the subsequent formation of light-scattering complexes are affected by the nature and concentration of the ions present in solution, pH, and the presence of hydrophilic molecules (Marrack and Richards, 1971).

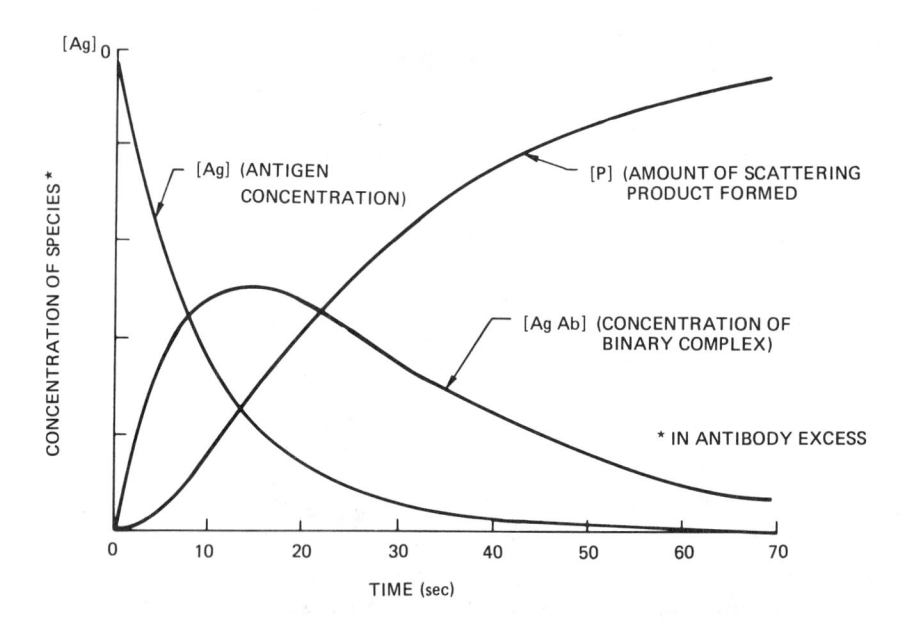

FIGURE 4 The time-dependence of concentrations of antigen, binary antigen-antibody complex, and large light-scattering complexes in an immunoprecipitin reaction. (From Anderson and Sternberg, 1978.)

The primary reaction is particularly important in determining the overall rate of reaction and this may be most usefully manipulated by altering the milieu. Ions form a spectrum from those that encourage macromolecular unfolding or dissociation (chaotropic) to those that encourage association (antichaotropic) (Dandliker et al., 1967). Ions with a low charge density such as thiocyanate, perchlorate, and chloride encourage dissociation and are termed chaotropic (Buffone et al., 1975). Phosphate, sulfate, and fluoride promote association and increase the rate of primary antibody-antigen interaction (Anderson and Sandberg, 1978). The antibody-antigen complexes have less associated solvent molecules than the primary reactants. Competition for solvent molecules, a feature of antichaotropic ions, results in an increased likelihood of antibody finding antigen and forming the less solvated complex. These effects are more marked at higher ionic concentrations. It is thus clear that sodium chloride has an inhibitory effect on the immunoprecipitation reaction, decreasing the number of antibody-antigen complexes formed per unit time but not affecting their character once formed (Savory et al., 1974). However, buffers containing phosphate enhance the reaction and are therefore more desirable for nephelometry. pH (Marrack and Grant, 1953; Dandliker et al., 1967; Marrack and Richards, 1971;

Killingsworth, 1978; Price and Spencer, 1981) and temperature (Price and Spencer, 1981) have little effect.

Similar effects probably give rise to the very marked enhancement of the antibody-antigen reaction brought about by nonionic hydrophilic polymers (Hellsing and Laurent, 1964). A wide range of such polymers has been used in nephelometry, all of which are characteristically molecules with a large hydrated radius (Hellsing, 1969). They compete for water molecules, decreasing their availability and thus encouraging association of other molecules in solution such as antibody or antigen (Laurent and Killander, 1964). The effect on the antibody-antigen reaction is to increase the effective concentration of antibody and antigen driving the primary reaction towards the formation of complexes. The result is an increase in slope of the antibody excess side of the precipitin curve and the displacement of equivalence towards higher antigen concentrations, together with a marked increase in the rate of reaction. In molecular terms there is an enhancement in the binding of both low- and high-affinity antibody molecules, dramatically increasing the avidity of the antiserum and thus the reaction rate. The increase in avidity also results in more antibody molecules reacting during the time of the assay thus increasing the effective antibody titer (dilution of antibody required to produce equivalence).

Hellsing studied the enhancing effect of 24 different polymers on the precipitin reaction using the Technicon automated immunoprecipitation system, a "steady state" nephelometric system (see Sec. VII. E) (Hellsing, 1974). Many of the polymers, with the exception of the polyethylene glycols, posed problems due to their inherent light-scattering properties or their viscosity which prevented filtration of solutions to remove particulate light-scattering materials. Most of the polymers tested enhanced reaction rate at low concentrations, increasing the sensitivity of the early part of the precipitin curve. At high concentrations a few of them moved the equivalence point towards higher antigen levels. Overall the most suitable was polyethylene glycol (PEG) of molecular weight 6000 which, at a concentration of 20 - 40 g/liter, increases reaction rate and sensitivity while at concentrations of greater than 60 g/liter it lengthens the preequivalence part of the precipitin curve (Fig. 5). PEG of MW 4000 has proven effective but requires slightly higher concentrations. The almost universal use of PEG 6000 has revolutionized nephelometry by dramatically decreasing the reaction time in end-point assays and increasing peak rate in kinetic assays by as much as 10-fold (Hellsing, 1978). PEG 6000 also has the advantage that addition to the antiserum solution results in the precipitation of lipoproteins which may be filtered out with a decrease in endogenous light scattering and thus of the reagent blank. It does however have the disadvantage of increasing the light scatter of diluted serum (see Sec. V.A.2) as a result of precipitation of high-molecular-weight proteins such as immune complexes and lipoproteins and also heparin. This may be satisfactorily avoided by predilution of the serum samples in PEG containing buffer followed by centrifugation to remove the precipitated proteins.

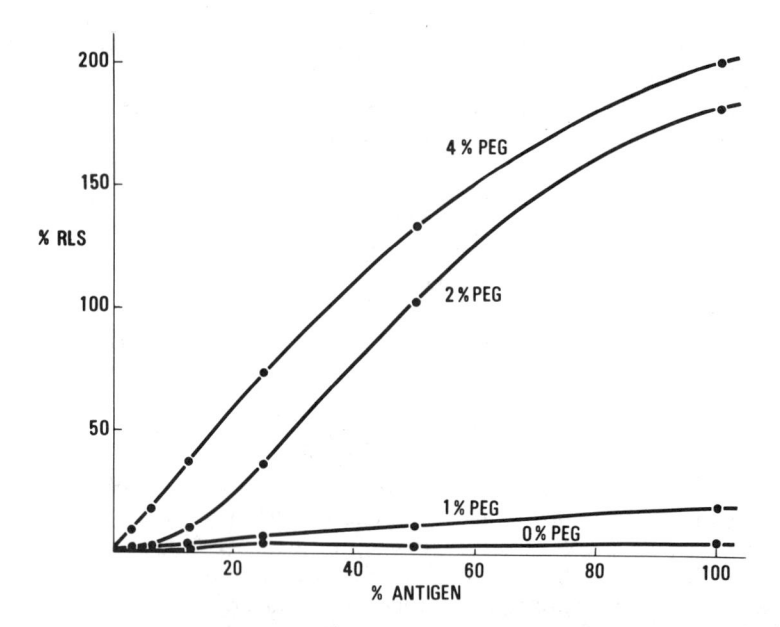

FIGURE 5 The effect on the production of light-scattering (% RLS) complexes generated by adding increasing amounts of antigen to a solution of antibody in the presence of different concentrations of polyethylene glycol (PEG). The assay was for IgA and was carried out in the Hyland Laser Nephelometer PDQ. (From Whicher and Blow, 1980.)

V. END-POINT NEPHELOMETRIC ASSAYS

A. Immunochemical Aspects

As has already been described, the number of light-scattering antigen-antibody complexes formed after reaction has reached completion is directly related to antigen concentration, providing that the complexes are homogenous with respect to both size and shape. The time course of an immunochemical reaction, monitored by nephelometry, is shown in Fig. 21. The "plateau" represents a pseudoequilibrium state, not a true completion of reaction. It is the result of a decreasing rate of light-scattering-complex formation balanced by the formation of larger complexes which scatter less light and tend to precipitate from solution. Under some circumstances "plateau" conditions may be very short-lived, particularly in the absence of polymers such as PEG. It is desirable for end-point nephelometry that measurements be made during the plateau; if this is not the case incubation time becomes the critical determinant in assay precision. This problem may be avoided by automation in which case short incubation times away from plateau conditions may be used.

1. Antiserum Requirements

The antiserum requirements for nephelometry are more rigorous than for the gel-based techniques. Monospecificity is essential and any contaminating antibodies present must be at a level at which they will not form complexes during the reaction time chosen. Unfortunately, with end-point measurements it is likely that contaminating antibodies, even if they are of low affinity, will form complexes during the rather long incubation times used (Fig. 6) (Ritchie, 1975). In many gel-based assays contaminating antibodies form predictable minor precipitin lines which may be recognized

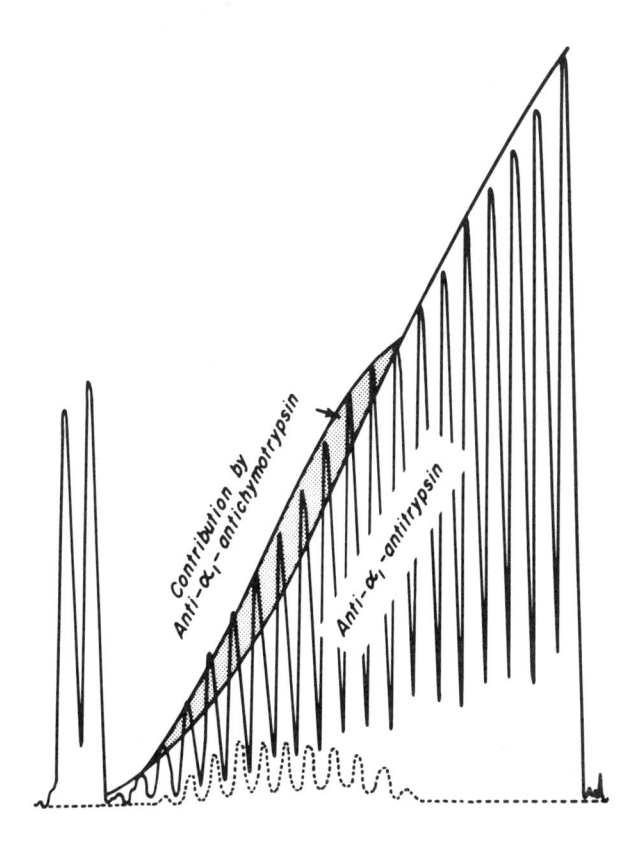

FIGURE 6 Automated Immunoprecipitation System recorder tracing showing the series of peaks of light scattering (y axis) produced by increasing antigen concentrations in an assay for α_1-antitrypsin. Note that a contaminating antibody to α_1-antichymotrypsin exerts a significant effect on the midportion of the calibration curve where normal serum samples will fall. The hatched area represents the contribution of the contaminating antibody if it were separated from the primary antibody. (From Ritchie, 1975.)

and ignored. It may even be possible to measure two antigens simultaneously, for example, by Laurell rockets using mixtures of antisera (Laurell, 1972). The avidity of the antiserum is very important for nephelometry as it is the major factor influencing reaction rate and thus incubation time. The titer of the antiserum for nephelometry is best thought of as the concentration of specific antibody molecules within the antiserum that are able to react and form light-scattering complexes by the time plateau conditions are reached. It is thus clear that avidity influences titer as it affects the speed with which the overall population of antibodies will react. The titer dictates the dilution at which the antiserum will give a sensitive calibration curve and the amount of antigen required to produce equivalence. The titer can be manipulated after the antiserum has been produced by concentration techniques such as ammonium sulphate precipitation or by dilution. Avidity can be manipulated by the method and duration of immunization and the type of animal used. It may also be effectively enhanced by the use of polymers in the reaction mixture (Hellsing, 1978). This is relatively more effective with low-avidity antisera. Gel techniques are much less dependent on avidity as very long incubation times (24-48 hr) can be used, thus allowing even very low affinity antibodies within the antiserum to react. Prolonged incubation times (12 hr) have been used for end-point nephelometry but they are undesirable owing to the inconstant formation, from the high-affinity antibodies present, of large flocculant particles of variable size which tend to precipitate from solution. Even if these are resuspended prior to assay, their variable size and light-scattering properties result in assays with poor precision. High avidity and titer are thus very desirable and antisera must be evaluated carefully by nephelometry before one chooses the most appropriate since there is a poor correlation with behavior in gel systems (Fig. 7) (Ritchie, 1975). The assessment of antisera is described in Sec. VIII.A.

2. Sample Blanks

The major factor limiting the sensitivity of end-point nephelometry is the presence of inherent light-scattering material in human serum. This gives rise to a background or blank light-scattering signal which necessitates a decreased instrument sensitivity to encompass the signal and may result in a very small measured increment due to specific antibody-antigen complexes (Anthony et al. 1980; Whicher and Blow, 1980). It is therefore undesirable to use neat or low dilutions of serum, particularly if it contains high levels of chylomicrons, very low density lipoproteins, low-density lipoproteins, endogenous immune complexes, or aggregated immunoglobulins. The latter are particularly produced by freezing and thawing of samples. Anything less than a final dilution of 1:50 of serum is likely to produce an undesirably high blank value. Unfortunately all those macromolecules are rendered less soluble in polymers such as PEG thus aggravating the problem. Fluids such as cerebrospinal fluid and urine which contain less endogenous light-scattering material are amenable to nephelo-

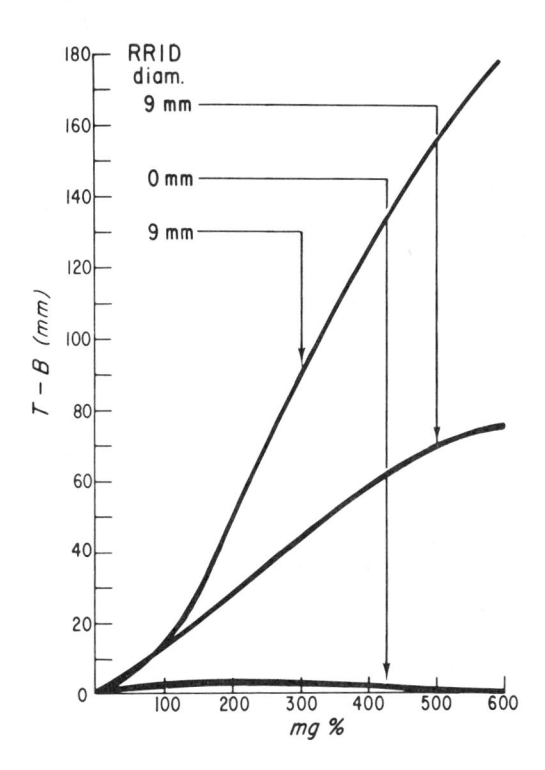

FIGURE 7 The nephelometric response (T-B) for three commercial antisera to α_1-antitrypsin in the Automated Immunoprecipitation System. The reversed radial immunodiffusion (RRID) ring diameters are shown in millimeters. It is clear that two antisera with ring diameters of 9 mm show very different nephelometric responses, thus reflecting the different requirements of the two types of technique. (From Ritchie, 1975.)

metric assays of considerably greater sensitivity than serum. The problem may be minimized by using high dilutions of serum (which necessarily limits the sensitivity of the assay) or by pretreatment of the sample dilutions to remove the offending particles (Whicher and Blow, 1980; Spencer and Price, 1981). Two approaches have been used. The first is to use high concentrations of polymers such as PEG (Hellsing and Enstrom, 1977), protamine sulfate (Wood et al., 1978), or dextran sulfate (Kallner, 1977), to precipitate macromolecules with centrifugation prior to the addition of antibody. This is very effective and the most generally applicable method (see Sec. VIII.A) though care must be taken to avoid precipitation of individual proteins, especially immunoglobulins.

The other approach is to remove lipoproteins specifically, either by solubilization in lipid solvents such as Arklone P (Whicher and Blow, 1980) (ICI, Macclesfield, U.K.) or by precipitation with calcium chloride. Filtration of samples should be avoided because of the possibility of protein absorption on filters (Walsh and Coles, 1980). It is, however, desirable to filter buffers and diluted antibody to minimize reagent blanks (Whicher and Blow, 1980). The presence of dust particles in cuvettes does present problems for some systems but this effect may be minimized by electronic filtration of high-amplitude scatter or by the use of two balanced photomultipliers which compensate for asymmetrical scattering of the sort produced by such large particles. Despite these precautions to minimize sample and reagent blanks, the measurement of blank values is highly desirable in end-point nephelometry as considerable variation may occur between serum samples.

3. Antigen Excess Detection

It is very important that nephelometric assays be designed in such a way that the antigen concentration in the test samples does not enter the antigen excess side of the precipitin curve (see Fig. 3). Clearly there are two potential antigen concentrations which may give rise to a given light-scattering signal. For most proteins an adequate assay range is available from the antibody excess side of the precipitin curve with antisera of good titer and avidity (Whicher & Blow, 1980). However, in the case of monoclonal immunoglobulins, serious problems arise due both to the very high concentrations that may be reached in serum and to their restricted antigenicity resulting in selective consumption of antibodies within the antiserum. This is due to the fact that monoclonal immunoglobulins are restricted in subclass and GM type in relation to the antibodies used to measure them which are raised against polyclonal immunoglobulins (Franklin and Buxbaum, 1978). It is also important to appreciate that the dose-response curves produced by such immunoglobulins may differ in shape from those of their polyclonal counterparts (Fig. 8) (Whicher, 1979). Their measurement by nephelometry is thus unsatisfactory as their lack of parallel behavior with reference materials means that a variable degree of inaccuracy will occur at different levels, thus jeopardizing the use of measurements even for monitoring the progress of individual patients. All samples for immunoglobulin measurement should be examined by electrophoresis for the presence of paraproteins and these may then be quantitated by nonimmunochemical means. If this is not done, systems for detecting antigen excess must be incorporated into the assay procedure. A few other proteins also show a very large incremental change in concentration in pathological conditions where it is not possible for all levels encountered in test samples to lie on the antibody excess side of the precipitin curve. This is particularly true of C-reactive protein in serum and β_2-microglobulin in urine. A number of approaches have been used for antigen excess detection in end-point assays. The simplest is

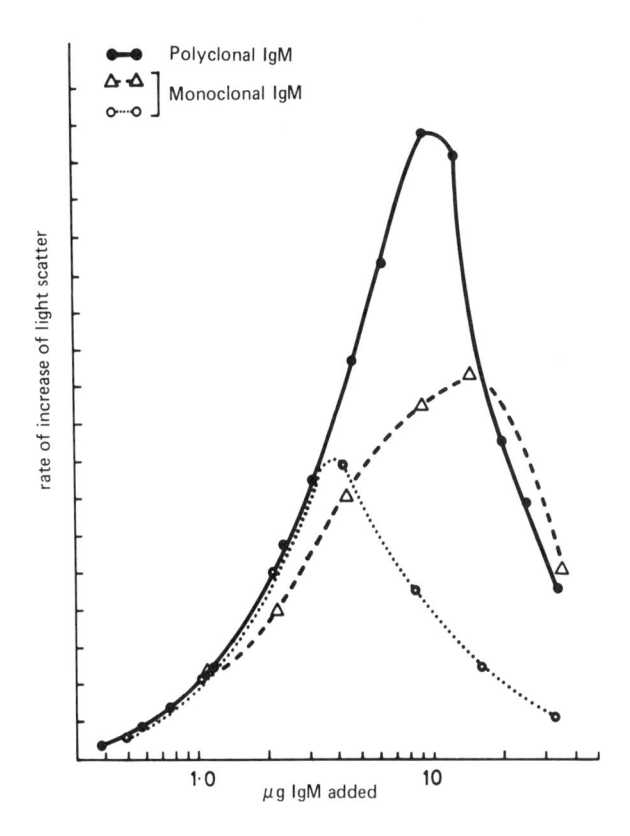

FIGURE 8 The dilution curves of two IgM paraproteins compared with that of a polyclonal standard in a rate nephelometric system. Both paraproteins pass into antigen excess at lower estimated levels of IgM than the polyclonal standard, suggesting selective consumption of antibody. It can be seen that paraprotein Δ shows a dilution curve that is not parallel to the other materials. (From Whicher, 1979.)

routinely to assay susceptible proteins such as β_2-microglobulin at two dilutions when the absence of a positive dose-response relationship, or a reversed one, indicates antigen excess. The sample is then reassayed at a greater dilution. Another commonly used method is to add more antibody or antigen to the reaction mixture, incubate for 5 - 10 min, and look for an increment in light scattering (Deaton et al., 1976). An increase following the addition of antibody implies antigen excess while an increase following addition of antigen demonstrates free antibody and discounts antigen excess. These approaches, while satisfactory, are time-consuming and may be expensive if antibody addition is employed. Antigen excess is directly recognizable only in the case of the Technicon automated immunoprecipitation system and other continuous-flow methods (Ritchie, 1975). This is due to the fact that the signal produced is recorded as a "peak" representing the amount of light-scattering immune complexes in a series of fluid segments of reaction mixture passing through the flow cell of the nephelometer. The central segments contain the highest antigen concentrations while those at the periphery are diluted by the neighboring wash segments. In antigen excess the redissolution of the precipitate in the center of the antigen "bolus" causes a characteristic notching of the peak profile (Fig. 9).

B. Instrumentation

The design of instruments for the immunonephelometric assay of proteins has depended very much on an understanding of the physics of light scattering. Very simple instruments have been used successfully, though with limited sensitivity, while optical and electronic refinements have allowed the measurement of proteins in serum down to concentrations of about 0.5 mg/liter. The absolute requirement for a nephelometer is to be able to measure scattered light without interference from transmitted light. Table 1 shows the factors affecting scattering and the properties of the scattered radiation.

The light source and optics are of considerable importance in nephelometer design since the proportion of incident light scattered is very small and any interference from stray light may significantly affect sensitivity. High-intensity light sources with good collimation of the beam are thus desirable. The light sources in use in commercial nephelometers are: Tungsten bulb: wavelength selection by filters; mercury discharge lamps: peak emission at 355 nm; xenon arc: peak emission at 467 nm; and, helium-neon laser: wavelength 632.8 nm.

Collimation is achieved by slits and lenses of varying complexity (Fig. 10). The helium-neon laser presents an attractive alternative as it is of high intensity and collimation and it is cheap, light, and does not require cooling. Unfortunately, however, the red light of 632.8 nm is not scattered as much as short-wavelength light, the optimal wavelength for scatter by antibody-antigen complexes being in the region of 460 nm. The helium-cadmium laser, with a wavelength of 441.6 nm, provides greater sensitivity but is much more expensive at the present time. Lasers produce light that is highly polarized, giving rise to different optimum scatter measurement angles for different-sized particles. This creates difficulty in kinetic nephelometry where immune complexes are rapidly increasing in size during the measurement period. In fact this is probably not a practical problem

Nephelometric Methods

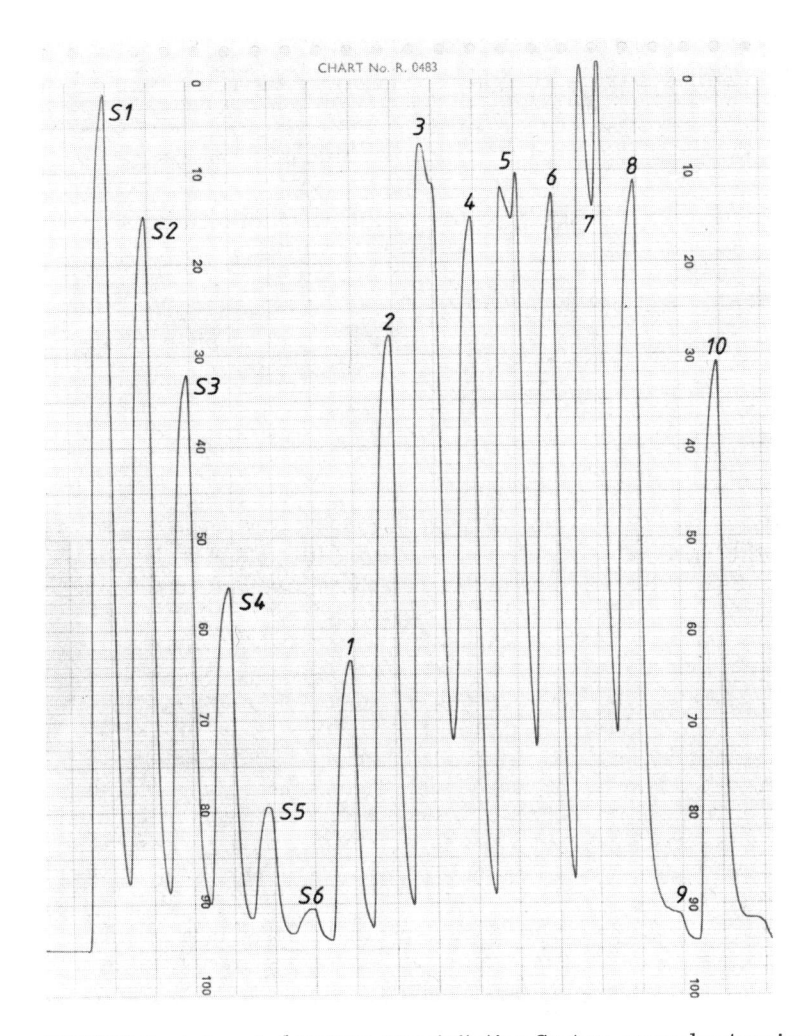

FIGURE 9 Automated Immunoprecipitation System recorder tracing showing peaks of light scattering for standards, S1 - S6 and test samples 1 - 10. Samples 3, 5, and 7 show the characteristic notching of antigen excess. (Albumin assay with samples prediluted 1/200 at a sampling rate of 120/hr; Reference curve from 5 to 40 g/liter.)

TABLE 1 Scattering and Scattered Radiation

Factors affecting scattering	Properties of scattered radiation
Wavelength of incident radiation	Majority is at same wavelength as incident light
Size of particles	
Shape of particles	Changes in polarization are common
Concentration of particles	Very low intensity
Anisotropy of particles	
Refractive index of particles and solvent	
Scattering properties of solvent	
Absorption of light by particles and solvent	

(see Sec. VI.B). Lasers are very satisfactory in end point nephelometry and the fact that stray light from the cuvette has a different polarity from the scattered light of the antibody-antigen complexes has been used in the Behring nephelometer to decrease background interference. This characteristic may also be used to advantage to minimize light scatter from smaller particles such as proteins present in serum.

The angle at which the scattered light is measured is important, as can be seen from Figure 1. A considerable increase in sensitivity can be obtained by measurement at forward angles near to 180° to the source of incident light. In practice nephelometers use anything between 90° and 5° to the axis of transmitted light. The highly collimated beam of the laser light source allows measurement at far forward angles as the narrow beam of 0.5 - 0.8-mm diameter can pass through the cuvette into a very small light trap (Fig. 10).

Attention to the optical design may minimize stray light arising from the cuvette and cuvette compartment and optical pathway of the instrument but spurious light scattering from the material of the cuvette and from dust particles in the reaction mixture will be sensed by the photodetector. A number of approaches have been used to minimize this problem. Two photodetectors with suitable electronics may be used to eliminate the measurement of asymmetrical light scatter arising from very large particles such as dust. These particles also produce scatter of much larger amplitude than that derived from antibody-antigen complexes which may be minimized by electrical filtration using suitable microprocessor circuits (Deaton et al., 1976). The differential polarity of light scattered from plastic cuvettes and from the assay mixture may be used to eliminate the effects of cuvette variation.

Scattered light is generally measured using photomultiplier tubes as these are the most sensitive form of photodetector. The output after

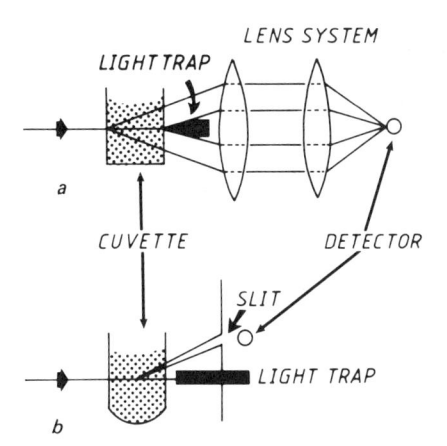

FIGURE 10 The two common light collection systems used with dedicated immunonephelometers. (a) Annular collection system (e.g., Behring). (b) Point collection system (e.g., Hyland).

suitable amplification is presented in most instruments as a digital display. Data handling varies considerably in sophistication while most instruments provide outputs for microcomputers.

VI. KINETIC NEPHELOMETRIC ASSAYS

A. Immunochemical Aspects

Kinetic enzyme and substrate assays have been available for many years with their advantages in speed, accuracy, and precision. However in recent years there has been an increasing interest in the kinetic monitoring of non-enzyme-mediated reactions (Pardue, 1977). This has been particularly aided by the incorporation into instruments of microprocessors to evaluate kinetic data.

As has been discussed in Sec. IV.A, the peak rate of reaction of antibody with antigen usually occurs within the first minute and is related, though not linearly, to antigen concentration if an excess of antibody is present. It is possible to derive a "kinetic precipitin curve" (Fig. 11) which is very similar to that of end-point nephelometric measurement, though the peak of equivalence is often narrower and sharper and the slope of the antigen excess side of the curve very steep. One of the difficulties of kinetic assays is that the interval before the peak rate occurs after initiating the reaction lengthens with decreasing antigen concentration (Fig. 12). Thus if a simple two-point kinetic measurement is made it is essential that the assay be established so that the peak rate is always in-

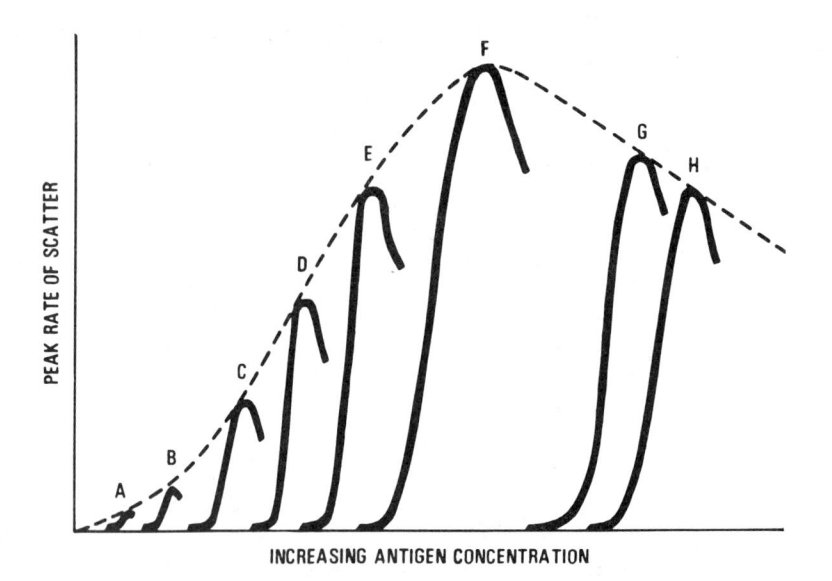

FIGURE 11 Kinetic precipitin curve. A - H represent the peak rate of scatter produced by adding increasing concentrations of antigen to antiserum in the Beckman ICS nephelometer. Kinetic equivalence is at antigen concentration F. (Courtesy of Beckman Instruments, Fullerton, California.)

cluded in the measurement interval for the lowest calibrator used on the reference curve. This results in a slow speed of assay as it may be necessary to use a 40-60-sec measurement interval. This problem may be obviated by the use of continuous electronic differentiation of the rate signal with "picking" of the peak rate. The overall assay rate is speeded up as high-concentration samples will have a much shorter measurement interval (10-15 sec) than low-concentration samples (Anderson and Sternberg, 1978).

1. Antiserum Requirements

Antiserum requirements for kinetic assay are more rigorous than for end-point systems because the measurement interval must be short and the majority of antibodies within the antiserum must have sufficient affinity to react within this time. The major difference between the two types of assay is that only a proportion of the antibody molecules present within the antiserum will react fast enough to contribute to the peak rate. Thus for a given antiserum kinetic equivalence usually occurs at lower antigen concentrations, since it does not include the low-affinity antibodies. The overall avidity of the antiserum is manifested as the time interval between mixing

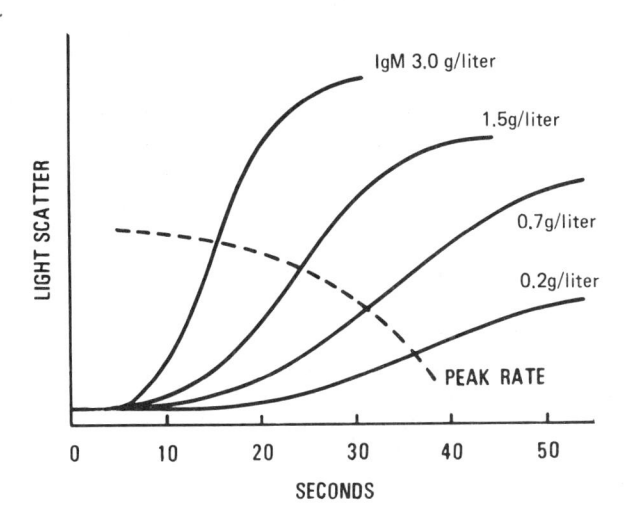

FIGURE 12 The light scattering produced by the addition of varying concentrations of serum to anti-IgM in a kinetic assay. Note the increasing lag phase with decreasing antigen concentrations with peak rate occurring after a longer interval for low antigen concentrations. (From "Immunonephelometric and Immunoturbidimetric Assays for Proteins," Whicher, J. T., Price, C. P., and Spencer, K. CRC Crit. Rev. Clin. Lab. Sci. 1983. 18; 213.)

antibody and antigen and the attainment of peak rate and also in the peak signal resulting from a particular antibody titer (Deverill, 1980). It is thus clear that high-avidity antisera are particularly important. The effect of modifying factors such as pH, ionic species, and the presence of polymers such as PEG is much more marked in kinetic assays (Anderson and Sternberg, 1978). The relative dependence of the assay on the subpopulation of high-affinity antibodies within the antiserum results in interesting differences in specificity between end-point and kinetic nephelometry (see Sec. VIII.B).

2. Sample Blanks

One of the major attractions of kinetic nephelometry is that it obviates the need to measure sample or reagent blanks. However, despite the absence of a problem due to endogenous light-scattering material it is important to appreciate that some macromolecules present in serum may interact with components in the reaction mixture other than the antiserum to produce light scatter. Thus lipoproteins; monoclonal immunoglobulins, especially of IgM class; and endogenous immune complexes may all be precipitated by

PEG to produce light scatter over a time course similar to that of the immunoprecipitin reaction. There may be a considerable peak rate produced by precipitation of the protein followed by a "negative rate" due to flocculation. It is therefore very important in kinetic assays to add the sample to PEG-containing buffer prior to the addition of antiserum and await zero-rate conditions before starting the reaction with antiserum addition.

3. Antigen Excess Detection

Antigen excess presents similar problems in kinetic and end-point nephelometry; however it is more easily detected in kinetic assay. Further addition of antiserum or antigen to the reaction mixture with a new peak rate measurement is a rapid approach to antigen excess detection but may be expensive in antiserum (Anderson and Sternberg, 1978). More interesting are attempts to recognize antigen excess directly by computer analysis of the rate curve. Anderson and Sternberg have examined the kinetics of the antibody-antigen reaction in great detail using a prototype of the Beckman Kinetic Nephelometer. They found both the time taken to reach peak rate and the shape of the rate curve depend on antigen concentration and may be used to detect antigen excess. Figure 13 shows the relationship of "time to peak rate" and peak rate to antigen concentration. It can be seen that the minimum of the time curve occurs near the maximum peak rate with respect to antigen concentration but there is a difference, ΔC, in antigen concentration which

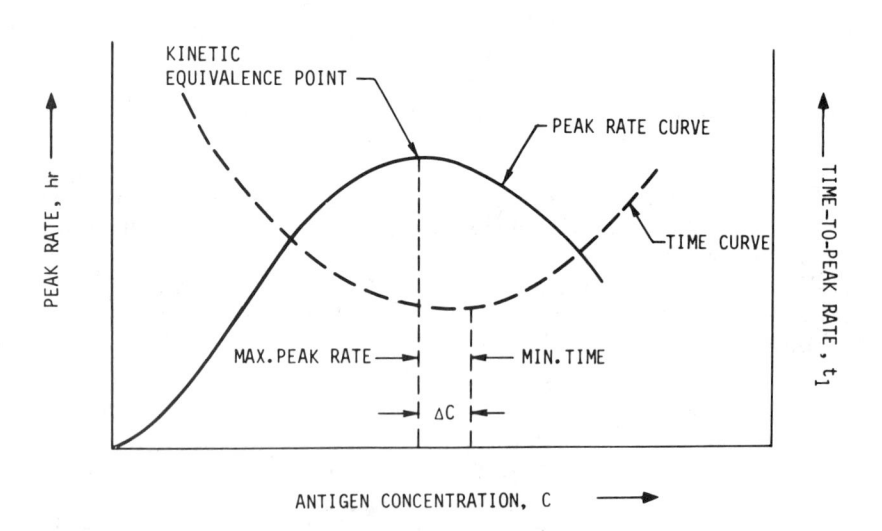

FIGURE 13 Peak rate and time-to-peak rate as functions of antigen concentration in a kinetic nephelometric assay. (From Anderson and Sternberg, 1978).

Nephelometric Methods 139

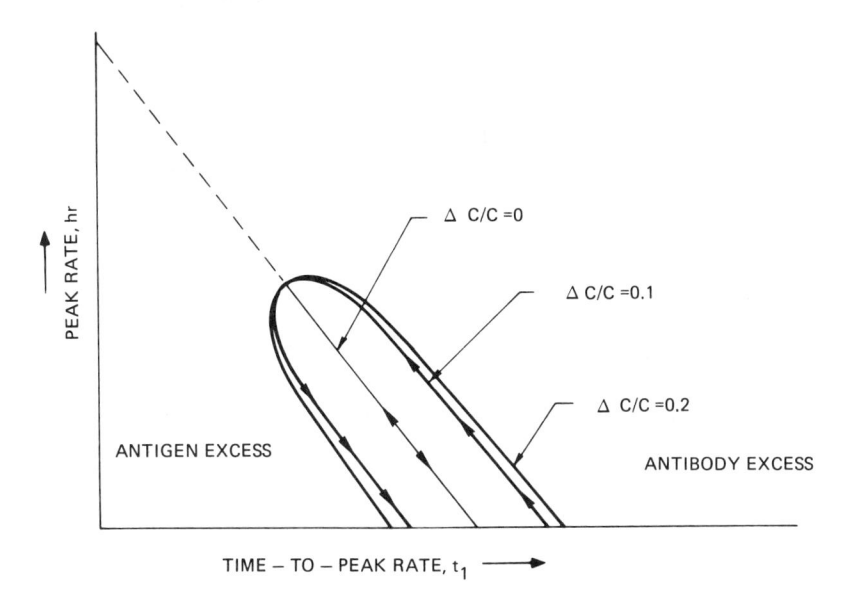

FIGURE 14 Idealized representation of peak rate as a function of time-to-peak rate. (From Anderson and Sternberg, 1978.)

separates them. Figure 14 shows the relationship of peak rate to time to peak rate for three ΔC conditions: $\Delta C/C = 0$ where time minimum and rate maximum occur at the same antigen concentration, $\Delta C/C = 0.1$ and $\Delta C/C = 0.2$. For all except $\Delta C/C = 0$ line, which represent antibody and antigen excess. It is therefore possible using simple algorithms to relate H to t_1 in such a way that the two sides of the parabolic curves may be distinguished. This is thus a way of quantifying the observations of Savory et al. (1974) that reactions in antigen excess show a rapid rate soon falling off while in antibody excess they proceed more slowly initially but continue for a longer time.

Figure 15 shows the shape of rate curves in antibody and antigen excess which reflect the characteristics already discussed. A number of workers, in particular Hills and Tiffany (1980), have used these characteristics to detect antigen excess. However, despite the potential for mathematical approaches these are still not generally reliable enough to be used routinely and postaddition of antibody or antigen remains the most widely used method. This is particularly applicable to kinetic assays as the reaction time is very short and an immediate subsequent rate can be measured (Fig. 16). Anderson and Sternberg (1978) have shown nonparallelism for dilution curves of the same antigen from different patient samples at antigen concentrations above about one-half of the kinetic equivalence concentration.

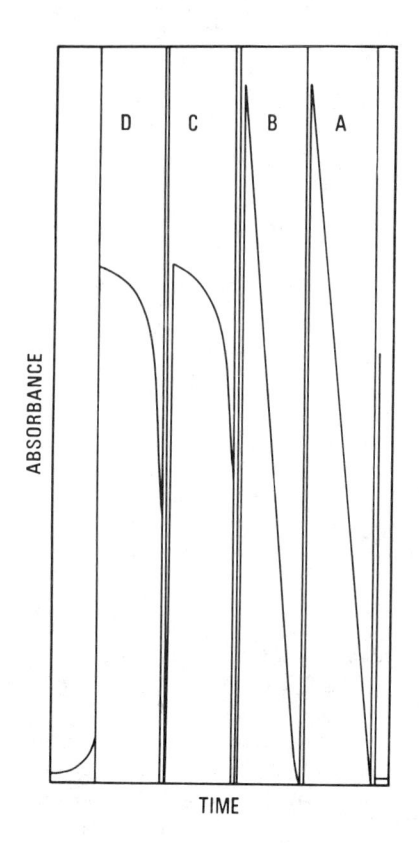

FIGURE 15 Reaction kinetics in antibody and antigen excess for a kinetic turbidimetric assay of IgG, A, B, -antigen excess, C, D, - antibody excess. The same characteristics are seen in nephelometric assays. (From "Immunonephelometric and Immunoturbidimetric Assays for Proteins," Whicher, J. T., Price, C. P., and Spencer, K. CRC Crit. Rev. Clin. Lab. Sci. 1983. 18; 213.)

This is an interesting phenomenon which we have not been able to confirm but is analogous to the problem we have seen with paraproteins. However, these authors have suggested that assays should be designed so that peak rates representing antigen concentrations above half the equivalence concentration are considered as out of range and reassayed with postaddition of antiserum or antigen. This is applicable to most antigens except immunoglobulins where paraproteins in gross antigen excess may produce peak rates lower than this. In this case either other limits can be defined, as is the case in the Beckman nephelometric system, or all samples are checked for antigen excess, as in the Baker system (see Sec. VII).

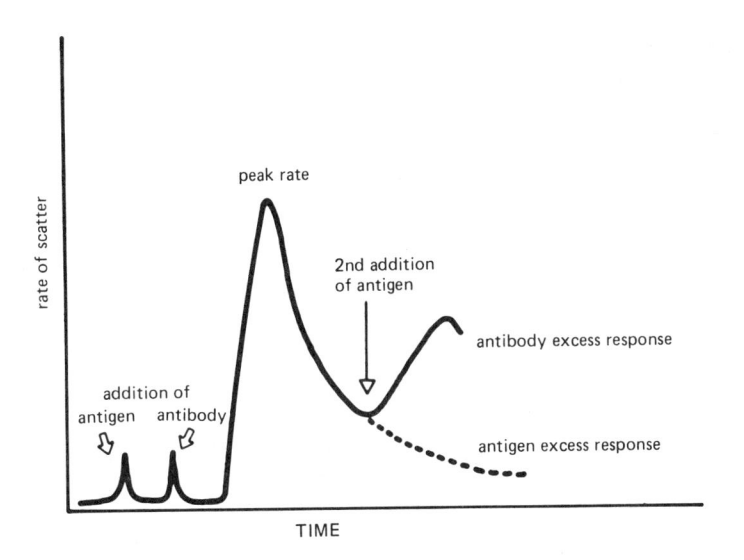

FIGURE 16 The method used for detecting antigen excess in the Beckman ICS kinetic nephelometer. Antigen and antibody are added to the cuvette and a peak rate of increase of light scattering is developed. This falls off to leave a constant amount of light scattering. If further antigen is now added a new rate will be seen if residual antibody is present; if however the initial rate was in antigen excess there will be no subsequent rate. (From Whicher, 1979.)

B. Instrumentation

Much of what has been said about instrument design for end-point nephelometry applies to kinetic measurement. There are, however, a number of fundamental differences which derive from the fact that antibody-antigen complex size changes with time and that peak rate occurs at varying times for different antigen concentrations.

Figure 17 shows the angular dependence of the antibody-antigen complex light-scattering signal together with that of the serum background. To obtain maximum differentiation between the two, Anderson and Sternberg (1978) found that with the cuvette design employed in the Beckman nephelometer angles between 20° and 40° were optimal and they chose 20°. These workers also found that the rate of change of scatter using monochromatic light provided by either a helium-neon laser at 633 nm or a helium-cadmium laser at 442 nm changed erratically during the initial phase of the antibody-antigen reaction whereas light over a broad waveband produced reproducible peak rates. This was presumed to be due to the fact that the optimal scattered wavelength is dependent on particle size which in the early phase of the antibody-antigen reaction is rapidly changing.

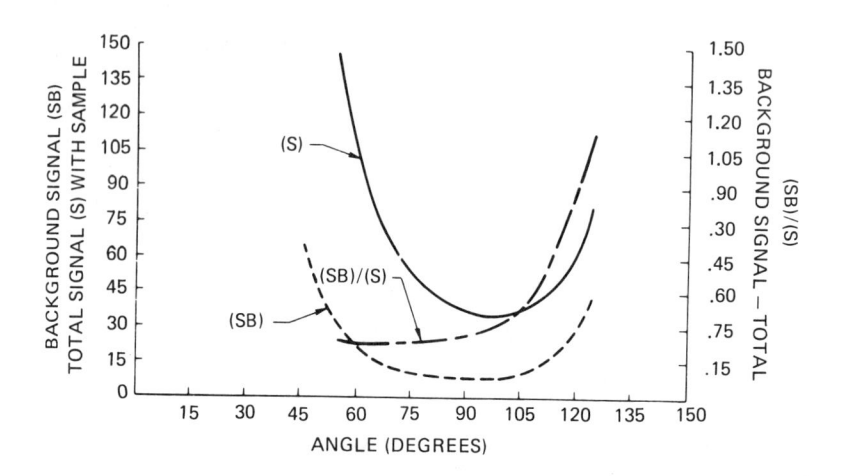

FIGURE 17 Dependence of scatter signal for an immunoprecipitin reaction and background scatter on angle of observation. (From Anderson and Sternberg, 1978.)

Short-wavelength light is still desirable since scattering increases with the fourth power of the wavelength. Interestingly the helium-cadmium laser presents problems due to both fluorescence and the presence of a serum absorption peak near to its wavelength of 422 nm (Anderson and Sternberg, 1978). The Beckman instrument was thus designed using a tungsten light source filtered to pass light from 400 to 550 nm. Despite these considerations the Baker Instruments Series 420 nephelometer which is designed to operate in the kinetic mode, uses a helium-neon laser, apparently quite satisfactorily.

Unlike end-point assay, kinetic measurement requires continuous stirring of the reaction mixture during the measurement interval in order to provide constant reaction kinetics. This requires a magnetic stirring bar which operates without producing bubbles.

An important data acquisition problem arises for kinetic measurement as a result of the time dependence of peak rate. Several instruments employ microprocessors to differentiate the scatter signal and thus to "pick" the peak rate. This approach is probably the most desirable in terms of producing maximum sensitivity. It is, however, possible to take multiple readings and use the linear part of the rate curve as is done for most kinetic enzyme assays. More simply "two-point" kinetic procedures have been used but it is essential to ensure that peak rate occurs within the time window alloted for the measurement. Such an approach usually results in slow assay rates. Another important aspect of kinetic measurement is the need to initiate measurements at the time of addition of antibody to the sample, since peak rate may occur very rapidly.

Kinetic nephelometry thus requires rather more sophisticated equipment than that employed for end-point assays, particularly in terms of data collection. For large batches of assays this mode usually proves rather slow particularly for low-antigen concentrations, owing to the delay before peak rate occurs.

VII. COMMERCIAL IMMUNONEPHELOMETRIC ANALYSIS SYSTEMS

Analysis systems for immunonephelometry include simple nephelometers, laser nephelometers, continuous-flow nephelometers, and centrifugal analyzers. They range in complexity from simple manual instruments to complete automated packages with total sample handling and data processing facilities. A brief description will be given of the currently available commercial systems in the United Kingdom at the time of writing (Table 2).

A. Baker Instruments*

Two immunonephelometric systems are available.

1. Baker Series 410 Immunonephelometer

This instrument was previously marketed by Kallestad as the LSA 290 nephelometer. It is of considerable interest as it is a simple manual nephelometer, designed for end-point analysis, which is less expensive than others on the market. The nephelometer consists of a control and display section housed above the radiation source and optics section containing the cuvette compartment and photodiodes.

The radiation source is a tungsten-filament bulb focused by an integral lens through a 610-nm filter. The light passes through the bottom of the cuvette, the resultant light scatter being detected by two silicon photodiodes placed at 90° to the light source. The light detected by the photodiodes, after compensation for asymmetrical light scattering such as would be produced by bubbles or dust, is converted to light-scattering units (LSU) on the digital display. Controls consist of an auto-zero button, a measure button, a sample blank button, and two adjustments: zero and span.

Sample handling is discrete, each cuvette being placed in the light path by hand. The cuvettes are disposable 10 x 75-mm round-bottomed glass test tubes with no special optical properties but quality controlled to be free of scratches and dust. The tubes must contain a minimum volume of 1 ml. Various types of tissue culture tubes may thus be used as cuvettes.

---

*(Baker Instruments Ltd., Rusham Park, Whitehall Lane, Egham, Surrey, United Kingdom; Baker Instruments Corporation, PO Box 2168, Bethlehem, Pennsylvania 18001)

TABLE 2 Comparison of Commercially Available Nephelometers

Manufacturer/model	Light source	nm	Mode	Automation available
Baker Series 410	Tungsten bulb	610	End point	No
Baker Series 420	He-Ne laser	633	End point/kinetic	Yes
Beckman ICS	Tungsten bulb	400-550	Kinetic	Yes
Behring Laser Nephelometer	He-Ne laser	633	End point	Yes
IL Multistat III	Xenon arc	467	End point/kinetic	Yes
Technicon AIP System	Mercury arc	355	End point	Yes
Hyland Laser Nephelometer PDQ	He-Ne laser	633	End point	Yes

The instrument uses samples at relatively high dilution of the order of 1:100 or greater and they are mixed with dilute antiserum containing PEG. Light scatter is read at the end point of the reaction normally after incubation for about 1 hr at room temperature. The operation of the nephelometer itself is simple; after an initial warm-up period of about 15 min the machine is electronically zeroed against a diluent blank. The light scattering of the antibody solution is measured and stored in memory. The highest reference blank and test value are measured. The span is adjusted by setting the instrument sensitivity to a maximum reading of 2000 light-scattering units using the span knob. Tests and blanks are read; the final scattering value displayed in this mode being the actual light scattering corrected for both antibody blank and individual test blank, both of which are stored in memory and automatically subtracted. This approach gives considerable versatility in sample handling since the span of the assay can be altered simply by setting the maximum reading of 2000 LSU on whichever of the references in the standard curve is chosen.

Optional antigen excess checking is performed by adding a further aliquot of antibody and reading the scatter again after a second short incubation period. Limited data processing is possible using an optional online microcomputer. The complete 410 system comprises the nephelometer, reference sera, reagents including polyethylene glycol and buffer, mixing rack, and a pipettor/dilutor. Assay kits are available for a number of proteins including C3, C4, IgA, IgM, IgG, α_1-antitrypsin, haptoglobin, transferrin, albumin, hemopexin, α_2-macroglobulin, ceruloplasmin, and orosomucoid; IgG and albumin can be measured in CSF. It is not, however, necessary to use the reagents supplied by Baker as, if it is appreciated that this instrument requires a relatively high volume (about 1 ml) of reaction mixture using dilute sample and antiserum, it is possible to develop assays using the operator's own reagents (see Sec. VIII.A).

On evaluation we found the nephelometer quick, reliable, and easy to use, though it was necessary to allow between 5 and 10 sec before a stable reading could be obtained on each tube. The response is linear over the full range of sensitivity settings on the instrument and blank values were not found to affect the accuracy of the test result up to blanks of about 40% of the test light-scattering units. The precision of the instrument for reading a single test is excellent with a coefficient of variation of about 0.65%. Most of the imprecision in analyses is thus inherent in dilution systems used, in dispensing the reagents, and in the calculation of results. The use of the two photodiodes as a method of compensating for scatter from dust and bubbles is not entirely satisfactory and it is important that reagents and cuvettes be completely dust-free. If this is not the case considerable imprecision results and it may be impossible to zero the instrument during the set-up procedure. Thus we felt that while the two photodiodes may compensate for scratched cuvettes, dust does in fact remain a problem.

2. Baker Series 420 Laser Nephelometer

The nephelometer consists of a base containing the cuvette compartment, sample turntable, and laser optics, above which is placed a keyboard and digital display together with the antiserum-dispensing syringe and a paper printout. Functionally the instrument is designed as a partly automated rate nephelometric system which can, however, be used quite satisfactorily in the end-point mode. The reaction takes place within a square glass flowthrough cell fitted with a small magnetic stirrer. The light source is a polarized helium-neon laser, the beam of which is split. One-half enters the flow cell where the scattered light is measured by a photomultiplier at a forward angle of $38.8°$ while the other half of the beam falls on a second photomultiplier used to monitor the laser output. The photomultiplier outputs are analyzed by a microprocessor but the data can be transferred to other systems by an RS232C interface. Predituted calibrants and samples are held on a 25-position turntable from which they are automatically aspirated into the flow cell. The assay reagent (antiserum diluted in buffered PEG solution) enters the flow cell directly from a motor-driven syringe. The sample is added to the cuvette first while the reaction is initiated by the addition of the diluted antiserum. The volume of reactant added (sample typically 3 µl, antibody 10-50 µl) and other reaction parameters, for example, any lag phase required before measurements are initiated, antigen excess parameters, etc., are entered by the analyst via a keypad. Ten assay protocols can be held within the nephelometer memory at any one time.

After the reaction is initiated in the flow cell, the light-scattering signal is automatically measured every 50 msec. The instrument takes the mean over a period of each 250 msec and this provides a data point, 24 of which are collected automatically. Rate data are obtained from the differential of this information though it is possible to print out all 24 data items to check whether peak rate did in fact occur during this period. The standard programs for the instrument automatically conduct an antigen excess check by adding a second aliquot of antibody into the cuvette and measuring four further data points. The criteria for the rate required by this second reaction to indicate antigen excess are set on the keypad by the operator prior to initiating the assay.

Calibration is achieved with a six-point standard curve obtained by taking varying volumes of a predituted calibrant. The microprocessors of the instrument calculate the assay values from this curve using a fourth-order polynomial. Samples are assayed sequentially from the turntable and the results printed out. Considerable versatility is available from the microprocessor and printout. It is possible to flag high and low results, and antigen excess situations, and automatically reanalyze high samples at greater dilutions. It is also possible to eliminate the automatic antigen excess check thus economizing on antiserum and speeding up the rate of assay. The nominal sampling rate is 25 samples/hr; the operator is only required to enter the assay protocol at the outset, initiate the assay sequence, and fill the antibody syringe.

The instrument is marketed together with reagent kits in much the same way as the 410 system but again it is quite possible to use the operator's own reagents. The versatility of the microprocessor data handling in this instrument allows the operator to have considerable control over the assay parameters and also to eliminate much of the automation, allowing this instrument to be used for experimental work in both the kinetic and end-point modes.

B. Beckman Instruments*

The Beckman nephelometric systems are based on the immunochemistry analyzer, a microprocessor-controlled rate nephelometer (Sternberg, 1977; Anderson and Sternberg, 1978). In this system the reaction takes place in a round glass cuvette containing a nickel bar acting as a magnetic stirrer. The cuvette is illuminated by filtered white light (400-550-nm) from a tungsten lamp while scattering is measured at a forward angle of $20°$ using a photomultiplier tube. The scatter signal is subject to peak height analysis to eliminate spurious peaks of high amplitude and is monitored continuously by the instrument microprocessors to determine peak rate. Data output is provided by a digital display while recorder outlets or an RS232C interface is available. Control of the instrument gain or sensitivity is achieved using optically read cards (Fig. 18).

There are two modes of operation for the basic nephelometer. In the manual mode, in which many of the features of the instrument are disabled, the operator sets the gain and performs a manual rate nephelometric assay utilizing the digital readout of peak rate data. Any antisera may be used in this mode, not only those supplied by Beckman. It is very convenient to attach the instrument to a two-pen recorder which will provide a continuous monitor of scatter and rate of change of scatter.

In the automatic mode the nephelometer becomes considerably more sophisticated in terms of data handling but much more difficult to use with non-Beckman reagents. Assays are based on a series of sixfold dilutions of calibrant and test samples. For a given antiserum and calibrant the standard curve characteristics are provided on instrument-read cards attached to the reagent containers (Fig. 18). The reading of these allows the instrument to select the correct gain, generate operate commands, and calculate data from a single point calibration using a predetermined curve-fit program. The cuvette containing buffered PEG solution is placed in the nephelometer and

---

*Beckman RIIC Ltd., Turnpike Road, Crassocks Industrial Estate, High Wycombe HP12 3NR, Buckinghamshire, United Kingdom; Beckman Instruments Inc., 2500 Harbor Boulevard, Fullerton, California.

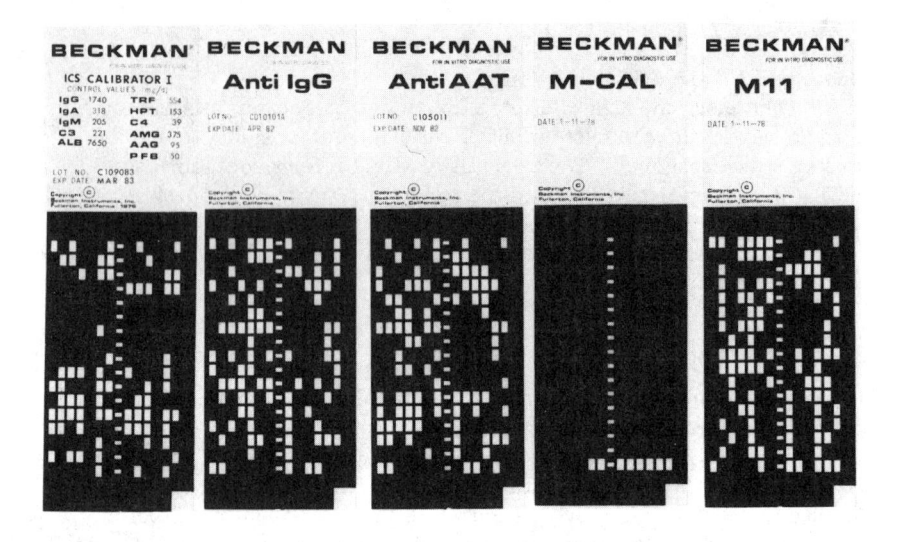

FIGURE 18   Optically read cards for the Beckman ICS System, showing two cards for the manual system (M-Cal and Mll), and cards for the automated system(s).

42 µl of a suitable sample dilution or reference is added. The light scattering of this mixture is monitored until a stable scatter level has been achieved and air bubbles have risen to the surface. The instrument then automatically commands the operator via the digital display to add 42 µl of antibody which automatically triggers the measuring sequence through a dye present within the antibody solution which is detected by a separate photodetector within the instrument. The reaction is monitored until peak rate is obtained (typically 25-90 sec). Calibration is achieved by replicate analysis of a calibrant at a single concentration point; the instrument accepts the calibrant's value once a duplicate response has been obtained within 3%. The data on the instrument-read card for a particular assay allow the instrument to set the gain so that the standard curve span is appropriate for the reading obtained of the single point calibration. Test samples are treated in exactly the same way as calibrants: test light-scattering measurements are automatically converted to concentration using the predetermined calibration curve present on the instrument-read card. The instrument informs the operator via the digital display of samples that are out of range, high or low, with instructions for redilution. Samples that fall into a range of potential antigen excess are reassayed using a second addition of antigen.

The Beckman immunochemical system is thus a very highly microprocessor-controlled kinetic nephelometric analyzer. Its advantages are good precision with a single-point calibration and economy of sample. The disadvantages are that in the automatic mode a rather large proportion of samples, particularly for some assays such as IgM, may have to be reassayed because of the potential of antigen excess. This results in a slow rate of assay of 15 - 30 samples/hr. It is also very difficult to use the instrument in the automated mode without using Beckman reagents because the generation of the instrument-read card with the characteristics for a particular antigen-antibody system is difficult for the operator to undertake.

Full sample-handling automation of the system is now possible using a microprocessor-controlled sample-handling and dilution system which includes a turntable for samples and a fluid transfer device for dilution and reagents. In place of the disposable glass round cuvettes, within the instrument is a glass flowthrough cuvette fitted with a magnetic stirrer. A series of sample dilutions is prepared on the turntable and appropriate volumes of reagent, diluted sample, and antibody are added to the flow cell by the fluid-handling system, which is a series of motor-driven syringes. Under these circumstances the measuring of the reaction is not initiated by dye-triggering within the reaction cell but purely by the microprocessor based upon the time of antibody addition.

The turntable holds 40 samples and the sample-handling module microprocessor is capable of holding data on assay characteristics for up to six assays together with patient information and test requests. It is thus possible to use the instrument in either a discretionary mode, where samples may be assayed for whichever protein is desired, or as a batch machine. The average sampling rate is 27 - 30 samples/hr so that if the instrument were filled with 40 samples and programmed to carry out six assays on each it could run unattended for 8 hr or more.

Despite the fact that this instrument was designed as a kinetic nephelometer for protein assay it is possible to use it for end-point analysis. More recently, assays have been developed for the measurement of drugs by kinetic hapten inhibition in both the manual and fully automated modes. Kits are now available from Beckman for a number of drugs (see Sec. IX).

C. Behringwerke Ag*

The Behring laser nephelometer system is modular comprising the nephelometer itself, sample and cuvette transporter, and dilutor. The system may thus be extended as required. The nephelometer (Sieber and Gross, 1976) is a relatively simple instrument using a helium-neon laser the light from which passes through the two slits of an optical gate and a polarizing filter prior to passing through the reaction mixture contained within plastic disposable cuvettes. The laser beam, after passing through the cuvette, enters a light trap allowing light to be measured at a far forward angle over a circular

---

*Hoechst Pharmaceuticals, Hoechst House, Salisbury Road, Hounslow, Middlesex GW4 6JH, United Kingdom; Behringwerke Ag, Maarburg, Lahn 3550, West Germany.

area between 5 and 12° from the optical path. The use of a polarizing filter in the laser beam eliminates problems due to spurious light scattering from the plastic cuvettes. The use of a relatively powerful laser (4 mW) together with the high intensity of the forward scattered light which is focused by a lens system provides enough energy to allow the use of a simple silicon photodiode as a photodetector. Electronic output from this passes directly without amplification to a digital voltmeter reading from 0 to 19.999 mV. The instrument is self-zeroing but it is not possible to alter the sensitivity or span due to the lack of amplification of the photodetector signal. The voltage output from the photodiodes can be passed to a chart recorder for kinetic measurements or via an RS232C output to a calculator or microcomputer.

In the manual mode diluted samples are added to antibody in microcuvettes containing a total volume of about 250 µl, incubated, and placed in the instrument manually for reading. Relatively concentrated antibody in small volumes without PEG is recommended by the manufacturer. This eliminates the need to measure blanks for many assays. Typically 40 µl of antiserum at a concentration of between neat and 1:10 dilution is required per assay. It is, however, possible to use more dilute antiserum in the presence of PEG if this is required. The system may be used with the operator's own reagents or with Behring kits. Full automation may be achieved by addition of module 2 which is an automatic cuvette-feeding system that sequentially places plastic cuvettes into the nephelometer optical housing. Module 3 is a sample handling system that performs dilutions on sera and dispenses these and antiserum into the cuvettes prior to incubation during the period of travel on the second module. The machine operates in a batch system with samples held in racks coded for the type of analysis and the sample number. Preprogrammed assay protocols may be held in an interfaced microcomputer, allowing complete automation. The incubation time is usually about 30 min in this system. On the whole there is no requirement for blank measurement because blanks are much less in the absence of enhancing polymers such as polyethylene glycol. However, automatic blank correction is possible if required. Calibration is usually performed with a six-point reference curve and a polynomial curve fit. There is no facility for automatic antigen excess check as there is in both the Beckman and the Baker instruments. This may be performed, however, either by reassaying samples on the automated machine or by the manual addition of further antiserum or antigen and the manual rereading of the scatter signal after a second short incubation. The system allows five proteins to be analyzed in one run with a maximum sample throughput of about 240 samples/hr.

This is thus a potentially completely automated end-point nephelometric system based upon one of the simpler and earlier of the laser nephelometers. It has proved very successful in operation, particularly in Germany where many laboratories are using the instrument. It has adequate sensitivity for most proteins analyzed in human serum and provides an instrument with a high throughput. As can be seen, it is possible to analyze samples at a far greater rate using end-point nephelometry than using kinetic measurements owing to the rather long measurement interval in the latter. The system may be used with the operator's own reagents or with specific reagents supplied by Behring.

D. Instrumentation Laboratory*

The IL Multistat III Centrifugal Analyzer has now been fitted with a fluorescent/nephelometric facility. Cuvettes are contained within a disposable acrylic rotor and have a reactant volume of about 150 µl. Antiserum and sample are placed in different compartments in the rotor and mixed in the cuvette by the rapid acceleration commencing the spin. The rotor is illuminated at $90°$ to the rotation axis with light from a xenon arc and light scattered at $90°$ to the optical axis is measured after passing through a holographic grating set at the desired wavelength. The instrument takes a measurement from each cuvette in the disc once in every revolution (1000 rpm). The information from 32 readings constitutes one data point and 24 data points are collected for each cuvette in an assay. The time from commencing the reaction to the first reading is variable. The analytic parameters and data handling are entirely at the discretion of the user but in general a multipoint standard curve is used employing one of a variety of curve-fitting options available from the online computer (PDP 11). The instrument thus represents a versatile centrifugal nephelometer which may be used for kinetic and end-point assays, with the advantages of continuous kinetic monitoring, small sample and antiserum requirement, and ability to run several samples simultaneously in a single rotor (Hills and Tiffany, 1980). It may also be used, of course, for the turbidimetric measurement of proteins (Spencer and Price, 1980) and has allowed a comparison of sensitivity for turbidimetry and nephelometry using an identical instrument under the same conditions (Tiffany, 1980).

E. Technicon Instruments[+]

The Technicon Automated Immunoprecipitation system was the first commercial nephelometer designed for measuring proteins by immunoassay. It arose from the work of a number of researchers in the late 1960s. Alper

---

*Instrumentation Laboratory UK Ltd., Station House, Stamford New Road, Altrincham, Cheshire, United Kingdom; Instrumentation Laboratory Inc., 113 Hartwell Avenue, Lexington, Massachusetts 02173
[+]Technicon Instruments Co. Ltd., Evans House, Hamilton Close, Basingstoke, Hampshire; United Kingdom; Technicon Instruments Corp., Tarrytown, New York 10591

and Propp (1968) first modified a fluorimeter for end-point nephelometric analysis in 1968 and were rapidly followed by Ritchie et al. in 1969 who used a continuous-flow system available from Technicon for developing an automated nephelometer. This was taken up by a number of other workers who developed assays for specific proteins using such a system and this was very soon marketed as the AIP system by Technicon (Alper et al., 1970; Eckman et al., 1970; Ritchie and Graves, 1970; Killingsworth and Savory, 1971; Killingsworth et al., 1971; Larson et al. 1971; Halberstam et al., 1972; Markowitz and Tschida, 1972; Aitken, 1973; Ritchie and Clark, 1973). There is no doubt that the commercial availability of this system provided a large impetus to the measurement and study of serum proteins in clinical medicine and the Technicon AIP system provided the mainstay of immuno-nephelometric assay for 10 years.

The AIP system is based on the conventional continuous-flow technology developed over many years by Technicon. It comprises a sample turntable and a proportioning pump with a manifold capable of diluting samples in a continuously flowing stream of air-segmented fluid with subsequent introduction of antibody into the flow. Incubation is performed in a coil while the final reaction mixture passes to the cuvette of a fluorimeter where light scatter is measured at $90°$. The tubular flow cell eliminates cuvette variability and has a low volume of 150 µl. The light source is a mercury arc lamp filtered at 357 nm and focused by a lens before passing through a narrow aperture into the flow cell. Scattered light exits through another aperture at $90°$ and is measured by a photomultiplier tube. The output after amplification is recorded as a series of peaks on a chart recorder. The fluorimeter contains a reference detector to compensate for fluctuations in light intensity.

The major advantage of the continuous flow system is a small sample and reagent volume requirement together with a very tightly controlled reaction and incubation time. The use of polyethylene glycol in the system has allowed incubation time to be shortened to between 3 and 5 min for most proteins with final readings being taken before completion of antibody-antigen reaction is reached (Hellsing, 1973). The sample dilution and volume and the volume of antiserum used is determined by the tube sizes used in the manifold of the peristaltic pump. Samples automatically diluted in the instrument are typically assayed at a rate of about 60/hr whereas prediluted samples can be assayed at a rate of about 100/hr (Markcroft and Newbanks, 1973). One of the major advantages of the AIP system is the ability to recognize antigen excess as a result of characteristic notching of the peak profile (Ritchie, 1975) (Fig. 9). The system does, however, have the limitation of all end-point nephelometric assays of a sample blank requirement for proteins at low concentration in the serum. Thus if the final sample dilution is less than about 1:600, it is important to reassay the diluted samples without antiserum in the presence of PEG to ascertain the blank value. It is essential in all such systems that blanks be performed in the presence of polymer as this often increases the blank value due to precipitation of proteins present within serum. The sensitivity of the system is more limited than that of some of the other nephelometers discussed, probably due to the nature of the optical system and the need to measure blank values. It is excellent for proteins present in plasma or serum above concentrations of about 2 g/liter but becomes progressively more imprecise below these levels.

F. Hyland*

The Hyland Laser Nephelometer PDQ consists of a control and display section which houses the instrument's electronics with the laser radiation source and an optics section containing the cuvette compartment and photomultiplier tube. The laser is a low-powered (0.5 Mw) helium-neon laser producing a highly collimated beam of light 0.99-mm in diameter at 632.8 nm. The light passes through the cuvette and scattered light is collected by a photomultiplier tube at forward angle of $30°$. The collection angle is set by two fixed slits. The scatter radiation is collected over a variable operator-controlled compute period of between 5 and 90 sec. During this period a pulse analyzer detects and rejects spurious large-amplitude scattering pulses from falling dust particles. The final relative light scatter is displayed on a four-digit digital display correct to one decimal place. The amplified signal is, however, available via recorder outputs (to allow monitoring of at least part of the reaction) and it is also possible to use an RS232C interface. In this basic form the instrument is used for manual end-point nephelometric assays. The instrument has a memory capable of holding the reagent blank and sample blank information and subtracting these from the test scatter to produce the final scatter results. In this mode the antibody-antigen reaction is carried out in disposable 10 x 75-mm glass test tubes in racks supplied by the manufacturer. The reaction tubes hold a volume of about 1 ml of reagent mixture. Typically 0.5 ml of diluted sample and 0.5 ml of diluted antiserum are added together, mixed by inversion, and incubated for a variable period of between 30 min and 1 hr prior to being placed in the light path of the instrument by hand.

The system is designed, unlike the rather similar Behring nephelometer but like the Baker 410, to use a large volume of dilute antiserum in the presence of polyethylene glycol, typically at a concentration of about 40 g/l. Owing to the lack of continuous mixing in the cuvette compartment it is not possible to use this instrument satisfactorily for kinetic measurements. In the manual mode it is thus similar to both the Baker 410 and 420 and also the Behring nephelometer but the use of polyethylene glycol invariably necessitates blank measurement.

---

*Travenol Laboratories Ltd., Caxton Way, Thetford, Norfolk IP24 3SE, United Kingdom; Hyland Division of Travenol Laboratories Inc., Costa Mesa, California.

The instrument can be upgraded by the use of an online microcomputer, such as a Hewlett Packard 9815 and the manufacturers provide suitable software for such an instrument. Typically the programs collect data straight from the instrument and process to produce hard copy of test protein concentrations. A six-point calibration curve is used with a quadratic curve fit. Results outside the reference curve limits are flagged as high or low.

Semiautomation of the system is possible with a sample transporter and a flow cell which essentially incubates the samples and reads them after a fixed period of time but does not perform sample dilution and antibody addition automatically.

Fully automatic sample handling is now possible using the DISC 120 system. This is a discrete analysis system based on reaction cups and sample cups held and transported in two plastic chains. A series of primary dilutions is automatically made from calibrants and samples; the samples are diluted 1:20 and a series of six different dilutions is made from the calibrant to give a standard curve. Different volumes of this primary dilution sequence are taken according to the assay characteristics of the protein being measured. These are added to aliquots of buffered PEG-containing antiserum in the test reaction cups while blanks are prepared from diluted sample with PEG alone. The machine operates in a batch mode assaying each protein sequentially, though the machine may be programmed in a discretionary fashion to analyze proteins selectively from the samples within the sample cups. Incubation is carried out for 1 hr at room temperature; blanks and samples are then read sequentially in the nephelometer by aspiration through a tubular flow cell placed in the cuvette compartment of the instrument. A single blank assay can be used for a number of different protein assays with an automatic correction for the sample volume used. This assumes that there is a linear relationship between blank value and sample concentration. The sample cup chain will hold up to 150 samples with a choice of 20 preprogrammed assay protocols. The sampling rate is 160/hr and results are printed out as they are obtained but they may also be printed out by patient number for a variety of different proteins. It is possible to perform antigen excess checks on a whole batch of samples by the further addition of calibrant and remeasurement of the whole batch after 10 min. This is not, however, an automatic facility.

G. Other Instruments

Many other instruments are adaptable for immunonephelometry. A number of commercial fluorimeters have been used for making measurements at 90° to the incident light path (Killingsworth and Savory; 1972; Savory et al., 1972; Durrington et al., 1976). A number of systems already described can be used in different ways from those for which they were originally marketed. For example, the Hyland PDQ nephelometer has been used as a detector system for continuous-flow analysis. We are bound to point out that there is now a swing towards the use of turbidimetry for specific protein measurements and this technique may be performed on a wide variety of instruments such as the discrete enzyme analyzers and the centrifugal analyzers. It is very probable that these instruments with their versatility and capability of being used for a wide variety of other assays will assume greater importance and popularity in the immunological measurement of proteins in future. Despite this, laser nephelometers are now being used for the measurement of drugs by hapten inhibition. It is, however, likely that in the long run lasers will prove more sensitive than turbidimetry, particularly with the use of such future developments as ultraviolet lasers, pulsed lasers, double-beam lasers, and instruments using differential polarization, all serving to improve the signal-to-noise ratio at low antigen levels and thus improving precision.

VIII. PRACTICALITIES OF IMMUNONEPHELOMETRY

A. Establishing Nephelometric Assays

Assay kits are available for all the commercial nephelometers for a very wide range of clinically valuable proteins. There are, however, many situations in which it is desirable to establish assays for proteins for which it is possible either to raise or purchase antisera. This is particularly true in using nephelometric assay for research purposes and in new areas of clinical practice. The general principles of establishing a nephelometric method, whether end-point or kinetic, are straightforward provided certain important criteria are understood and it should be possible for any laboratory to initiate its own nephelometric methods. It is in many ways unfortunate that there are many descriptions in the literature of nephelometric assays for measuring a wide variety of proteins. These simple developments are not, on the whole worthy of publication and the details of a method developed with a particular antibody are usually not applicable to another worker using a different antiserum from a different source. Whicher and Blow (1980) have provided a comprehensive protocol for the development of nephelometric assays using discrete nephelometers. Similar principles, however, apply to the automated immunoprecipitation system (Ritchie, 1975).

Kinetic nephelometry, and turbidimetry, are based on analogous concepts but require a slightly different type of evaluation; this has now been reviewed in detail by Anderson and Sternberg (1978) and Deverill (1980). The factors to be considered when establishing a nephelometric assay are shown in Table 3. In this section the method of establishing the reaction conditions for both endpoint and kinetic nephelometric assay will be outlined.

TABLE 3 Factors to be Considered when Establishing a Nephelometric Assay

---

Position on the precipitin curve of the antigen-antibody rate range used in the assay

Assay sensitivity and range
  Instrument sensitivity
  Problem of sample blanks
  Values likely to be encountered in patient samples

Time course of the reaction

Need for polymer enhancement

Suitability of the antiserum for nephelometry

Economy of antiserum consumption

Source: Whicher, J. T., Price, C. P., and Spencer, K. (1983). Immunonephelometric and immunoturbidimetric assays for proteins, CRC Crit. Rev. Clin. Lab. Sci. 18: 213.

1. Antigen-Antibody Ratio

The shape of the precipitin curve produced by the reaction of antibody with antigen varies considerably depending upon the nature of the antigen, its antibody, the animal origin of the antiserum, and the presence or absence of polymer. In general, polymer enhancement is desirable as it shifts the equivalence point towards high antigen concentrations and thus allows a greater economy of antiserum consumption while also allowing rather low-affinity antibodies to be effective. As a general guide, it is desirable to have the highest reference standard at no greater antigen concentration than half that of equivalence. The first step is to establish the range of antibody-antigen ratios that can usefully be used, and the second is to decide what total mass of antibody and antigen to introduce into the system. The latter dictates the amount of immunoprecipitate formed and thus the sensitivity of the assay. This will vary of course from instrument to instrument, and an assay developed on one instrument cannot necessarily be transferred to another. In order to establish the optimal antibody-antigen ratio, a series of precipitin curves must be produced for a wide range of antigen dilutions at different antibody concentrations. For instruments using high volumes of reactant with relatively dilute antibody it is essential that these experiments be carried out in the presence of an enhancing polymer, such as 40 g/liter of PEG. Only in the case of the Behring nephelometer is it possible to establish assays in the absence of polymer using small volumes of concentrated antiserum.

The mass of reactants used is dictated by the cost of antiserum and the dilution at which it can be used while producing a sensitive calibration curve. It is usually reasonable to consider using antiserum at dilutions of between 1:20 and 1:200. A series of standard curves is then produced from antigen dilutions at each antiserum concentration. For example, in the case of the Hyland and the Baker instruments, it is convenient for serum proteins to use a range of dilutions of whole serum calibrant from about 1:10 through to 1:100. The volume of these dilutions to be added to the reaction mixture again dictates the amount of antigen added. It is convenient in the first instance to take an arbitrary volume such as 100 µl, which is appropriate for most proteins where a total reaction volume of 1 ml is being used. Indications can be obtained from looking at the recommendations supplied with the instruments for the proteins for which the kits are available.

Figure 19 shows the family of curves obtained for a series of antigen concentrations over a range of antiserum dilutions using an end-point nephelometric system. A similar family of curves could be obtained for a kinetic assay, though the peak of equivalence would tend to be narrower and

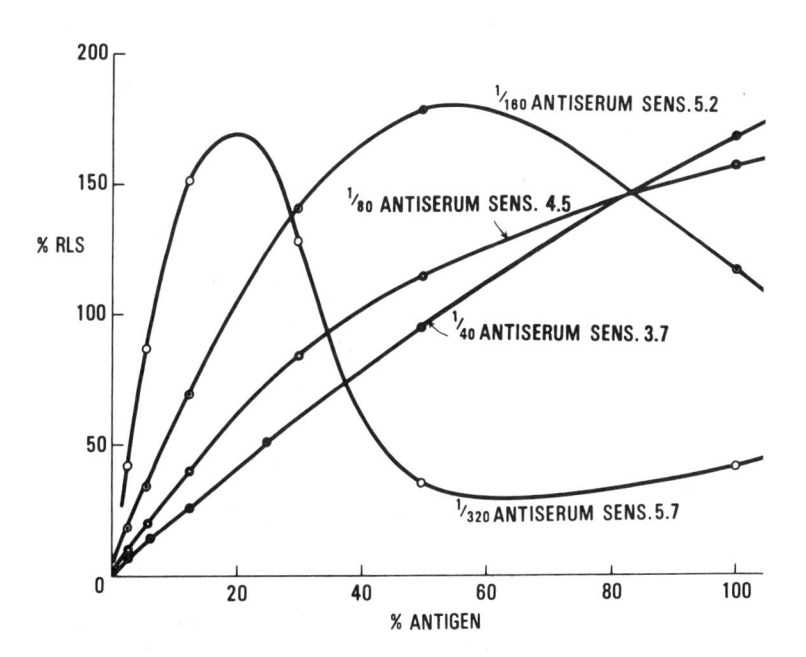

FIGURE 19 Assay of prealbumin on the Hyland nephelometer. 100% antigen concentration is a 1:12.5 dilution of reference serum. Calibration curves have been produced for different antiserum dilutions. Sensitivity setting (Sens) on the instrument is shown. 1:40 and 1:80 dilutions of antiserum are usable but 1:160 and 1:320 are in antigen excess. (From Whicher and Blow, 1980).

at a rather lower antigen concentration. The curves obtained are examined for linearity, sensitivity, and signs of antigen excess at high antigen concentration. It can be seen from Fig. 19 that this assay for prealbumin rapidly passes into antigen excess with the antiserum diluted 1:160 at about 50% of antigen added. The equivalence point has thus been established and it is possible to calculate for a given antiserum dilution the number of grams of antigen required to produce equivalence. It is desirable to operate at less than half this antigen-antibody ratio. Providing the antibody-antigen ratio remains the same, the amount of antibody and antigen present in the system can now be altered. Within the limits prescribed by antiserum economy and instrument sensitivity, the range of antibody-antigen ratios finally chosen should provide a linear or slightly convex dilution curve. In general it is desirable to aim for as low a dilution of serum as possible and for as high a concentration of antiserum as is economical. Figure 20 shows a modification of the assay conditions used in Figure 19 and it can be seen that by decreasing the amount of antigen used and increasing the instrument sensitivity slightly (from 3.7 at 1:40 antiserum dilution to 4.7), a new series of curves can be produced.

The assay range can easily be adjusted by altering the ratio of the serum sample dilutions to those of the reference materials, allowing the reference curve to encompass as much of the normal and pathological range of concentration of the protein being measured as possible. It is most convenient to establish a reference curve from a series of dilutions of reference material and then to alter the sample dilution according to the biological fluid being used. For example, from Fig. 19 the prealbumin assay was established using 100 µl of 1:25 serum sample dilution and antiserum was diluted at 1:80. On the basis of the prealbumin content of the reference serum the protein content for the highest reference was 1.18 µg of antigen in the reaction mixture. At the lowest reference, 3.125% of the top reference, this was 0.037 µg of antigen. It is thus clear that using these assay conditions with 100 µl of undiluted urine as the test sample prealbumin would be measurable at a concentration of 0.0015 µg/µl or 1.5 mg/liter. If, under these conditions, the blank values for the urine test sample were found to be low the assay sensitivity could be increased by increasing the instrument sensitivity setting. However, serum-based reference materials could not be used as they would provide too high a blank value.

2. Incubation Time

It is important to establish the optimal time course for the reaction. For end-point assays the time course of the reaction for the highest and lowest reference should be monitored for about 300 min. Figure 21 shows the reaction of serum with anti-IgG in the Hyland nephelometer with 40 g/liter polyethylene glycol 4000 in the reaction mixture. The plateau represents a stable solution of antibody-antigen complexes and has been reached by about 20 min. The lowest reference shows an initial long lag phase and

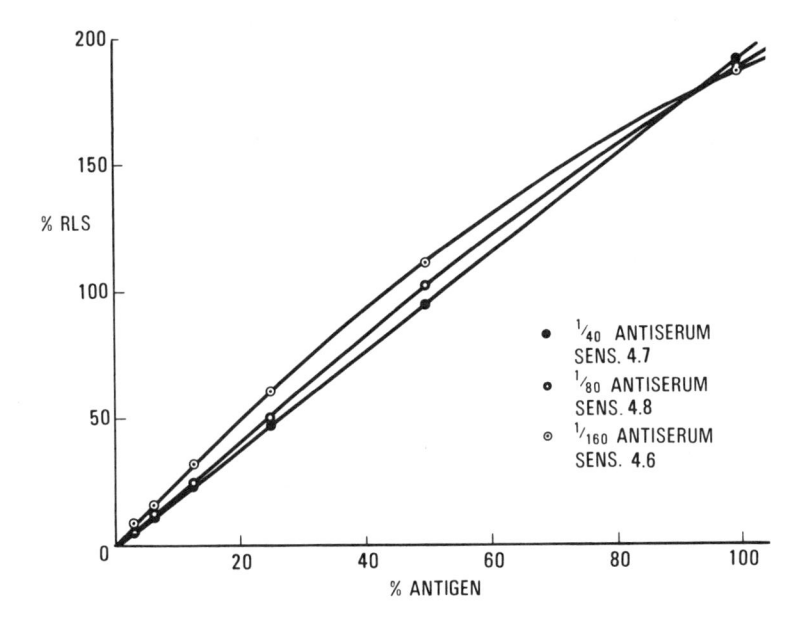

FIGURE 20 Assay of prealbumin on the Hyland nephelometer. 100% antigen concentration is a 1:25 dilution of reference serum. Calibration curves have been produced for different antiserum dilutions. Sensitivity setting (Sens) on the instrument is shown. (From Whicher and Blow, 1980.)

therefore reaches plateau more slowly. In the absence of polymer enhancement, antibody-antigen complexes tend to aggregate and precipitate more easily, particularly at high antigen concentrations. Figure 22 shows the time course of the IgG/anti-IgG reaction without polymer in the Behring nephelometer. As can be seen from these figures it is important that the time course is actually examined before the incubation time is chosen. For kinetic assays peak rate will occur later after the addition of antigen at low antigen concentration and it is important to establish a series of rate curves such as is shown in Fig. 12 so that the reaction can be monitored for an adequate length of time always to include peak rate. It is clear from this figure that the use of a two-point "kinetic" assay is not entirely desirable: A large part of the lag phase will be included for samples of low concentration giving rise to poorly reproducible reference curves which are very far from linear in shape.

3. Polymer Enhancement

The use of polymers such as PEG of 4000 or 6000 daltons is almost universal in both end-point and kinetic nephelometric systems. The increase

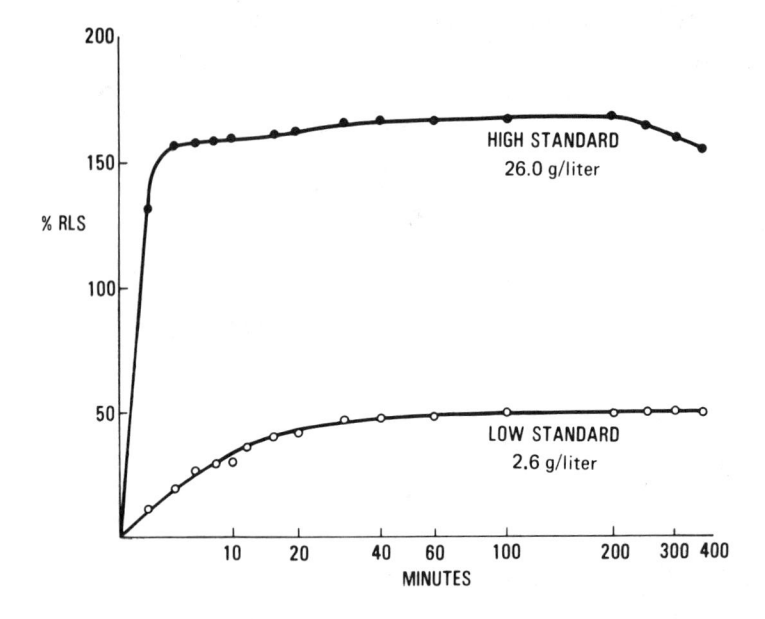

FIGURE 21 Time course for an IgG assay using 40 g/l PEG in the Hyland nephelometer. (From Whicher and Blow, 1980.)

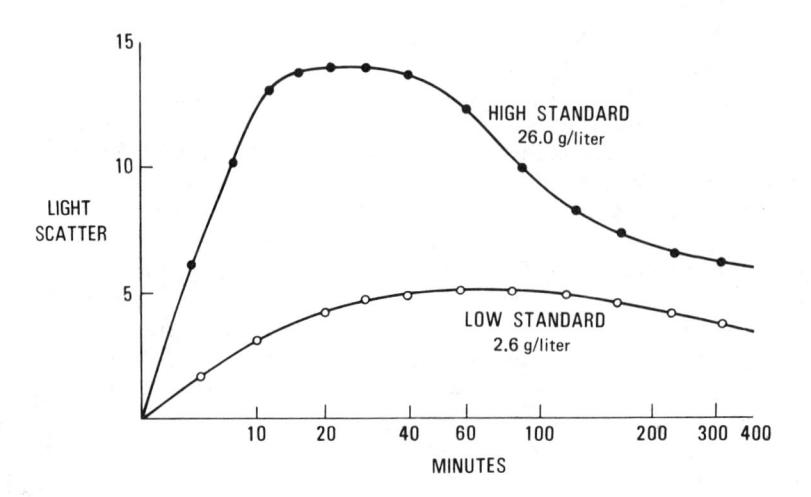

FIGURE 22 Time course for an IgG assay in the absence of PEG in the Behring nephelometer. (From Whicher and Blow, 1980.)

in reaction rate and shifting of equivalence towards higher antigen concentrations gives rise to faster and more economical assays. Polyethylene glycol 4000 is slightly less effective than PEG 6000 (0.75 w/w). The optimal concentration is usually about 40 g/liter but by no means invariably. While PEG has these advantages, it also decreases the solubility of large proteins and lipoproteins in serum, causing a considerably higher blank light-scattering than the same dilution of serum in buffer. It is therefore desirable to use PEG at the lowest concentration capable of producing an adequate calibration curve. Figures 5 and 23 show the effect of PEG in two different assays and it can be clearly seen that they are by no means the same: the lower-affinity antiserum in Fig. 5 shows greater enhancement.

4. Sensitivity

The sensitivity of immunonephelometric assays is limited by the presence of light-scattering particles in serum and also by the noise produced by dust and particulate matter in reagents and on the surfaces of cuvettes. Decreasing the total mass of antibody and antigen while increasing economy

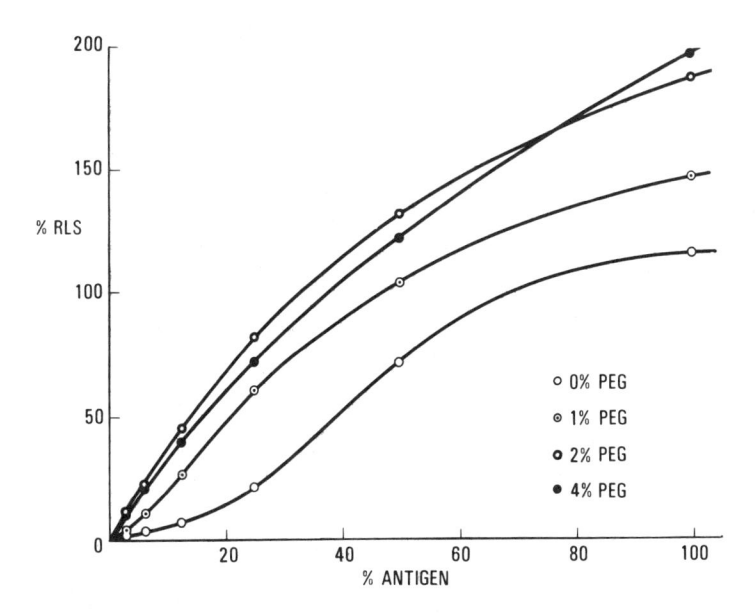

FIGURE 23 The effect on the production of light scattering (% RLS) from the addition of increasing amounts of antigen to a solution of anti-IgG in the presence of different concentrations of PEG in the Hyland laser nephelometer PDS. (From Whicher and Blow, 1980.)

will also have the advantage of providing a greater dilution of serum within the assay mixture and decreasing the contribution from endogenous light-scattering material present. This must be balanced against the fact that it may be necessary to increase the instrument sensitivity setting to provide an adequate light-scattering signal from the top standard on the reference curve, which may result in noise due to dust particles. The balance of these two factors dictates the sensitivity setting required and the optimal antigen dilution to provide the most sensitive assay possible within serum. The addition of PEG to the reaction mixtures will increase blank light-scattering from lipoproteins and proteins present in serum. This may to some extent be avoided by preprecipitation with PEG prior to assay. This is usually performed with concentrations of about 10% PEG 6000 which are then diluted to 4% by the addition diluted antiserum. Inherent light-scattering material does not present a problem for kinetic assay as only a rate of change is being measured after addition of antiserum. It is, however, important to be aware that it is possible for materials within the sample, such as high concentrations of IgM, to produce an apparent rate of scatter when polymeric buffers are added.

5. Antisera

Antisera vary enormously in both titer and affinity even when obtained from the same manufacturer directed against the same antigen over a period of years. It is essential when establishing nephelometric assays to evaluate antisera in the same measuring system in which they are to be used. So important is this that many laboratories evaluate samples of a series of antisera of a particular specificity in the nephelometric system which they are proposing to use prior to purchase. Commercial antisera from reputable companies may vary in affinity in nephelometry by as much as fourfold thus enormously influencing the final cost. Figures 24 and 25 show two different antisera tested on the Hyland nephelometer for the measurement of IgG. Clearly antiserum A is more satisfactory for nephelometry, being usable at a dilution of 1:320 whereas antiserum B is only suitable at a dilution of 1:40. They behave in a very similar fashion in gel techniques such as radial immunodiffusion. The importance of high-affinity antisera should not be underestimated, particularly for kinetic nephelometry, and poor antisera are the most common cause of failure to establish a satisfactory assay.

6. Antigen Excess Detection

It is important to establish the assay range in such a way that the concentrations of the protein being measured that are likely to be encountered in biological fluids do not pass beyond half of the equivalence point in terms of antigen concentration. This may not be possible for some proteins which have enormously wide ranges of biological variation, notably C-reactive

Nephelometric Methods

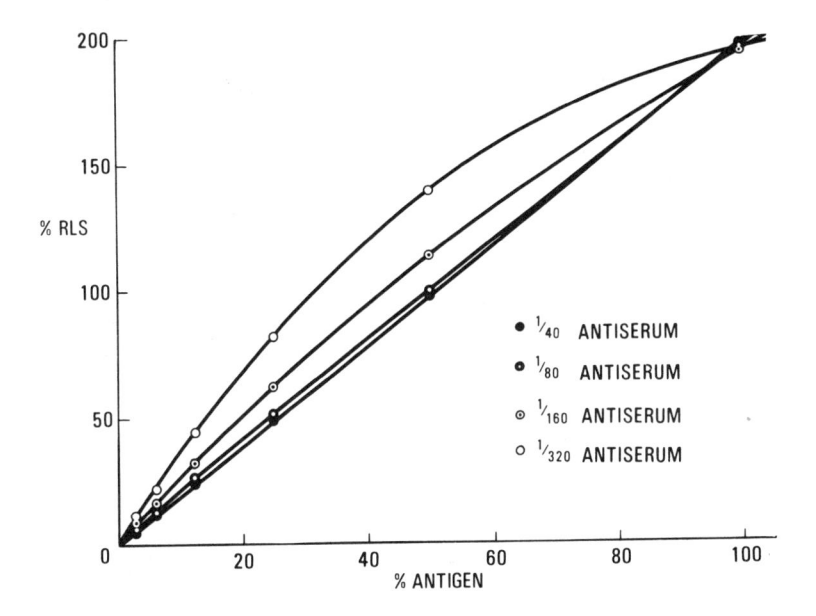

```
200 ┤
         
150 ┤
%RLS     
100 ┤         • 1/40  ANTISERUM
                 ◉ 1/80  ANTISERUM
 50 ┤         ⊙ 1/160 ANTISERUM
                 ○ 1/320 ANTISERUM
  0 └────┬────┬────┬────┬────┬
        20   40   60   80  100
              % ANTIGEN
```

FIGURE 24 IgG assay on the Hyland nephelometer. A series of curves produced for various dilutions of a commercial antiserum (A). (From Whicher and Blow, 1980.)

protein and β_2-microglobulin. In such cases it may be necessary to build an antigen excess detection method into the assay. This can be done in end-point assays by adding a further aliquot of antigen or antibody, incubating for a further period of time, and remeasuring light scatter or in kinetic assays by looking for an increment in rate. Paraproteins can largely be avoided by prior electrophoresis of serum samples. Another approach is to assay samples routinely at two or more dilutions and look for a linear dose-response. The use of computer or microprocessor analysis of the rate curves in kinetic nephelometry to identify curves characteristic of antigen excess has not been particularly successful to date though it may prove possible in the future.

B. Accuracy

Accuracy is a problem in the measurement of proteins as there is only a very limited number of purified proteins available in sufficient quantity to provide primary standards (Ritchie, 1978). The problem is further accentuated by the fact that many proteins are altered functionally and antigenically during purification and storage. They thus differ from the proteins present in biological fluids and also from the antigens that may have been used for

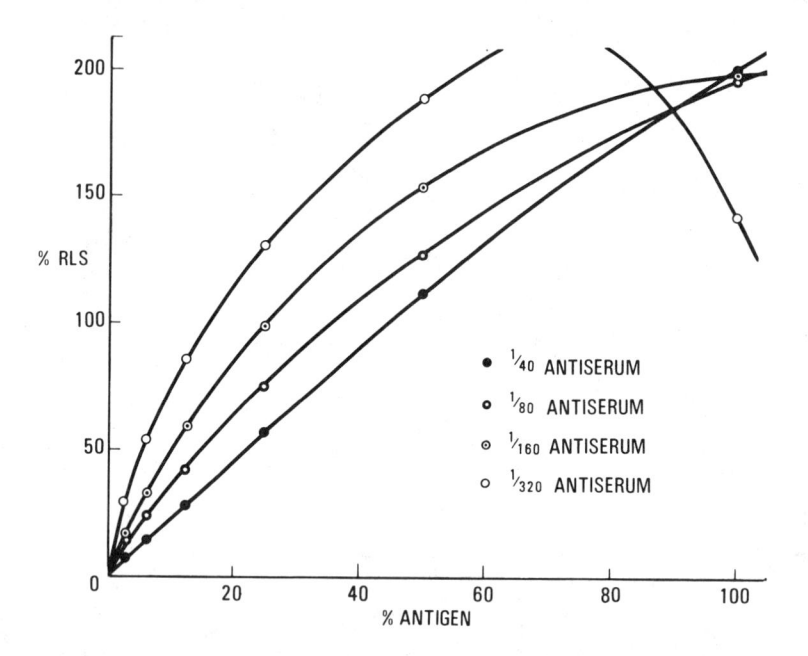

FIGURE 25 IgG assay on the Hyland nephelometer. A series of curves produced for various dilutions of a commercial antiserum (B). (From Whicher and Blow, 1980.)

raising antisera. Accuracy is thus dependent on the molecular characteristics of the antigen, the specificity of the antiserum, and the physical principles of the method (Whicher, 1979).

For these reasons it is important to assess the accuracy of immunonephelometric methods only with reference to a specific analyte and a specific method. With a few exceptions, such as immunoglobulins, internationally agreed standards are not available and the calibration of protein methods is largely arbitrary with normal ranges being related to a particular calibrant. It is thus not surprising that there is an enormous variation between methods and between laboratories for most protein analyses and this is a state of affairs that we have to accept at the present time. The most important aspect of accuracy, however, is an understanding of the reasons for this heterogeneity of results and an awareness of the factors that may alter accuracy in a particular assay.

The main problem of accuracy in specific protein measurement is that the reference and test materials are not identical in physical or chemical characteristics and thus show dissimilar comparative behavior under different assay conditions (Grubb, 1973). Gel techniques are particularly vulnerable to alteration in molecular size of antigen (Ritchie and Clark, 1973;

Alper and Rosen, 1975; Perrin et al., 1975) while free fluid phase systems such as nephelometry are very much less affected. However all immunoassays are affected by alterations in the antigenicity of the test protein and variations in antiserum specificity. Major discrepancies exist in the case of monoclonal immunoglobulins due to their altered and restricted antigenicity. It is generally accepted that accurate quantitation of such proteins annot be performed with immunochemical techniques due to the differing specificities of antiserum (Markowitz and Tschida, 1972; Havez, 1972; Whicher, 1979, 1980; Seneviratne and Moores, 1980). Genetic phenotypes in certain proteins such as haptoglobin may also cause problems (Van Lenti et al., 1979).

Interfering antibodies present in human serum may theoretically affect all immunochemical assays. Thus patients with IgA deficiency occasionally have antibodies to serum proteins of the family Bovidae. If such animals, including sheep and goats, are used for antiserum production complexes may be formed between the human antibody and the animal serum protein, thus obscuring the diagnosis of IgA deficiency (Gilliland, 1977). In practice this has never been demonstrated as a problem in nephelometry (Ritchie, 1975). It has, however, been suggested that IgM rheumatoid factors in human serum may react with the IgG of the antiserum (Ritchie, 1975).

Fluid phase immunoassay is susceptible to the influence of chemical factors affecting the rate of the antibody-antigen reaction. Thus urea is urine at physiological concentrations may inhibit nephelometric assays by as much as 20% (Anthony et al., 1980). This does, however, appear to be avoidable by adequate sample dilution or by the use of 80 g/liter polyethylene glycol in the reagent mixture. Heparin in samples has been shown to affect nephelometric measurements in the Technicon automated immunoprecipitation system (Shenkin et al., 1977).

There are, as has already been discussed, important differences between the specificity of kinetic and end-point nephelometric assays. This is due to the fact that in kinetic assays only a proportion of the high-affinity antibodies present within the antiserum will react within the time course of the assay. They may thus recognize a different aspect of the antigen than does endpoint nephelometry or gel techniques.

C. Precision

Much of the impetus towards the use of nephelometric techniques has been the improvement in precision that automation can bring. Typical data for within batch precision for a variety of techniques are given in Table 4. In general terms, end-point systems tend to have poorer precision than kinetic systems owing to the need for blank measurement. Nephelometry is particularly susceptible to imprecision arising from spurious light scattering due to particulate matter in cuvettes and reagents. Scrupulous care thus

TABLE 4 Within-Batch Precision for Various Specific Protein Assay Techniques

Method	Protein	Concentration (g/liter)	Coefficient of variation (%)	Reference
Radial immunodiffusion	IgG	14.0	4.7	Grabner et al., 1972
Electroimmunoassay	AFP	0.014	2.92	Borel-Giraud et al., 1976
Turbidimetry				
LKB	IgM	1.73	2.08	Dandliker et al., 1967
Multistat	Haptoglobin	2.40	1.27	Dandliker et al., 1967
Nephelometry				
End-point	IgG	10.0	3.2	Laurent and Killander, 1964
Rate	CRP	0.102	1.8	Gilletal, 1981
Radioimmunoassay	TBG	0.230	6.6	Kagedal and Kallberg, 1977
Enzyme immunoassay	HPL	0.195	5.1	Macdonald et al., 1980
Fluoroimmunoassay	HPL	$2\text{-}10 \times 10^{-3}$	3.9	Chard and Sykes, 1979

Source: Whicher, J. T., Price, C. P., and Spencer, K. (1982). Immunonephelometric and immunoturbidimetric assays for proteins. CRC Crit. Rev. Clin. Lab. Sci., 18: 213.

needs to be taken in filtering reagents and avoiding dirty or scratched cuvettes. In our experience these are major determinants of assay precision.

Automated systems avoid some of these problems with a single flowthrough cuvette though coating with protein may result in drift during an assay run. Automated sample dilution substantially improves the precision obtained.

Contrary to expectations the findings of the first year's survey of the U.K. National Quality Control Scheme for immunoglobulins suggests that between laboratories nephelometric systems are no more precise than radial immunodiffusion.

D. Sensitivity

Sensitivity is probably best defined as the smallest test value that can be discriminated with a stated probability from a blank. Many of the factors affecting sensitivity in nephelometry have already been discussed in some detail. The most important factors are antiserum titer and affinity which allow reaction of a larger proportion of the antiserum within the time course of the assay, thus producing a greater mass of antibody-antigen complex to be detected within the measurement system. The other major limitation of nephelometry is the presence of endogenous light-scattering material within samples. This is not a problem for kinetic assays but in end-point assay it severely limits the sensitivity of measurements in biological materials such as serum. The importance of blank interference in serum samples can be minimized by treatment with polyethylene glycol prior to assay; this does, however, lengthen the time needed for the measurements.

IX. APPLICATIONS

A. Serum Proteins

Both kinetic and end-point nephelometry have, to a large extent, replaced the gel-based techniques of radial immunodiffusion and Laurell rocket immunoelectrophoresis for the measurement of the clinically important proteins present in serum. In general terms these methods are extremely suitable for measuring proteins present in concentrations down to about 1 mg/liter with a between-batch precision of between 5 and 10%. Instrumentation for protein assay appeals to many clinical chemists who are not used to the skill-dependent techniques of the gel-based assay. Commercial manufacturers have understandably priced kits for nephelometric systems in such a way as to encourage their use instead of commercial radial diffusion plates. However there are probably no proteins which can be measured by nephelometry which cannot be measured by the gel-based techniques and the reverse is probably true. The most "difficult" proteins to assay by nephelometry are such analytes as C-reactive protein and β₂-microglobulin which are present at rather low concentrations. The advantages of nephelometry are speed; precision, particularly if they are automated; and cost-effectiveness compared with commercial gel-based systems. It suffers from less inaccuracy resulting from molecular heterogeneity of antigens and probably has greater user acceptability. It is, however, important to realize that the capital cost of these instruments is high and with the increasing use of immunoturbidimetry it is probable that many laboratories will move from nephelometry to immunoturbidimetry using centrifugal analyzers or enzyme analyzers. Most of the evidence to date suggests that such techniques are as sensitive and economical as nephelometry.

B. Other Biological Fluids

There are certain attractions in the nephelometric measurement of proteins in urine and cerebrospinal fluid as they contain much lower concentrations of endogenous light-scattering materials. More sensitive assays are thus possible and a nephelometric assay for β₂-microglobulin in urine has recently been developed in one of our laboratories.

C. Immunoinhibition Assays for Haptens

It is not possible, when using straightforward nephelometric techniques, to measure haptens with a single antigenic site because macromolecular antibody-antigen complexes which scatter light will not form. In the same way single monoclonal antibodies are not appropriate to use in these systems. However recent work suggests that it may be possible to assay low-molecular-weight haptens such as steroids or drugs using a technique in which the molecular size of the hapten is increased by coupling it to a high-molecular-weight carrier. The coupled hapten may then be used in a competitive assay in much the same way as a radiolabeled antigen is used in radioimmunoassay. Nephelometric immunoinhibition assay depends upon the fact that antisera contain rather few high-avidity antibodies to low-molecular-weight haptens and that only a small amount of hapten is required to occupy and bind that antibody. Pauling in the 1940s (Pauling et al., 1944) showed that small amounts of hapten cause large reductions in the amount of precipitate produced by interaction of an antihapten with a hapten conjugated to a protein. The "carrier" protein molecule had exposed on its surface several hapten molecules thus rendering the antihapten-hapten conjugate complex precipitable. The idea was taken up by Riccomi et al. in 1972 who described an assay using a Technicon AIP system. An antiserum was raised against the hapten conjugated to an immunogenic carrier protein. The antiserum was then mixed with a so-called developer antigen comprising several hapten molecules complexed to a high-molecular-weight carrier such as a large protein or latex particle. At equivalence, appropriate ratios of antibody and developer antigen resulted in cross-linking and antibody-antigen complex formation with the development of maximal light scattering. The addition of free hapten resulted in the breakdown of light-scattering complexes by competition with the developer antigen for antibody molecules. There was thus a decrease in light scattering as free hapten was added. Preliminary assays for haptens in pure solution were described by Riccomi et al. (1972) for DNP-lysine, Cambiaso et al. (1974b) for progesterone, and Gauldie and co-workers (1975) for digoxin and morphine. In 1979 Nishikawa and colleagues (1979a) described the first successful direct assay in human serum for theophylline. This was performed on the Hyland laser PDQ nephelometer and the drug was assayed in 10 µl of serum to a sensitivity of 10^{-7} m/liter with a coefficient of variation of 6%. Sample background scatter was compensated for by measuring a serum blank. Further assays were described for phenobarbital and phenytoin (Nishikawa et al., 1979b). Kinetic measurements have subsequently been developed for phenytoin, phenobarbital (Finley et al., 1981) and gentamicin (Cheng et al., 1979) on the Beckman ICS.

There seems little doubt that this technique will constitute a serious competitor to radioimmunoassay, fluorescent immunoassay, and enzyme immunoassay for haptens present in low concentrations in biological fluids. At the present time both Beckman and Baker are marketing kits for the measurement of several drugs using this method.

D. Other Assays

Outside the range of this review is a variety of other types of nonimmunochemical assays using nephelometry where the interaction of a ligand with an analyte results in a light-scattering precipitate. Such assays have been described using simple chemical precipitation techniques for substances such as total protein in urine (Instrument Information, Hyland Division, Travenol Labs., Costa Mesa, California) and CSF (Lafay et al., 1979). Lectin-protein interactions have been investigated using nephelometry and methods for the measurement of glycoproteins in serum have been described using concanavalin A (Kottgen et al., 1979; Warren et al., 1980; Martin et al., 1982). Bacterial endotoxin has been measured in biological fluids using increase in light scattering produced by proteins present in the lysate of horseshoe crab amebocytes (Bang, 1956; Galanos et al., 1979; Laurent and Zeller, 1979; Nishi et al., 1979). Mucopolysaccharides have also been measured by their interaction with complexing agents such as cetyl pyridinium chloride (Warren and Manley, 1981).

X. CONCLUSION

End-point and kinetic nephelometry have become important methods for the measurement of specific proteins in the last 10 years. In common with other immunological assay techniques they produce nonlinear dose-response curves requiring multiple calibrants or sophisticated computer-based curve linearization. Their main advantages lie in their ability to be automated and thus to save work time. They have not, in general, proved significantly more sensitive or precise than their predecessors, the gel-based techniques. The detection of antigen excess, while in theory a problem in such systems, is not of great significance in practice and though techniques for detection are at present rather tedious there is every prospect for computer analysis of rate curves as a means of antigen excess detection. The appearance of nephelometric inhibition assays has come at an opportune time to save the nephelometer from extinction as the specific protein assays can be as successfully performed by either end-point or kinetic immunoturbidimetry.

REFERENCES

Aitken, J. (1973). Automated analysis of α_1-antitrypsin and immunoglobulins. Colloquium University of Louvain, Belgium, 1972, Technicon AIP Monograph, 33.

Alper, C. A. and Propp, R. P. (1968). Genetic polymorphism of the third component of human complement (C3). J. Clin. Invest. 47: 2181.

Alper, C. A., and Rosen, F. S. (1975). Clinical applications of complement assays. Adv. Intern. Med. 20: 61.

Alper, C. A., Abramson, N., Johnston, R. B., Janek, J. H., and Rosen, F. B. (1970). Studies in vivo and in vitro on an abnormality in the metabolism of C3 in a patient with increased susceptibility to infection. J. Clin. Invest. 49: 1975.

Anderson, R. J., and Sternberg, J. C. (1978). A rate nephelometer for immunoprecipitin measurement of serum proteins. In Automated Immunoanalysis, Part 2, R. F. Ritchie (ed.), Marcel Dekker, New York, p. 409.

Anthony, F., Spencer, K., Mason, P., Massom, G. M., Price, C. P., and Wood, P. J. (1980). The variable influence of an α-component in pregnancy plasma on four different assay systems for the measurement of pregnancy specific β_1-glycoprotein. Clin. Chim. Acta 105: 287.

Bang, F. B. (1956). A bacterial disease of Limulus polyphemus. Bull. Johns Hopkins Hosp. 98: 325.

Borel-Giraud, N., Maillawin, A., and Vernet-Nyssen, M. (1976). Dosage de l'alpha-1-foetoproteine serique par electroimmunodiffusion sur agarose. Clin. Chim. Acta 71: 117.

Buffone, G. J., Savory, J., and Hermans, J. (1975). Evaluation of kinetic light scattering as an approach for the measurement of specific proteins using the centrifugal fast analyser. II. Theoretical considerations. Clin. Chem. 21: 1735.

Cambiaso, C. L., Masson, P. L., Vaerman, J. P., and Heremans, J. F. (1974a). Automated nephelometric immunoassay (ANIA) I. Importance of antibody affinity. J. Immunol. Methods 5: 153.

Cambiaso, C. L., Riccomi, H. A., Masson, P. L., and Heremans, J. F. (1974b). Automated nephelometric immunoassay II. Its application to the determination of hapten. J. Immunol. Methods 5: 293.

Chard, T., and Sykes, A. (1979). Fluoroimmunoassay for human choriomammotrophin. Clin. Chem. 25: 973.

Cheng, A., Bray, K., and Polito, A. (1979). Nephelometric inhibition assay for gentamycin. Clin. Chem. 25: 1095.

Chow, B. F. (1947). The determination of plasma or serum albumin by means of a precipitin reaction. J. Biol. Chem. 167: 757.

Dandliker, W. B., Alonso, R., de Saussere, V. A., Kierzenbaum, F., Levison, S. A., and Schapiro, H. C. (1967). The effect of chaotropic ions on the dissociation of antigen-antibody complexes. Biochemistry 6: 1460.

Deaton, C. O., Maxwell, K. W., Smith, R. S., and Creveling, R. L. (1976). Use of laser nephelometry in the measurement of serum proteins. Clin. Chem. 22: 1.

Debye, P. (1915). Zerstreung Von Rongerstrahen. Ann. Physik. 46: 809.

Deverill, I. (1980). Kinetic measurements of the immunoprecipitin reaction using the centrifugal analyzer. In Centrifugal Analyses in Clinical Chemistry, C.P. Price and K. Spencer (eds.), Praeger, New York, p. 109.

Durrington, P. N., Whicher, J. T., Warren, C., Bolton, C. H., and Hartog, M. (1976). A comparison of methods for the immunoassay of apolipoprotein B in man. Clin. Chim. Acta 71: 95.

Eckman, I., Robbins, J. B., Van den Hammer, C. J., Lentz, J., and Scheinberg, H. I. (1970). Automation of quantitative microanalysis of human serum transferrin: A model system. Clin. Chem. 16: 558.

Finley, P. R., Dye, J. A., Williams, J., and Lichti, D. A. (1981). Rate nephelometric inhibition immunoassay of phenytoin and phenobarbitol. Clin. Chem. 27: 405.

Franklin, E. C., and Buxbaum, J. (1978). Immunoglobulins in the normal state and in nepolasms of B cells. In Comprehensive Immunology, Vol. 4, R. A. Good and S. B. Day (eds.), Plenum Medical Book Company, New York, p. 361.

Galanos, C., Freudenberg, M. A., Luderitz, O., Rietschel, E. T., and Vestphal, O. (1979). Chemical, physicochemical, and biological properties of bacterial lipopolysaccharides. Prog. Clin. Biol. Res. 29: 321.

Gauldie, J., Sherrington, E., and Sircar, P. K. (1975). Automated nephelometric inhibition immunoassay of digoxin and morphine. Fed. Proc. 34: 104.

Gill, C. W., Bush, W., Burleigh, W. M., and Fischer, C. L. (1981). An evaluation of a C-reactive protein assay using a rate immunonephelometric procedure. Am. J. Clin. Pathol. 75: 50.

Gilliland, B. C. (1977). Immunological quantitation of serum immunoglobulins. Am. J. Clin. Pathol. 68: 664.

Gitlin, G., and Edelhoch, H. (1951). A study of the reaction between human serum albumin and its homologous equine antibody through the measurement of light scattering. J. Immunol. 66: 67.

Grabner, W., Bergner, D., Sailer, D., and Berg, G. (1972). Untersuchungen zur zuverlassigkent quantitativer Immunoglobulibestimmugen (IgG, IgA, IgM) Durche einfache radiale Immunodiffusion. Clin. Chim. Acta 39: 59.

Grubb, A. (1973). Immunochemical quantitation of IgG. Influences of the antiserum and of the antigenic population. Scand. J. Clin. Lab. Invest. 31: 465.

Halberstam, D., Diamond, H. S., and Derrico, E. (1972). Analysis of serum a_i-antitrypsin, Technicon International Congress, 1972. Adv. Autom. Anal. 4: 23.

Havez, R. (1972). Laboratory experience and results in the determination of immunoglobulins and haptoglobin in an automated system. In Automated Immunoprecipitation Reactions, J. P. Ham (ed.), Technicon Instrument Corp., New York, p. 45.

Heidelberger, M., and Kendall, F. E. (1935). Quantitative theory of precipitin reaction: Study of cryoprotein: Antibody system, J. Exp. Med. 62: 467.

Hellsing, K. (1969). Immune reactions in polysaccharide media. The effect of hyaluronate, chondroitin sulphate, and chondroitin sulphate-protein complex on the precipitin reaction. Biochem. J. 112: 475.

Hellsing, K. (1973). Influence of polymers on the antigen-antibody reaction in a continuous flow system. Colloquium, University of Louvain, Belgium 1972. Technicon AIP monograph 17.

Hellsing, K. (1974). The effects of different polymers for the enhancement of the antibody-antigen reaction as measured with nephelometry. In Protides of the Biological Fluids, Vol. 21, H. Peters (ed.), Pergamon, Oxford, p. 579.

Hellsing, K. (1978). Enhancing effects of non-ionic polymers on immunochemical reactions. In Automated Immunoanalysis, Part I, R. F. Ritchie (ed.), Marcel Dekker, New York, p. 67.

Hellsing, K., and Enstrom, H. (1977). Pre-treatment of serum samples for immunonephelometric analysis by precipitation by polyethylene glycol. Scand. J. Clin. Lab. Invest. 37: 529.

Hellsing, K., and Laurent, T. C. (1964). The influence of dextran on the precipitin reaction, Acta Chem. Scand., 18: 1303.

Hills, L. P., and Tiffany, T. O. (1980). Comparison of turbidimetric and light scattering measurements of immunoglobulins by use of a centrifugal analyser with absorbance/fluorescence light scattering optics. Clin. Chem. 26: 1459.

Kagedal, B., and Kallberg, M. (1977). Determination of thyroxine binding globulin in human serum by single radial immunodiffusion and radioimmunoassay. Clin. Chem. 23: 1694.

Kallner, A. (1977). Removal of background interference in nephelometric determination of serum proteins. Clin. Chim. Acta 80: 293.

Killingsworth, L. M. (1978). Analytical variables for specific protein analysis. In Automated Immunoanalysis, Part I, R. F. Ritchie (ed.), Marcel Dekker, New York, p. 113.

Killingsworth, L. M., and Savory, J. (1971). Automated immunochemical procedure for the measurement of immunoglobulin IgG, IgA and IgM in human serum. Clin. Chem. 17: 936.

Killingsworth, L. M., and Savory, J. (1972). Manual nephelometric methods for the determination of IgG, IgA and IgM in human serum. Clin. Chem. 18: 335.

Killingsworth, L. M. and Savory, J. (1973). Nephelometric studies of the precipitin reaction: A model system for specific protein measurements. Clin. Chem. 19: 403.

Killingsworth, L. M., Savory, J., and Teague, P. O. (1971). Automated immunoprecipitation technique for the analysis of the third component of complement in human serum. Clin. Chem. 17: 374.

Kottgen, E., Bauer, C., and Gerok, W. (1979). Laser nephelometry of lectin-glycoprotein complexes: A new method for their quantitation and differentiation. Anal. Biochem. 96: 391.

Kusnetz, J., and Mansberg, H. P. (1978). Optical considerations: Nephelometry. In Automated Immunoanalysis, Part I, R. F. Ritchie (ed.), Marcel Dekker, New York, p. 1.

Lafay, S., Diemert, M. C., and Galli, J. (1979). Etude comparative de deux methodes de dosage des proteins dans le urine et le LCR. In Compte Rendu de la Deuxieme Reunion Nationale du Group d'Etudes sur les Applications de la Nephelometrique Laser, Hoechst-Behring, Paris, p. 23.

Larson, C., Orenstein, P., and Ritchie, R. F. (1971). An automated method for the quantitation of proteins in body fluids, Technicon International Congress, 1970. Adv. Autom. Anal. 1: 104.

Laurell, C-B. (1972). Electroimmunoassay. Scand. J. Clin. Lab. Invest. 29 (Suppl. 124): 21.

Laurent, P. H., and Zeller, H. (1979). Formation des gels: Etude nephelometrique—Laser Application aux Endotoxines Bacteriennes. In Compte Rendu de la Deuxieme Reunion Nationale du Groupe d'Etudes sur les Applications de la Nephelometrique Laser, Hoechest-Behring, Paris, p. 25.

Laurent, T. C., and Killander, J. (1964). A theory of gel filtration and its experimental verification. J. Chromatogr. 14: 317.

Libby, R. L. (1938c). The photonreflector: An instrument for the measurement of turbid systems. J. Immunol. 34: 71.

Libby, R. L. (1938b). A new and rapid quantitation technique for the determination of the potency of Type I and Type II anti-pneumococcal serum. J. Immunol. 34: 269.

Macdonald, D. J., Nicol, K. M., Belfield, A., Shah, M. M., and Mack, D. S. (1980). Enzyme linked immunoassay for placental lactogen in human serum. Clin. Chem. 26: 745.

Marcroft, J., and Newbanks, I. M. (1973). Evaluation of a specific protein analyser. Clin. Chim. Acta 46: 399.

Markowitz, H., and Tschida, A. R. (1972). Automated quantitative immunochemical analysis of human immunoglobulins. Clin. Chem. 18: 1364.

Marrack, J. R., Grant, R. A. (1953). The interaction of antigen and antibody in low concentrations of salt. Br. J. Exp. Pathol. 34: 263.

Marrack, J. R., and Richards, C. B. (1971). Light scattering studies of the formation of aggregates in mixtures of antigen and antibody. Immunology 20: 1019.

Martin, M. F. R., Dieppe, P. A., Jones, H. E., Warren, C., Whicher, J. T., and Kohn, J. (1982). Serum concanavalin A binding in rheumatoid arthritis. Ann. Rheum. Dis. 41: 133.

Mie, G. (1908). Beiträge zur optik trüber Medien, speziell kolloidaler Metallosungen. Annalen der Physik, 25: No. 3, 4th series: 377.

Nishi, H. H., Kestner, J., and Elin, R. J. (1979). A new semi-automated nephelometric procedure for the determination of bacterial endotoxin. Clin. Chem. 25: 1106.

Nishikawa, T., Kubo, H., and Saito, M. (1979a). Competitive nephelometric immunoassay for theophylline in plasma. Clin. Chim. Acta 91: 59.

Nishikawa, T., Kubo, H., and Saito, M. (1979b). Competitive nephelometric immunoassay for antiepileptic drugs in patient blood. J. Immunol. Methods 29: 85.

Pardue, H. L. (1977). A comprehensive classification of kinetic methods of analysis used in clinical chemistry. Clin. Chem. 23: 2189.

Pauling, L., Pressman, D., and Gronberg, A. (1944). Serological properties of simple substances: VII. Quantitative theory of inhibition by haptens of precipitation of heterogeneous antisera with antigens, and comparison with experimental results for polyhapten in simple substances and for azoproteins. J. Am. Chem. Soc. 66: 784.

Perrin, L. H., Lambert, P. H., and Miescher, P. A. (1975). Complement breakdown products in plasma from patients with systemic lupus erythromatosus and patients with membranoproliferative glomerulonephritis. J. Clin. Invest. 56: 165.

Price, C. P., and Spencer, K. (1981). Kinetic immunoturbidimetry of human choriomammotrophin in serum. Clin. Chem. 27: 882.

Lord Rayleigh. (1871a). On the light from the sky, its polarisation and colour. Phil. Mag. 41: 107.

Lord Rayleigh. (1871b). On the scattering of light by small particles. Phil. Mag. 41: 447.

Lord Rayleigh. (1881). On the electromagnetic theory of light. Phil. Mag. 12: 81.

Lord Rayleigh. (1910). The incidence of light upon a transparent sphere of dimensions comparable with wavelength. Proc. R. Soc. A84: 25.

Riccomi, H., Masson, P. L., Vaerman, J. P., and Heremans, J. F. (1972). An automated nephelometric inhibition immunoassay (NINIA) for haptens. Colloq. AIP, Brussels 9.

Ritchie, R. F. (1972). Evaluation of antibody preparations for the AIP system. Colloq. AIP, Brussels 59.

Ritchie, R. F. (1975). Automated immunoprecipitation analysis of serum proteins. In The Plasma Proteins, Vol. 2, F. W. Putnam (ed.), Academic Press, New York, p. 375.

Ritchie, R. F. (1978). Reference materials for plasma protein analysis. In Automated Immunoanalysis, Part I, R. F. Ritchie (ed.), Marcel Dekker, New York, p. 159.

Ritchie, R. F. and Clark, C. (1973). Fully automated analysis of low density lipoproteins. Colloquium, University of Louvain, Belgium 1972, Technicon AIP monograph, 33.

Ritchie, R. F., and Graves, J. A. (1970). Automated analysis of cerebrospinal fluid proteins, Technicon International Congress. Adv. Autom. Anal. 1: 117.

Ritchie, R. F., Alper, C. A., and Graves, J. A. (1969). Experience with a fully automated system for immunoassay of specific serum proteins. Arthritis Rheum. 12: 693.

Savory, J., Buffone, G. J., and Reich, R. (1974). Kinetics of the IgG anti-IgG reaction as evaluated by conventional and stopped flow nephelometry. Clin. Chem. 20: 1071.

Savory, J., Heintges, M. G., Killingsworth, L. M., and Potter, J. M. (1972). Manual and automated determination of immunoglobulins in unconcentrated cerebrospinal fluid. Clin. Chem. 18: 37.

Schultz, J. E., and Schwick, G. (1959). Quantitative Immunologische bestimmung von plasmaproteinen. Clin.Chim. Acta 4: 15.

Seneviratne, C. S., and Moores, S. (1980). Kinetic turbidimetric determination of serum immunoglobulins using a Multistat III centrifugal analyser. In Centrifugal Analysers in Clinical Chemistry, C. P. Price and K. Spencer (eds.), Praeger, Eastbourne, England, p. 449.

Shenkin, A., Morrison, B., and Robertson, D. A. (1977). Some problems in the correction for intrinsic light scattering in samples in specific protein analysis by automated immunoprecipitation. Ann. Clin. Biochem. 14: 163.

Sieber, A., and Gross, J. (1976). Determination of serum proteins by laser nephelometry. Protides Biol. Fluids 23: 295.

Spencer, K., and Price, C. P. (1980). The use of the IL Multistat III centrifugal analyser for kinetic immunoturbidimetry. The measurement of serum transfer. In Centrifugal Analysers in Clinical Chemistry, C. P. Price and K. Spencer (eds.), Praeger, New York, p. 457.

Spencer, K., and Price, C. (1981). Clinical chemistry instrumentation and light scattering measurement. U. V. Spectro. Group Bull. 8: 38.

Sternberg, J. C. (1977). A rate nephelometer for measuring specific proteins by immunoprecipitin reactions. Clin. Chem. 23: 1456.

Steward, M. W. (1977). Affinity of the antibody-antigen reaction and its biological significance. In Structure and Function of Antibodies, L. E. Glyn and M. W. Steward (eds.), John Wiley, New York, p. 233.

Tanford, C. (1961). Light scattering. In Physical Chemistry of Macromolecules, John Wiley, New York, p. 275.

Tiffany, T. O. The concepts of centrifugal analysers. In Centrifugal Analysers in Clinical Chemistry, C. P. Price and K. Spencer (eds.), Praeger, Eastbourne and New York, p. 2.

Tyndall, J. (1854). On some phenomena connected with the motion of liquids. Proc. R. Inst. 1: 446.

Van Holde, K. E. (1971). Scattering in Physical Biochemistry, Vol. 9, Prentice-Hall, Englewood Cliffs, New Jersey.

Van Lenti, F., Marchand, A., and Galen, R. S. (1979). Evaluation of a nephelometric assay for haptoglobin and its clinical usefulness. Clin. Chem. 25: 2007.

Walsh, R. L., and Coles, M. E. (1980). Binding of IgG and other proteins to microfilters. Clin. Chem. 26: 496.

Warren, C., and Manley, G. (1981). Measurement of serum glycosoaminoglycans by laser nephelometry. Clin. Chim. Acta 116: 369.

Warren, C., Whicher, J. T., and Kohn, J. (1980). The use of concanavalin A to measure acute phase proteins by laser nephelometry. J. Immunol. Methods 32: 141.

Whicher, J. T. (1979). Problems encountered in immunochemical technique: Methodology. In Immunochemistry in Clinical Laboratory Medicine, A. Milford Ward and J. T. Whicher (eds.), MTP Press, Lancaster, England, p. 51.

Whicher, J. T. (1980). Consideratione della immunoglobulini. Estratto da Clinica et Laboratorio 4: 295.

Whicher, J., and Blow, C. (1980). Formulation of optimal conditions for an immunonephelometric assay. Ann. Clin. Biochem. 7: 170.

Wood, J. P., Cockett, D., and Mason, P. (1978). A rapid and inexpensive laser nephelometric assay for plasma pregnancy specific β_1-glycoprotein levels. Clin. Chim. Acta 90: 87.

7
Labeled Antibody Immunoassays

HEATHER A. KEMP AND J. STUART WOODHEAD
Welsh National School of Medicine, Cardiff, Wales

RHYS JOHN
University Hospital of Wales, Cardiff, Wales

I. INTRODUCTION

During the last 20 or more years, labeled antibodies have been used as tools for the investigation of many aspects of biochemistry and medicine, including antibody structure, kinetics of immune reactions, protein synthesis, and cancer therapy. In 1968 their use was extended to the measurement of soluble antigens by a technique called the immunoradiometric assay. The technique was introduced by Miles and Hales (1968a, b) in an attempt to overcome some of the disadvantages that they perceived in conventional radioimmunoassay procedures. They argued that an immunoassay should possess the following fundamental characteristics. Firstly, the signal response measured should be detectable at very low concentrations of the product. Secondly, all the unknown compound should be reacted at least once in the assay. Thirdly, the procedure should give a low background and show changes in signal response in direct proportion to the changes in product concentration. Radioimmunoassay, a technique already well-established in 1968, satisfies the first of these requirements but not the last two. Miles and Hales (1968b) argued that by using excess labeled antibodies to convert antigen to a labeled derivative which was measured directly, these requirements could be satisfied. In addition, it was suggested that antibody labeling would facilitate the development of assays for peptides which were difficult to label because of a low content or absence of tyrosine.

The first immunoradiometric assay developed by Miles and Hales was for plasma insulin. The procedure involved affinity purification of antibodies from antiserum using a solid-phase insulin preparation (immunoadsorbent). The antibodies were labeled with ^{125}I while bound to the immunoadsorbent and then dissociated by pH reduction. In the assay procedure, insulin standards or unknown samples were reacted with excess labeled antibody until equilibrium was reached, then unreacted antibodies were removed by reacting them with an excess of immunoadsorbent for a short period of time (Fig. 1). The excess labeled antibody was separated from that bound to antigen by centrifugation. The latter fraction, representing supernatant radioactivity, was counted and yielded a direct measurement of antigen concentration. The performance of the technique appeared optimal when the labeled antibodies were of high specific activity and used in only moderately excess concentrations. Immunoradiometric assays were subsequently developed for a range of polypeptides including growth hormone, luteinizing hormone (LH), follicle-stimulating hormone (FSH), parathyroid hormone (PTH), calcitonin, angiotensin, and immunoglobulin G (IgG) (Addison and Hales, 1971a; Woodhead et al., 1974).

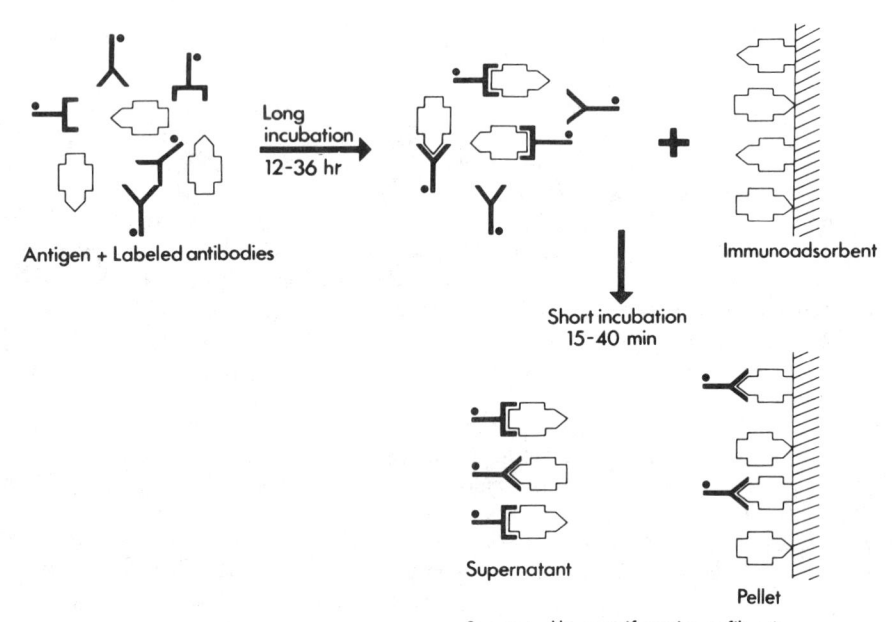

FIGURE 1 The immunoradiometric assay procedure. The antigen (⌂) is reacted with excess labeled antibody (⊣[⊢<) until equilibrium is reached (12-36 hr). Unreacted antibodies are removed by reaction with excess immunoadsorbent. The radioactivity of the supernatant from centrifugation gives a direct measure of the antigen concentration.

An interesting development demonstrated by Miles (1971) was the ability to use solid-phase antigens to remove cross-reacting antibodies from antisera to gonadotrophins. In this way a specific assay was produced for FSH from a relatively nonspecific antiserum. A further practical advantage of using ^{125}I-labeled antibodies as opposed to ^{125}I-labeled antigens has been their increased stability; their shelf-life is up to 2 months or more (Kemp et al., 1981). However, the wider application of the immunoradiometric technique was hampered by three main factors. First was the requirement for relatively large quantities of purified antigen, although it was found that partially purified material could be used satisfactorily for separation purposes (Addison et al., 1971). Second was the requirement for large quantities of antiserum. In addition to the use of antibodies in a reagent-excess rather than reagent-limiting mode, has been the problem of variable recovery from the affinity chromatography and labeling stages. Finally, in the majority of published assays, backgrounds were rarely lower than 20% of the added radioactivity and occasionally as high as 40% thus limiting the potential sensitivity of the methodology.

These disadvantages have been largely overcome by the development of two-site assays. In this procedure, the labeled antibody-antigen complex is separated from excess labeled antibody by reaction with solid-phase antibody (Fig. 2). In the first assay described using this procedure to measure growth hormone (Addison and Hales, 1971b), antigen was extracted from samples onto antibody linked to filter paper discs and, after the discs were washed, a second reaction was carried out with ^{125}I-labeled antibodies. High sensitivity could be obtained by extracting relatively large volumes of plasma and washing the final complex extensively to reduce background radioactivity. Readhead et al. (1973) developed an assay for FSH using the modification of antibody-coated plastic tubes as a means of extracting antigen from samples.

The increasing availability of antibodies on a large scale coupled with the simplicity of two-site assay procedures has led to the introduction of a range of polypeptide assays, several from commercial kit manufacturers. Substances frequently measured using two-site assays include α-fetoprotein (AFP), thyroid-stimulating hormone (TSH), ferritin, and hepatitis B surface antigen.

There have been many modifications of the basic technique during the last 10 years. A method designed to overcome the problem of iodinating small quantities of valuable antibody which utilized an indirect ^{125}I-labeled (anti-IgG) antibody was developed by Beck and Hales (1975). In addition to providing "universal" reagents, labeled (anti-IgG) antibodies offer the potential for increasing specific activity, up to three molecules of labeled reagent being bound for each molecule of first antibody. Initially the labeled (anti-IgG) antibody was intended for use after combination with

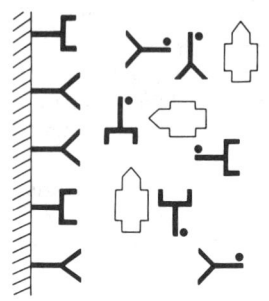

Antigen + excess solid-phase antibody
+ excess labeled antibody

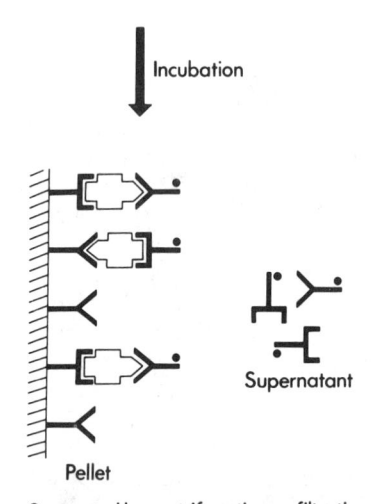

↓ Incubation

Supernatant

Pellet

Separated by centrifugation or filtration

FIGURE 2 The two-site assay procedure. In this procedure the labeled antibody-antigen complex is separated from excess labeled antibody by reaction with solid-phase antibody.

first antibody (Beck and Hales, 1975) though successful procedures for measuring human (Rainbow et al., 1979) and rat (Yue et al., 1979) proinsulins have subsequently been developed involving sequential addition of first antibody and labeled (anti-IgG) antibody.

The use of solid-phase immunoglobulin "arms" has also been described (Miles et al., 1974a). These have been used as a means of increasing sensitivity and reducing undesirable plasma effects. It has thus been possible to develop an assay for glial fibrillary acid protein in which the guinea pig antibody was attached to plastic tubes coated with rabbit (anti-guinea-pig IgG) antibody.

Two different antibody populations have been used to increase the specificity of two-site assays for polypeptides. Using selected N- and C-terminal antibody populations, it was possible to develop an assay for bovine PTH in which the intact (1-84) sequence of the hormone could be measured without interference from molecular fragments (Woodhead et al., 1977). A similar principle was employed for the measurement of proinsulin (Rainbow et al., 1979) involving sequential reaction of sample with antibodies to insulin and then to the C-peptide.

The more recent applications of two-site assay technology have exploited the sensitivity, speed, and convenience of the technique. In this chapter we describe some of these developments by reference to specific assays where the immunoradiometric procedure has produced marked improvements over conventional radioimmunoassay (RIA) methodology.

II. METHODOLOGY

A. Solid-Phase Derivatives

Immunoradiometric assay procedures may utilize solid-phase antigen derivatives (immunoadsorbents) and solid-phase antibodies. The former are required for affinity purification of antibodies from antisera and the latter provide a means of precipitating immune complexes. The matrix for such derivatives should exhibit a high binding capacity, low nonspecific protein absorption, and provide a stable conjugate. In addition, the matrix used for affinity purification should yield high recoveries of active antibody, present the antigen in a variety of steric configurations, and not take part in iodination reactions. Thus while polymerized proteins have been used to affinity purify antibodies to human serum albumin and ACTH (Girard et al., 1971), such immunoadsorbents would preclude the possibility of iodination of the antibody while complexed with the solid phase.

B. Solid-Phase Antigens (Immunoadsorbents)

Cyanogen-bromide-activated cellulose (Homburger et al., 1981) and Sepharose (Jones and Worwood, 1978; Eber et al., 1980; Muller et al., 1980) have been used widely to prepare solid-phase derivatives for immunoassay purposes; coupling is achieved via an amino derivative (Axén et al., 1967). Protein uptake on to these matrices is frequently high, up to 1000 µmol Gly-Leu/g Sephadex G-25 and 225 µmol Gly-Leu/g cellulose, and there is generally negligible nonspecific binding. However, cyanogen bromide activation is a potentially hazardous procedure requiring care and Sepharose is not an easy matrix to handle as it packs down poorly on centrifugation. Agarose and polyacrylamide beads have also been used as solid supports (Langone, 1978). These can be functionalized with free carboxyl and amino groups respectively and coupled using carbodiimides. The performance of these matrices in labeled antibody immunoassay systems has not been assessed.

For several years we have favored the use of solid-phase reagents based on the diazonium salt of reprecipitated powdered cellulose (Gurvich et al., 1961). This derivative reacts through primary amines and also tyrosine and histidine residues so that an antigen may be coupled in several steric configurations. Details of this coupling procedure have been described previously (Kemp et al., 1981; O'Riordan and Woodhead, 1976). Briefly, it involves the reprecipitation of an amino benzyl-O-methyl cellulose derivative from solution in ammoniacal cupric hydroxide (Schweitzer's reagent). This is followed by diazotization of the amino group using nitrous acid and reaction of the diazonium salt with protein at pH 8 for 18-48 hr. The solid-phase derivative can be washed vigorously to remove noncovalently bound protein. In the case of peptides such as insulin or PTH it may be necessary to wash with acid solutions to achieve this. Solid-phase antigens and antibodies prepared in this way are stable for up to 2 years when stored in buffer at 4°C in the presence of sodium azide to inhibit bacterial growth. The cellulose matrix fulfils the requirements listed above. Protein uptakes vary from 0.09 μmol/g cellulose for human myeloma IgM to 40 μmol/g for oxytocin (Addison, 1971). Nonspecific absorption is less than 6.6 nmol/g (Miles and Hales, 1968a). Uptakes are generally higher with low-molecular-weight proteins, though the apparent immunoreactivity of the immunoadsorbent is correspondingly decreased; as little as 20% of coupled insulin is immunoreactive. It is possible that small peptides are coupled at reactive sites of the cellulose which are not readily accessible to the relatively large antibody molecules.

For affinity purification of antibodies it is standard practice to react 25-250 μl of undiluted antiserum with 50-100 μg of immunoadsorbent for several hours. Antibody uptakes vary depending upon the nature of the antigen and the antibody titer, being typically 0.3 mol/mol antigen for peptides such as insulin and 13 mol/mol antigen for large proteins such as ferritin (Addison, 1971).

C. Solid-Phase Antibodies

Solid-phase antibodies are used in two-site assays to precipitate labelled antibody-antigen complexes (Fig. 2). In addition to the properties listed above, the matrix should also be convenient to handle in separation systems and have standardized and uniform characteristics (Hales and Woodhead, 1980). One of the first solid-phase systems described was that using antibody-coated plastic tubes (Catt and Tregear, 1967). The tubes are convenient to handle and store but have the disadvantages of variable degrees of coating, low density of antibody, and slow reaction times due to the relatively low surface-to-volume ratio. Readhead et al. (1973) described a two-site assay for FSH based on antibody-coated tubes. More recently the facility for carrying out sequential reactions in antibody-coated tubes has been exploited in the measurement of PTH (Woodhead et al., 1977) and proinsulin (Rainbow et al., 1979; Yue et al., 1979) where intact peptide sequences can be measured without interference from cross-reacting molecular fragments.

Several radioimmunoassays and two-site assays using antibodies coupled to porous glass are presently marketed by Corning Medical Products. While this solid phase is easily derivatized to yield high-capacity matrices, the high nonspecific binding of haptens and peptides to glass surfaces restricts the usefulness of these derivatives in immunoassay systems.

As in the case of immunoadsorbents we prefer reprecipitated cellulose to other matrices for the preparation of solid-phase antibodies. In general, IgG precipitates from antisera can be coupled to the diazonium salt of cellulose to yield protein uptakes of the order of 300-400 µg/mg cellulose. In some cases losses in antibody activity may result. For example one preparation of antibody to ACTH (1-24) showed a binding capacity only 2% of the original antiserum while the apparent affinity constant showed no significant change. Solid-phase antibodies based on cellulose have the advantages that they stay in suspension for up to 1 hr without agitation, reactions are relatively rapid, and they are easily removed by centrifugation or filtration.

Recently, antibodies linked to magnetic solid supports have been described (Smith and Gehle, 1980). Separation by this means is cheap, simple, and rapid and enables complex washing cycles to be carried out with maximum precision and minimum operator inconvenience.

D. Labeled Antibodies

Antibodies to large molecules such as hemophilic factor can be affinity purified prior to labeling without the need for solid-phase antigen since they can be precipitated by antigen in 25% ammonium sulfate, dissociated from the antigen, and then labeled in solution (Addison, 1974). There are, however, potential advantages to labeling antibodies while they are bound to immunoadsorbent in that the antigen-binding site may be protected from radiolytic damage, and the iodination reagents can be removed rapidly following termination of the reaction.

After initial extraction of the antiserum, the immunoadsorbent-antibody complex is washed free of other protein and iodinated using chloramine-T (Greenwood et al., 1963). In cases where the antibody titer is low, iodination using the lactoperoxidase procedure (Thorell and Johannsen, 1971) may yield more satisfactory preparations of labeled antibody. After the iodinated complex is washed at pH 8 to remove unreacted iodine and other reagents, low-affinity antibodies can be dissociated using dilute HCl (pH 3.0). Subsequent washing with dilute HCl (pH 2.0) dissociates high-affinity antibodies. These washes are collected directly into strongly buffered solutions to minimize exposure of the antibodies to the low pH conditions. The elution of steroid antibodies where interaction with immunoadsorbent may be particularly hydrophobic in nature is facilitated by the incorporation of a solvent such as dimethyl formamide in the acid wash (Woodhead et al., 1975). Recently, it has been demonstrated that recovery of peptide antibodies may be improved using aqueous acetonitrile at pH 4.0 (Hodgkinson and Lowry, 1982). Each system may require optimization, however. For example, high-affinity antibodies to ferritin may be eluted from immunoadsorbent at pH 2.6, elution at pH 2.0 actually being disadvantageous since it results in dissociation and consequent elution of the ferritin molecular subunits.

Recoveries from immunoadsorbent may vary considerably from one system to another, and may even depend on the ratio of immunoadsorbent to antiserum used for extraction (Kemp et al., 1982). Of the label recovered in the washes, that in the pH 2.0 fraction should constitute 15-25%. In excess of 20% of the radioiodine used may remain irreversibly bound to the immunoadsorbent as a result of labeling of the antigen.

In reagent excess systems, sensitivity may be improved by increasing the specific activity of the labeled antibodies, thereby allowing low uptakes to be detected. In practice, however, the incorporation of more than one atom of ^{125}I per molecule of antibody invariably results in reduced immunoreactivity. This is presumably a result of radiolytic damage, since as many as seven atoms of ^{127}I can be incorporated per molecule of antibody without significant effect on insulin assays (Miles and Hales, 1970). In general, an incorporation of one or two atoms per antibody molecule is achieved by the use of 1-2 mCi ^{125}I to label 50 μg antibody protein; this yields specific activities of 20-30 μCi/μg (Addison and Hales, 1971a). Miles and Hartree (1971) have claimed improved sensitivity in the measurement of LH using iodinated Fab fragments which may derive from the higher specific activity of these monovalent molecules. It is likely, however, that the potential improvements deriving from higher-specific-activity labeled antibodies will only be realized through the use of nonisotopic compounds such as chemiluminescent molecules (see Chapter 5) which can be detected with equal sensitivity to that of ^{125}I but do not reduce immunoreactivity.

In general, labeled antibodies can be stored in solution at 4°C for several weeks without obvious deterioration in performance. Long-term performance is usually improved, however, if the labeled antibody is stored complexed with immunoadsorbent. This procedure has been routinely used in a conventional immunoradiometric assay of PTH (Addison et al., 1971). If necessary the immunoreactivity of the labeled antibody can be assessed before use by incubating with an excess of immunoadsorbent. If more than 80% of the label binds at this stage, it should then be satisfactory for assay purposes.

E. Assay Methodology

The immunoradiometric assay outlined in Fig. 1 requires careful optimization. One critical feature is the quantity of labeled antibody used. Though this reagent should be in excess in order to convert all the antigen present to a labeled derivative, performance may be adversely affected by the presence of a large antibody excess, since this may lead to an unacceptably high background radioactivity. A second consideration is the incubation time with immunoadsorbent where unreacted antibody is removed. If the dissociation rate for the antigen-antibody complex is rapid an extended reaction time with immunoadsorbent will produce a poor apparent binding and low sensitivity. Since the reaction time with immunoadsorbent must be kept constant for all assay tubes, the technique is demanding if carried out manually.

By comparison, the two-site assay is more robust and easier to perform. In this procedure, the reagents may be added sequentially or simultaneously. Because reaction with the solid phase is the rate-limiting step, some peptide assays such as TSH may be carried out most efficiently by first reacting the sample with labeled antibody and then reacting with solid-phase antibody. In addition to their greatly simplified procedure when compared with other immunological assay methods, two-site assays have the unique advantage of extremely low backgrounds particularly when cellulose is used as the solid phase. This, combined with the use of high-specific-activity labels, is responsible for the high potential sensitivity of many of these assay systems.

F. The High-Dosage Hook

A feature of many reported two-site assays is the paradoxical fall in bound label at extremely high antigen dosages (Miles et al., 1974b; Vyzantiadis et al., 1974). Despite attempts to analyze this phenomenon in theoretical terms (Rodbard et al., 1978) its nature remains somewhat obscure. Explanations include insufficient washing of solid-phase complexes and dissociation of antigen from low-affinity solid-phase antibodies (Rodbard et al., 1978). The dosage at which the hook appears relates to the quantity of solid-phase antibody used which implies that it is a feature of saturation of the solid phase. The situation may then be considered analogous to a saturation assay where antigen and ^{125}I-antibody-labeled antigen compete for binding sites on the solid-phase antibody when the antigen is present in large excess. In practice the high-dosage hook generally occurs well beyond the working range of most assays though it is important to note that in the case of ferritin, which may be present in serum at concentrations ranging from 1 to 10,000 ng/ml, it is necessary to assay samples at dilutions to prevent misinterpretation resulting from this phenomenon.

G. Automation

In general, automation in the immunoassay field has been slow by comparison with other areas of clinical biochemistry. This is due partly to the constantly evolving methodology and partly to the wide variety of separation techniques currently in use. Thus the majority of automated systems are adapted to individual methodologies and so lack the flexibility for general application. The Kemtek 3000 (Kemble Instrument Co., Burgess Hill, Sussex, U.K.) is a computer-controlled, fully automated system initially developed by Bagshawe and his colleagues at Charing Cross Hospital Medical School in London. This system comprises an input/dilution unit, a reagent addition unit, a filtration unit, and a five-head counting unit together with standard data-reduction facilities (Bagshawe et al., 1968). Separation by filtration allows for wide flexibility in assay procedures and we have successfully developed a range of labeled antigen and labeled antibody assays for routine use on this system (Woodhead et al., 1981). An important feature of the automated system is the ability to carry out nonequilibrium reactions since reagent addition and separation can be coordinated to yield constant reaction times. A major development has been the automation of α-fetoprotein (AFP) and TSH assays which are widely used for population screening. Two-site assays have been developed for both analytes (Kemp et al., 1981; John and Woodhead, 1982) which combine the speed of the analytical procedure with the precision and high throughput capacity of an automated system. These two procedures are discussed in detail below as examples of two-site assays providing improvements in speed and sensitivity when compared with conventional radioimmunoassays. The measurement of ACTH is used as an example of a two-site assay which offers considerably improved specificity over a conventional system.

1. Two-Site Assay of α-Fetoprotein

α-Fetoprotein (AFP) is frequently measured in the serum and amniotic fluid of pregnant women as an prenatal diagnostic test for neural tube defects. Because this is a screening assay, large numbers of samples must be processed quickly. We developed a two-site assay for AFP which is fully automated and requires a short reaction time (Kemp et al., 1981). Initially the solid-phase antibodies and labeled antibodies were both prepared from a polyclonal sheep antiserum (Dr. A. Munro, Scottish Antibody Production Unit, Law Hospital, Carluke, Lanarkshire, U.K.). AFP was purified from amniotic fluid by anion exchange chromatography and affinity chromatography on concanavalin-A-Sepharose 4-B and used to prepare an immunoadsorbent. Antibodies labeled on the immunoadsorbent usually show high incorporation (approximately 20%) of the eluted radioactivity. These antibodies can be stored either on immunoadsorbent or in buffer at 4°C for several weeks. In the assay procedure labeled antibodies and solid-phase antibodies are allowed to react simultaneously with serum

AFP. Because of the reagent excess, the reaction proceeds rapidly: uptake of labeled antibody onto solid-phase antibody in the presence of 44 and 88 kU/liter of AFP is considerable after only 30 min (Fig. 3). An incubation time of 1 hr was found to yield a satisfactory assay. This represented a considerable improvement over conventional radioimmunoassay and other two-site assays utilizing overnight reactions (Belanger et al., 1973; Grenier et al., 1978). The short incubation period means that the reaction is not at equilibrium when separated, so it is very important that the tubes are all incubated for exactly the same length of time to avoid assay drift. The use of the delayed reagent addition facility coupled with the appropriate filtration cycle on the Kemtek 3000 is invaluable in this respect, allowing large assays (100 or more samples) to be processed accurately.

The assay has a within-assay coefficient of variation of 4.8% and a between-assay coefficient of variation of 9.3% at a dosage of 65 kU/liter. It gives good correlation on serum samples ($r = 0.99$) with a double antibody radioimmunoassay in use in our laboratory. The sensitivity achieved is 6 kU/liter using 50 μl of serum which is adequate for detection of AFP in maternal serum; the median values for weeks 15-18 of gestation range from 34 to 51 kU/liter in normal pregnancies.

Recently, we have modified the procedure by the use of a monoclonal antibody as the label (Weeks et al., 1981). The use of ^{125}I-labeled antibody (Hybritech, La Jolla, California) with an affinity constant of 2×10^9 liter/mol yielded a satisfactory assay in a 1-hr reaction with no detectable bias. It is likely that the relative ease of preparing labeled monoclonal antibodies which do not require affinity purification will extend the use of immunoradiometric technology in the measurement of polypeptides.

2. Two-Site Assay of Thyroid-Stimulating Hormone

Thyroid stimulating hormone (TSH) assays account for a large workload in most endocrine laboratories and form an essential part of the thyroid profile. The short incubation times and high sensitivity that two-site TSH assays exhibit compared to conventional radioimmunoassays are of particular advantage, for instance, when assaying for TSH as part of a screening program for neonatal hypothyroidism. We routinely measure TSH in dried blood filter paper spots (John and Woodhead, 1982) which contain the equivalent of 5 μl of serum using an assay which has a sensitivity of 3.4 mU/liter (Fig. 4). This is achieved by reacting the blood spots in an overnight incubation with rabbit anti-TSH antibodies labeled with ^{125}I. After a further incubation for 2 hr with sheep anti-TSH antibodies coupled to cellulose, the insoluble labeled complex is separated from unreacted labeled antibodies by filtration on the Kemtek 3000. Good precision of this automated procedure has allowed us to use single samples and results are available within 24 hr of receipt of a sample. Of 59,106 samples so far screened, only 17 had values above 50 mU/liter. Of these, 15 were confirmed cases of neonatal

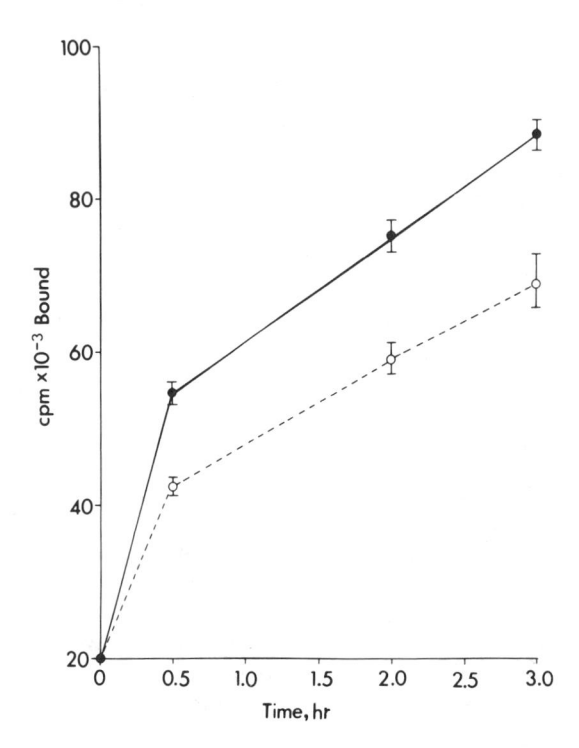

FIGURE 3 Uptake of labeled antibodies onto solid phase in the presence of 44 (O - - O) and 88 (● —— ●) kU/1 AFP.

hypothyroidism. One was a case of transient hypothyroidism and one was a false positive which resulted from the presence of a circulating TSH-binding immunoglobulin in both mother and child (Lazarus et al., 1983). The distribution of results obtained so far with this screening program is given in Table 1. Two cases of congenital hypothyroidism were undiagnosed initially as a result of error in the screening program which led to the samples not being assayed.

The assay is robust and reliable and is easily automated so that TSH is now the assay of choice when setting up a neonatal hypothyroid screening program and will replace the T_4/TSH approach still favored by a few American centers.

A serum assay based on the same reagents as the neonatal TSH assay and using the Kemtek automated immunoassay system, allows us to measure TSH with high precision and sensitivity with a total incubation time of 5 hr, so that results are available within the same working day. Even with this very fast assay time, the sensitivity achieved is better than 0.4 mU/liter with a sample size of 100 µl. The range of TSH values found in

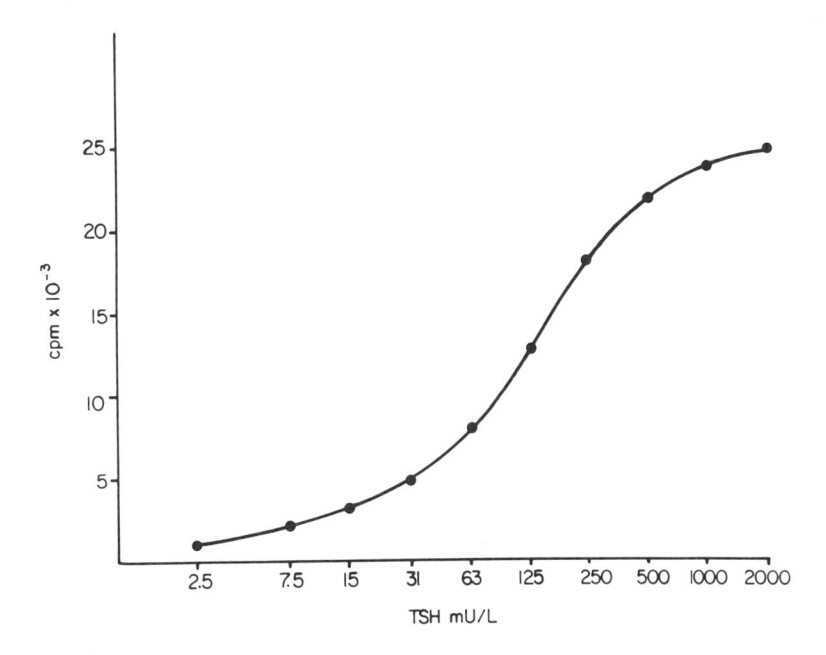

FIGURE 4 Standard curve for the assay of TSH in dried blood filter paper spots.

90 normal subjects varied from 0.4 up to 5 mU/liter. The assay is specific, showing no interference from polypeptides with a similar structure, samples from the first and second trimester of pregnancy and from postmenopausal women have TSH values within the normal range. Patients with rheumatoid factor present in their sera also do not show any interference in the assay, as has been suggested previously (Kato et al., 1979).

3. Two-Site Assay for Adrenocorticotrophin

There is considerable evidence that adrenocorticotrophin (ACTH) exists in the peripheral circulation as high-molecular-weight precursors and small fragments, as well as the intact peptide of 39 amino acids. Only the intact peptide seems to have full biological activity so that radioimmunoassays, which measure fragments, also may give values which are misleading. The development of a two-site assay using antibodies directed against two different parts of the peptide should not detect fragments that contain only one or other of these antigenic determinants, thus conferring an extra degree of specificity on the measurement of ACTH.

We have developed a two-site assay using an N-terminal antiserum raised in sheep against ACTH (1-24) linked to bovine thyroglobulin and a

TABLE 1 Distribution of TSH Results in Neonatal Hypothyroid Screening Program

	TSH (mU/l)				
	<15	16-30	31-50	51-100	101-250	>250
No. of samples	59,076	13	0	3	10	4

C-terminal antiserum raised in sheep against cross-linked ACTH (1-39). The cross-reactions of these two antisera are shown in Figs. 5 and 6. It can be seen that the N-terminal antiserum cross-reacts well with ACTH (1-39) and ACTH (1-24) but poorly with ACTH (1-13) and so is predominantly directed against the midportion of the peptide. The antiserum raised against cross-linked ACTH has the highest cross-reaction with ACTH (25-39) and so is predominantly directed against the extreme C-terminal.

The solid-phase antibodies were prepared by coupling an IgG precipitate of the anti-ACTH (1-24) antiserum to cellulose. This was used at a dilution of 1/20 in the assay. The labeled antibodies were prepared by

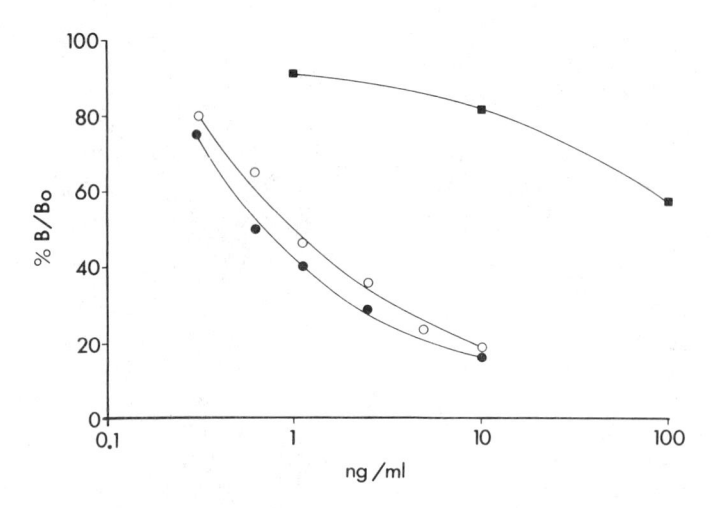

FIGURE 5 Cross-reactions in a conventional radioimmunoassay of ACTH sequences 1-39 (●——●), 1-24 (o——o), and 1-13 (■——■) with a sheep antiserum raised to ACTH (1-24) linked to bovine thyroglobulin.

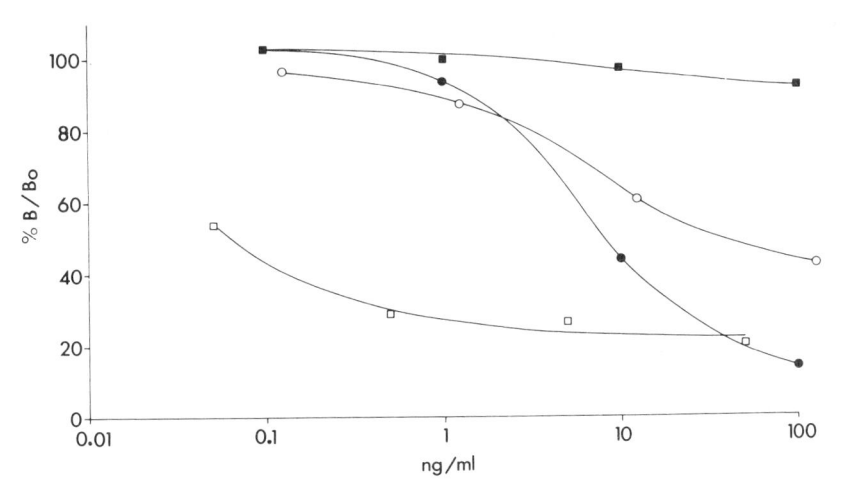

FIGURE 6 Cross-reactions in a conventional radioimmunoassay of ACTH sequences 1-39 (●——●), 1-24 (o——o), 1-13 (■——■), and 25-39 (□——□) with a sheep antiserum raised to cross-linked ACTH (1-39).

attaching antibodies onto an ACTH (1-39) immunoadsorbent, iodinating by the lactoperoxidase method, then removing the unwanted N-terminal labeled antibodies by incubating the label overnight with ACTH (1-24) immunoadsorbent. The remaining C-terminal-labeled antibodies can then be used in the assay diluted to give 60,000 cpm/100 µl. The final recovery of radioactivity is low (about 3%) and only 50% of the labeled C-terminal antibodies will bind to excess ACTH immunoadsorbent, indicating a certain degree of antibody damage during iodination. The assay is performed by incubating 100 µl of standard or sample with 100 µl of solid-phase antibody and 100 µl of labeled antibody at 4°C for 40 hr. Separation can be achieved either by centrifugation or filtration.

The assay shows no reaction with the fragments ACTH (1-24) or ACTH (25-39), as shown in Fig. 7.

III. CONCLUSION

The pioneering work of Miles and Hales in the development of labeled antibody techniques for the measurement of soluble antigens has now evolved to the stage where these techniques represent the logical methods of choice for a variety of polypeptide assays. The most important advantages over conventional radioimmunoassay procedures are those deriving from the use of excess binding reagent, namely speed and sensitivity. In addition,

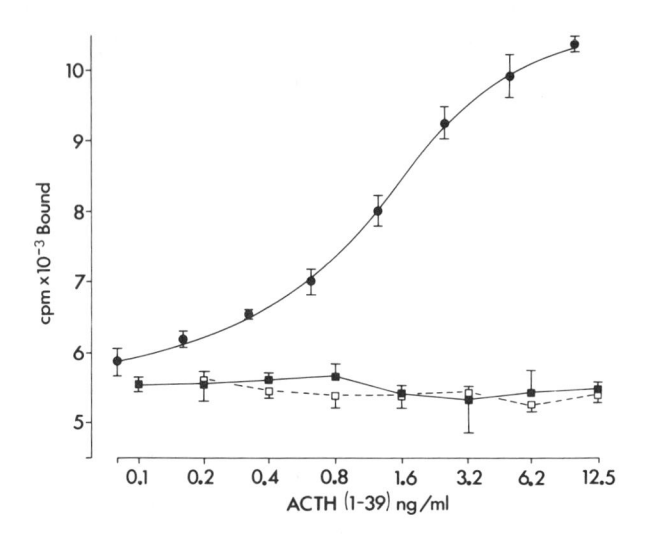

FIGURE 7 Two-site assay of ACTH (1-39) (●——●) showing lack of reactivity with either N-terminal (1-24) (o——o) or C-terminal (25-39) (□——□) fragments.

the increased specificity obtainable in a two-site system with antibodies directed at different epitopes can prove beneficial. The simplicity of the assay procedures makes two-site systems attractive to the producer and the consumer for control of performance; it also leads to ease of automation.

The main disadvantages, namely the need for relatively large amounts of antibody and the requirement for affinity purification, can be overcome by the use of suitable monoclonal antibodies. In the future, we expect to see these methods used to provide further improvements in assay sensitivity. At present a major constraint is the limited specific activity of ^{125}I-labeled antibodies if immunoreactivity is to be retained. This constraint does not apply to nonisotopic labels so that labeling of antibodies with compounds such as chemiluminescent molecules will lead to improvements in immunoassay performance in terms of stability and sensitivity.

ACKNOWLEDGMENTS

We gratefully acknowledge financial assistance from Ciba-Geigy Ltd. and the Welsh Office.

REFERENCES

Addison, G. M. (1971). Preparation and properties of labeled antibodies. Thesis, University of Cambridge.

Addison, G. M. (1974). New developments in the immunoradiometric assay. In Radioimmunoassay and Related Procedures in Medicine, International Atomic Energy Agency, Vienna, p. 131.

Addison, G. M., and Hales, C. N. (1971a). The immunoradiometric assay. In Radioimmunoassay Methods, K. E. Kirkham and W. M. Hunter (eds.), Churchill Livingstone, Edinburgh, p. 447.

Addison, G. M., and Hales, C. N. (1971b). Two-site assay of human growth hormone. Horm. Metab. Res. 3: 59.

Addison, G. M., Hales, C. N., Woodhead, J. S., and O'Riordan, J. L. H. (1971). Immunoradiometric assay of parathyroid hormone. J. Endocrinol. 49: 521.

Axèn, R., Porath, J., and Ernbach, S. (1967). Chemical coupling of peptides and proteins to polysaccharides by means of cyanogen halides. Nature 214: 1302.

Bagshawe, K. D., Harris, F. W., and Orr, A. H. (1968). Automated radioimmunoassay for human chorionic gonadotrophin and luteinizing hormone. In Automation in Analytical Chemistry, Vol. II. E. Kawerau (ed.). Technicon Symposia, Mediad, New York, p. 53.

Beck, P., and Hales, C. N. (1975). Immunoassay of serum polypeptide hormone by using ^{125}I-labelled anti-(immunoglobulin G) antibodies. Biochem. J. 145: 607.

Belanger, L., Sylvestre, C., and Dufour, D. (1973). Enzyme-linked immunoassay for alpha-fetoprotein by competitive and sandwich procedures. Clin. Chim. Acta 48: 15.

Catt, K., and Tregear, G. W. (1967). Solid-phase radioimmunoassay in antibody-coated tubes. Science 158: 1570.

Eber, M., Abecassis, J., Grob, J. C., Ott, G., and Methlin, G. (1980). Immunoradiometric assay for human thyroglobulin and variations in thyroid pathology. Clin. Chim. Acta 105: 51.

Girard, J., Baumann, J. G., and Hernandez, R. (1971). Attempts to use polymerized antigen for the isolation and labeling of specific antibodies. In Radioimmunoassay Methods, K. E. Kirkham and W. M. Hunter (eds.), Churchill Livingstone, Edinburgh, p. 461.

Greenwood, F. C., Hunter, W. M., and Glover, J. S. (1963). The preparation of ^{131}I-growth hormone of high specific activity. Biochem. J. 89: 114.

Grenier, A., Morissette, J., Valet, J-P., and Belanger, L. (1978). Polystyrene tube immunoradiometric assay for human α_1 fetoprotein and its use for mass screening. Clin. Chem. 24: 2158.

Gurvich, A. E., Kuzovleva, O. B., and Tumanova, A. E. (1961). Production of protein-cellulose complexes (immunoadsorbents) in the form of suspensions able to bind great amounts of antibodies. Biokhimiia 26: 803.

Hales, C. N., and Woodhead, J. S. (1980). Labeled antibodies and their use in the immunoradiometric assay. In Methods in Enzymology, Vol. 70, H. Van Vanukis and J. J. Langone (eds.), Academic Press, New York, p. 334.

Hodgkinson, S. C., and Lowry, P. J. (1982). Selective elution of immunoadsorbed anti-(human prolactin) immunoglobulins with enhanced immunochemical properties. Biochem. J. 205: 535.

Homburger, H. A., Smith, J. R., Jacob, G. L., Laschinger, C., Naylor, D. H., and Pineda, A. A. (1981). Measurement of anti-IgA antibodies by a two-site immunoradiometric assay. Transfusion 21: 38.

John, R., and Woodhead, J. S. (1982). An automated immunoradiometric assay of thyrotropin (TSH) in dried blood filter paper spots. Clin. Chim Acta 125: 329.

Jones, B. H., and Worwood, M. (1978). An immunoradiometric assay for the acid ferritin of human heart: Application to human tissue, cells and serum. Clin. Chim. Acta 85: 81.

Kato, K., Umeda, A., Suzuki, F., Hayashi, D., and Kasaka, A. (1979). Use of antibody Fab fragments to remove interference by rheumatoid factors with the enzyme-linked sandwich immunoassay. FEBS Letts. 102: 253.

Kemp, H. A., Simpson, J. S. A., and Woodhead, J. S. (1981). Automated two-site immunoradiometric assay of human alphafetoprotein in maternal serum. Clin. Chem. 27: 1388.

Kemp, H. A., John, R., and Woodhead, J. S. (1982). Labelled antibody immunoassays. Ligand Q. 5: 27.

Langone, J. J. (1978). ^{125}I-protein A: A tracer for general use in immunoassay. J. Immunol. Methods 24: 269.

Lazarus, J. H., John, R., Ginsberg, J., Hughes, I. A., Shewring, G., Rees Smith, B., Woodhead, J. S., and Hall, R. (1983). Transient neonatal hyperthyrotrophinaemia: A serum abnormality due to a transplacentally acquired antibody to thyroid stimulating hormone. Br. Med. J. 286: 592.

Miles, L. E. M. (1971). Isolation of specific high affinity antibodies from non-specific antisera. J. Clin. Endocrinol. Metab. 33: 399.

Miles, L. E. M., and Hales, C. N. (1968a). The preparation and properties of purified ^{125}I-labelled antibodies to insulin. Biochem. J. 108: 611.

Miles, L. E. M., and Hales, C. N. (1968b). Labelled antibodies and immunological assay systems. Nature 219: 186.

Miles, L. E. M., and Hales, C. N. (1970). Immunoradiometric assay procedures: New developments. In In vitro Procedures with Radioisotopes in Medicine. International Atomic Energy Agency, Vienna, p. 483.

Miles, L. E. M., and Hartree, A. S. (1971). Immunoassay of human luteinizing hormone using univalent radioactive antibodies. J. Endocrinol. 51: 91.

Miles, L. E. M., Bieber, C. P., and Eng, L. F. (1974a). Properties of two-site immunoradiometric (labelled antibody) assay systems. In Radioimmunoassay and Related Procedures in Medicine, Vol. 1. International Atomic Energy Agency, Vienna, p. 149.

Miles, L. E. M., Lipschitz, D. A., Bieber, C. P., and Cook, J. D. (1974b). Measurement of serum ferritin by a two-site immunoradiometric assay. Anal. Biochem. 61: 209.

Muller, H. P., Van Tilburg, N. H., Bertina, R. M., Terwiel, J. P., and Veltkamp, J. J. (1980). Immunoradiometric assay of procoagulant factor VIII antigen. Clin. Chim. Acta 107: 11.

O'Riordan, J. L. H., and Woodhead, J. S. (1976). Immunoradiometric assay of parathyroid hormone in serum. In Methods of Hormone Analysis, H. Breuer, D. Hamel, and H. L. Kruskemper (eds.), Thieme, Stuttgart, p. 66.

Rainbow, S. J., Woodhead, J. S., Yue, D. K., Luzio, S. D., and Hales, C. N. (1979). Measurement of human proinsulin by an indirect two-site immunoradiometric assay. Diabetologia 17: 229.

Readhead, C., Addison, G. M., Hales, C. N., and Lehmann, H. (1973). Immunoradiometric and two-site assay of human follicle-stimulating hormone. J. Endocrinol. 59: 313.

Rodbard, D., Feldman, Y., Jaffe, M. L., and Miles, L. E. M. (1978). Kinetics of two-site immunoradiometric ("sandwich") assays II: Studies on the nature of the high dose hook effect. Immunochemistry 15: 77.

Smith, K. O., and Gehle, W. B. (1980). Semiautomation of immunoassays by use of magnetic transfer devices. In Methods in Enzymology, Vol. 70, H. Van Vanukis and J. J. Langone (eds.), Academic Press, New York, p. 388.

Thorell, J. I., and Johannsen, B. G. (1971). Enzymatic iodination of polypeptides with ^{125}I to high specific activity. Biochim. Biophys. Acta 251: 363.

Vyzantiadis, A., Danielidis, B., Liatsis, J., and Valtis, D. (1974). Detection of Australia antigen by immunoradiometric assay and other methods. In Radioimmunoassay and Related Procedures in Medicine, Vol. 2. International Atomic Energy Agency, Vienna, p. 389.

Weeks, I., Kemp, H. A., and Woodhead, J. S. (1981). Two-site assay of human α_1-fetoprotein using ^{125}I-labelled monoclonal antibodies. Biosci. Rep. 1: 785.

Woodhead, J. S., Addison, G. M., and Hales, C. N. (1974). The immunoradiometric assay and related procedures. Br. Med. Bull. 30: 44.

Woodhead, J. S., Evans, M. J., Scarisbrick, J. J., Read, G., and Cameron, E. H. D. (1975). The preparation and properties of ^{125}I-labelled antibodies to progesterone. In Fifth Tenovus Workshop on Radioimmunoassay of Steroids, E. H. D. Cameron, S. G. Hillier, and K. Griffiths (eds.), Alpha Omega, Cardiff, p. 269.

Woodhead, J. S., Davies, S. J., and Lister, D. (1977). Two-site assay of bovine parathyroid hormone. J. Endocrinol. 73: 279.

Woodhead, J. S., Simpson, J. S. A., Davies, S. J., Foster, H., and Davies, C. J. (1981). Accuracy and precision in automated immunoassay systems. In Quality Control in Clinical Endocrinology, D. W. Wilson, S. Gaskell, and K. Griffiths (eds.), Alpha Omega, Cardiff, p. 117.

Yue, D. K., Gibby, O. M., Luzio, S. D., Yanaihara, H., and Hales, C. N. (1979). Indirect two-site immunoradiometric assay of rat and mouse proinsulin. Diabetologia 17: 235.

8
Monoclonal Antibodies

NOEL R. LING AND ROYSTON JEFFERIS
University of Birmingham School of Medicine,
Birmingham, England

I. INTRODUCTION

The complexity of a typical antibody response to an antigen derives from the many clones of antibody-secreting cells generated. The revolutionary new hybridoma technique of Kohler and Milstein (1975) allows the polyclonal response to be dissected into its monoclonal components. This is achieved by fusing the antibody-secreting cells to cells of a plasmacytoma line. Each of the resulting hybrid cells (after cloning) will give rise to immortal hybridoma cell lines secreting antibody of a single molecular species.

The monoclonal antibody (McAb) product of a hybridoma cell line will be specific for a single epitope (antigenic determinant) which may characterize or define a target molecule. Thus McAb may be produced following immunization with crude antigen preparations and each McAb may be employed in the isolation and purification of individual components. This application is exemplified in the purification of interferon (Secher and Burke, 1980) and an already burgeoning literature demonstrating the definition of individual biological macromolecules following immunization with whole cells, bacteria, or virus. When McAb of a required specificity, affinity etc., have been produced, their permanent availability can be exploited to establish standardized assay systems.

II. PRINCIPLES OF PRODUCTION

In principle the technique is simple; in practice, it is demanding and labor-intensive. Considerable expertise in the handling of cells is particularly needed. The technology of the production, growth, and cloning of hybridomas is similar whatever the nature of the antigen, but the immunization and selection techniques require the specialized knowledge which the group attempting to produce McAbs will usually have already acquired in the preparation of polyclonal antisera. The techniques described elsewhere in this volume can all be adapted as procedures for the selection and characterization of McAb. The following scheme is based on our own experience over the past 3 years in producing monoclonal antibodies to human immunoglobulins.

A. Choice of Plasmacytoma Line

A suitable plasmacytoma should (1) grow well in vitro, (2) fuse well with Ig-secreting cells, (3) produce stable hybridomas, (4) have some distinctive property such as an enzyme deficiency by which it can be selected against, and (5) ideally not produce Ig itself. The two mouse lines which have been most used are NSI/1-Ag 4.1 (Kohler and Milstein, 1976) which produces but does not secrete light-chain and Sp 2/0-Ag 14 (Shulman et al., 1978) which produces no Ig. A suitable rat cell line is Y3-Ag 1.2.3 which produces light-chain (Galfré et al., 1979). Cells of these lines lack the enzyme hypoxanthine phosphoribosyl transferase and when grown in a medium containing aminopterin (which blocks the main pathway of nucleic acid synthesis) are unable to utilize the salvage pathway whereas normal cells are able to do so provided thymidine and hypoxanthine are also available. By adding hypoxanthine, aminopterin, and thymidine to conventional media a selective (HAT) medium is produced. Conversely HAT-sensitive cells are unaffected by thioguanine whereas normal cells are killed by it. Growth of the plasmacytoma cell line cells in thioguanine-containing medium for 1 week prior to fusion ensures that HAT-resistant mutants are not present.

B. Immunization Schedules

In general, the first injection of a soluble antigen (routinely 1 mg/ml, but a much lower dosage may be tried for valuable antigens and will probably be effective) is given emulsified in incomplete or complete Freund's adjuvant as a subcutaneous injection (0.1 ml per mouse). It is important that the antigen be as pure as is reasonably possible, particularly if it is not very immunogenic; otherwise the majority of hybridomas eventually produced may be secreting antibody directed against impurities in the preparation. Subsequent boosts at intervals of approximately 2-4 weeks should be of either soluble antigen or antigen precipitated with Alhydrogel (aluminium hydroxide) and injected intraperitoneally (0.1 ml). After one or more boosts a test bleed from the tail is taken and the antibody content assessed by titration. The titer need not be high, but it is desirable that there should not be high titers of antibodies to irrelevant antigens. Mice that are considered to be satisfactorily immunized are allowed to rest for 2-4 weeks. They then receive a final boost 3-4 days before fusion. It has been our practice that this should take the form of an intraperitoneal injection (0.05 mg) in the morning followed by an intravenous injection (0.05 mg in 0.1 ml into the heart) in the afternoon.

C. Screening and Selection Procedures for Soluble and Cellular Antigens

For most antigens a binding assay is preferable; this is usually either an enzyme-linked immunosorbent assay (ELISA) or a radioimmunoassay. It is vital that the assay system chosen be simple, rapid, and sensitive as a very large number of assays must be performed over a short period of time once the hybridomas start to grow. An alternative, which is applicable to all antigens that may be attached to red cells (provided they are available in milligram quantities) is a passive hemagglutination system (Ling et al., 1977; Lowe et al., 1981). Protein antigens are coupled to sheep red cells with chromic chloride and are stored at 4°C until required. Provided sterility is maintained, the coated cells may be kept for several weeks. For a titration or screening all that is required is to dilute the contents of a tube of coated cells (4 ml of a 2.5% suspension) to 20 ml with saline and add one drop to 0.05 ml of culture supernate or serially diluted antibody in a U-bottom microtiter plate. If the dilute suspension is to be kept for some time it is better to dilute not with saline but with sterile Hepes-buffered RPMI medium containing 2% fetal calf serum.

Most binding assays for the screening of culture supernates containing McAbs to cell surface antigens are performed with glutaraldehyde-fixed target cells. Convenience outweighs the risk of loss of cell surface determinants by the aldehyde treatment. A variant of this procedure involves attachment of cells to polylysine-coated wells of microtiter plates; cells are sedimented onto the base, adhere as a monolayer, and are then permanently fixed with glutaraldehyde (Cobbold and Waldmann, 1981). The ELISA technique has also been successfully applied to unfixed cells (Posner et al., 1982).

D. Hybridization, Screening, and Cloning

The original fusogen, Sendai virus, has now been replaced by polyethylene glycol (PEG). We have routinely used PEG 1500 in a technique based on that of Galfré et al. (1977), but systematic studies have shown that PEG 4000 is superior (Fazekas de St Groth and Scheidegger, 1980). Spleen cells from the immune mouse and plasmacytoma cells are separately washed, then mixed, spun down together and the pelleted cells exposed to a high concentration of PEG under defined conditions. The cell suspension is then carefully diluted and the cells washed free of PEG and placed in growth medium. This is a critical stage in the experiment and the way it is performed will determine the number of hybridomas produced. There are many small variations of procedure in different laboratories. It is conventional to use a ratio of spleen cells to plasmacytoma cells of 10:1 or 5:1 but ratios of 2:1 and 1:1 have also been used successfully. Following fusion the cells are resuspended, the clumps lightly dispersed, and the suspension further diluted with growth medium. It is important that this be the best-quality medium available because it must promote the growth of single hybridomas into colonies.

Single-strength RPMI-1640 (purchased as such) is slightly superior to medium prepared from 10X concentrate or from powder and should be used at this stage. It should be supplemented (20%) with good-quality serum. This is usually fetal bovine, but sometimes horse serum has been used instead. Addition of 2-mercaptoethanol (to 60 μM) is also beneficial. It is usual to use "non-HAT" medium at this stage and change to HAT medium 1 day later. There appears to be no advantage in doing this and the full medium with HAT ingredients may be used for all postfusion cultures. The cell suspension must now be aliquoted into either 48 wells of two Linbro trays or 6 x 96 wells of six microtiter trays. All cultures are transferred to a gassed incubator (5% CO_2 in moist air, 37°C). If the larger wells are preferred, early cloning is desirable since, following a good fusion, each well is likely to contain a number of different hybridomas and if cultured collectively the strongest grower (which may well not be the producer of the desired antibody) will outgrow other cells in the well.

Hybridomas (looking very much like the NSI plasmacytoma cells) will first be observed under the inverted (Olympus) microscope as single cells. Small colonies may start to appear at 5 days postfusion, by which time other cell types, such as macrophages and fibroblasts, will already have spread out over the surface of the well and often continue to grow until outgrown by the hybridomas. Red cells will also be seen. Medium is now added to the cells at intervals, as needed to support growth. When cell multiplication reaches a stage when near confluence is reached and the medium shows signs of turning acid (yellow) after overnight incubation it is necessary to decide whether or not the cells are to be grown up in large volumes or to be cloned or discarded. This decision will obviously depend on the result of the screening test.

Cell colonies from microwells will probably be monoclonal by the time they have been grown up in large volumes and need not be cloned at this stage. Colonies from Linbro wells may be cloned in soft agar. In this case the resulting colonies (presumed to represent single clones) are picked off from the agar and the cells grown up first in microwells then in Linbro wells and finally grown in bulk, in bottles or flasks. Some cells are frozen and some injected into Pristane-primed mice (see later). The antibody content of the supernates should be checked systematically during the growing-up phase. Inevitably there will be some unpleasant surprises in that loss of antibody production may be encountered with some hybridomas even after cloning. This is due to the fact that some hybridomas are genetically unstable.

Once the bulk phase has been reached, however, it is unusual for loss of antibody production to occur, although recloning (to avoid overgrowth by nonproducing mutants which may have arisen) may be advisable at a later stage. Freezing of some of the cells at the earlier stages of growth leaves open the option of regrowth and cloning. It is important always to aim at maximal growth conditions and particularly not to allow overgrowth of cultures with extensive cell death as this seems to favor loss of antibody production. The cells at the culture stage are frozen down and more cells are frozen down after clones have been grown up in pristane-primed mice. The cells are spun down from the ascitic fluid and the clear supernate is tested for antibody content, aliquoted, and stored frozen. It is our policy to use stored cultured cells for future cultures in vitro and stored ascitic cells for future growth in mice.

1. Use of McAb to Select Antigens from a Crude Preparation for Use as Immunogens for Further Production of McAbs (Cascade Procedure)

Columns of Sepharose-bound McAbs have been successfully used for the purification of interferon (Secher, 1980) and rat lymphocyte surface antigens (Sunderland et al., 1979).

2. Large Scale Production of McAb in Vitro and in Vivo

It is reasonably easy to grow liter volumes of cell suspensions in stationary plastic or glass flasks. Cells will grow to a concentration of approximately 0.5×10^6/ml. Larger numbers of cells may be grown in "spinner" or "roller" cultures, but in our limited experience antibody levels achieved have been disappointing. Growth of lymphoid cells on a factory scale has been achieved for human lymphoblastoid cell lines and the same methods should be applicable to mouse hybridomas. The hybridomas may also be grown up as ascitic tumors in BALB/c mice injected intraperitoneally with 0.5 ml of Pristane (2, 6, 10, 14-tetramethylpentadecane; Koch Light Laboratories Ltd., Colnbrook, Buckinghamshire SL3 OBZ, U.K.) 3-60 days previously. Approximately $2.5-5 \times 10^6$ cells in 0.5 ml of serum-free medium are injected intraperitoneally. After approximately 10-15 days, when an obvious abdominal swelling appears, the mice are killed and the fluid drained off; 5-10 ml of fluid per mouse is obtained. The first passage yields enough tumor cells to inject a large number of other mice and large volumes of fluid are readily obtained. The antibody content is usually between 0.5 and 5 mg/ml. Larger volumes of fluid ($\simeq$ 50 ml) may be obtained by growing rat hybridomas in syngeneic rats.

It is often desirable to obtain pure antibody preparations; indeed that is one of the reasons for preferring McAb. In culture supernates McAbs are accompanied by large amounts of fetal bovine serum proteins. The antibodies may be removed by affinity-binding to an antigen column and recovered by elution with 3 $\underline{M}$ KCNS; or they may be affinity-purified on a column of staphylococcal protein A (if of the appropriate subclass of mouse

IgG; see Goding, 1978). Both these procedures, however, are liable to produce some aggregation of antibodies. Recovery from ascitic fluid (which is likely to contain a "background" of "natural" antibodies) is most simply achieved by preparation of a suitable fraction by DEAE chromatography. The fraction will inevitably be contaminated by polyclonal immunoglobulin of the same class as the McAb, or affinity chromatography on an antigen column may be used. None of these procedures is ideal when pure McAb, free of aggregates, is required. For this purpose it is preferable to grow the hybridoma in serum-free medium. Some hybridomas will grow in HAT medium lacking FCS. Others will grow in the presence of dialyzable serum components (Klinman and McKearn, 1981) or in Iscove's medium (Iscove and Melchers, 1978). The antibody in the culture supernates may usually be concentrated by dialyzing the fluid against 0.01 $\underline{M}$ phosphate buffer pH 6.5, running it through a CM-cellulose column and eluting with 0.15 $\underline{M}$ phosphate buffer pH 7.0. If labeled antibodies are required, a labeled amino-acid (e.g., ^{3}H-leucine or ^{35}S-methionine) is added to the medium. If high-activity labeling is required it is necessary to omit this particular (unlabeled) amino acid from the medium (Haas and Kennett, 1980).

McAb are susceptible to the same destructive effects (denaturation, proteolysis) as polyclonal antibodies but there is considerable individual variation between different McAbs.

Long-term storage of culture supernates at 4°C is not recommended because serum components present will eventually destroy antibody activity. Jonak (1980) recommends precipitation of McAb by adding an equal volume of saturated ammonium sulfate and storing the isolated redissolved McAb at -70°C.

E. Freezing Down and Recovery of Hybridomas

1. Freezing Down

Cultured cells are stored as a source material for future cultures and must be kept sterile. Ascites cells are needed for regrowth in mice and need not be kept sterile. In both cases spin down approximately 10^7 cells for each vial to be frozen. Discard supernate and resuspend cells to approximately 20×10^6/ml in H-RPMI (stage 1). Transfer 0.5 ml aliquots to each small polypropylene tube. Spin down cells and discard supernate completely. Resuspend cells in 7 drops of 5% dimethyl sulfoxide (DMSO) in 95% FCS (this mixture is made up in bulk and frozen in aliquots). Transfer tubes to a 50% glycerol bath at -32°C after marking top and rim in different colors with felt-tipped pens (Staedtler lumocolor 317 permanent, waterproof). Leave the tubes in bath for 40 min. Wipe tubes with paper towel and transfer directly to liquid nitrogen storage container.

2. Recovery

Thaw the tube quickly by holding it under the warm water faucet. Have ready some H-RPMI already warmed to 37°C. Add 1.7 ml to the thawed suspension in the polypropylene tube, the first few drops slowly, with mixing. Replace cap, mix by inversion. Spin for 5 min. at approximately 1400 rpm on a bench centrifuge. Suck off and discard supernate. Take up the cells in HAT medium (or PBS if to be injected into mice) and transfer to two Linbro wells. Incubate and expand cultures by adding HAT medium as necessary. A typical cell suspension (86% viable before freezing) gave viability values of 50-78% after recovery and culture for 24 hr. Higher cell concentrations may be frozen but care must be taken not to dilute the DMSO significantly with medium trapped by the cells. This may be avoided by using H-RPMI containing 5% DMSO at stage 1.

F. A Specific Fusion Experiment

The immunogen used is affinity-purified Bence-Jones kappa chain (Bo). Immunization consists of a first injection (day 0) of 0.1 ml of Bo (1 mg/ml) in incomplete Freund's adjuvant injected subcutaneously. An intraperitoneal boost is given on days 8, 15, 22, 29, and 39. The final boost is given intraperitoneally in the morning and intravenously in the afternoon (day 39).

Fusion is performed on day 42 as follows. Remove spleen, trim, place in Petri dish in sterile hood (in which all future manipulations will be performed). Two syringes each containing 10 ml of warm RPMI-1640 medium (no serum) are fitted with No. 26 needles. Cells are blown out of the spleen by multiple puncture and expulsion of small volumes of medium (Kennett et al., 1978). Filter the cell suspension through sterile nylon gauze into a universal container. The yield is 5×10^7 lymphocytes (about half the cells in the spleen) and the viability is 90% (as assessed by trypan blue exclusion). Sample the suspension of NSI cells (fed 24 hr previously) and perform a viable cell count (normally 90% viable) taking approximately 10^7 cells. Each cell preparation is spun separately for 5 min at 2000 rpm on a bench centrifuge and the supernate sucked off and discarded. Each cell pellet is suspended in RPMI-1640 (room temperature).

The two suspensions are now mixed, centrifuged, at 1850 rpm, the supernate discarded, and the cell pellet dispersed by flicking and adding medium. The pellet is warmed to 37°C (in the hand) for at least 1 min. PEG 1500 (40% in RPMI-1640, prewarmed to 37°C) is added over 1 min and the tube held in the hand for 1 min. Warm RPMI-1640 (1 ml) is then added over 1 min and a further 20 ml over 5 min. The tube is centrifuged at 1500 rpm, the supernate discarded, and the cells resuspended in 30 ml of RPMI-1640 supplemented with 20% FCS and 2ME (60 µ$\underline{M}$). One drop of cell suspension is added to each well of six labeled Costar microtrays and the trays then transferred to a gassed incubator (5% CO_2 in moist air).

After 24 hr some complete HAT medium is poured into a sterile aluminum dish. An eight-channel multiple diluter (8 MD) set at 0.05 ml is fitted with polypropylene tips. The tips are sterilized in boiling water in a stainless steel bath by sucking back and forth. The 8 MD is then used to add 0.05 ml of medium to each microwell. The trays are inspected for growth of hybridomas and any change of color noted. Six days after fusion 0.15 ml of HAT is added to each well using the 8 MD.

By day 9 some large colonies are present and the medium is turning yellow in a few wells. Six U-bottom microtiter trays are labeled to correspond to the culture trays. Using the sterilized 8 MD, 0.05 ml samples from the first line of eight wells of the culture tray are transferred to the corresponding U-bottom tray; the 8 MD is resterilized, the next eight wells are sampled, and so on. Again using the 8 MD, fresh HAT medium (0.05 ml) is added to each well of the sampled trays before returning them to the incubator. The samples removed are now tested for antibody by adding to each well (with the 8 MD) 0.05 ml of a 0.5% suspension of sheep red cells coated with a Bence-Jones kappa (NA) protein.

After 2 hr the settled pattern agglutination is read. Culture wells showing positive results in the supernate test are inspected for a hybridoma colony. If wells with positive results contain a good growth of hybridoma cells all is well. Otherwise it is necessary to suck off and replace all the medium in all wells. Medium is sucked off at a water pump through rubber tubing fitted with a very fine tipped bent glass tube (prepare by bending the narrow portion of a Pasteur pipette to make a right angle bend and drawing out the end to a fine tip). The medium is replaced using the 8 MD as usual. When growth is near confluence the cells from wells with positive results are transferred with a Pasteur pipette to a Linbro well containing 0.5 ml of HAT. The microwell is topped up with HAT. A second screening is performed 3 days later.

Fresh positives identified are also grown up in Linbro wells. When growth in the Linbro well is established fresh HAT medium is added and after further growth the colony is split into two wells and as growth continues cells are transferred to small plastic flasks. Samples are frozen as soon as 20×10^6 cells are available. Cells are also injected into pristane-primed mice: two mice initially for each colony. Culture supernates and ascitic fluids are monitored at all stages for antikappa antibody by passive hemagglutination.

The following reagents may be used:

1. RPMI-1640 medium single strength (Gibco Biocult Ltd., Washington Road, Paisley, Scotland).
2. HAT medium: RPMI-1640 + 20% FCS + hypoxanthine (10^{-4} M), thymidine (10^{-5} M), methotrexate (0.5×10^{-8} M), and 2 ME (6×10^{-5} M).

3. H-RPMI: RPMI-1640 medium prepared from 10x concentrated medium and buffered with Hepes instead of bicarbonate

The following stock solutions should be prepared:
1. Hypoxanthine: 408 mg in 100ml H_2O; add N NaOH until dissolved.
2. Thymidine: 114 mg in 100 ml H_2O.
3. Combine 1. and 2. and make up to 300 ml with H_2O. Adjust pH to 10.0. Filter to sterilize, aliquot, and store at $-20^{\circ}C$.
4. Methotrexate: Vials contain 2 ml at 25 mg/ml. Dilute 0.9 ml to 1 liter and adjust pH to 7.5 with N NaOH or HCl. Filter to sterilize.
5. 2 ME: Add 0.06 ml of concentrated 2-mercaptoethanol to 10 ml of saline. Filter to sterilize.
6. Fetal calf serum (FCS; Gibco, Flow, or Seward Laboratories) Test several batches by adding 10% (v/v) to stock NSI medium and testing the capacity to support growth of very dilute suspensions of NSI. Bottles stored at $-20^{\circ}C$ (or lower temperature if available). Quick freezing is preferable to slow freezing.
7. Gentamicin: 1 mg/ml.
8. Thioguanine solution ($2 \times 10^{-3}\underline{M}$): 33.4 mg of thioguanine in 100 ml of H_2O. Add N NaOH to bring pH to 9.5. Filter.

HAT medium is prepared from 160 ml of RPMI-1640, 40 ml FCS, 2 ml of 3., 2 ml of 4., 0.12 ml of 5., and 0.2 ml of 7.

NSI medium is prepared from powdered or 10X concentrated RPMI-1640 medium supplemented with penicillin (200 units/ml), streptomycin (100 units/ml), gentamicin (1 ug/ml), and FCS (10% v/v).

Thioguanine medium is made from 200 ml of NSI medium added to 2 ml of 8.

To prepare polyethylene glycol (PEG) 40%, melt 8 g of PEG in a water bath and add 12 ml of prewarmed RPMI-1640. Filter to sterilize and store at room temperature in aliquots. All sterile filtration is done through membranes of pore size 0.2 µ.

G. Cloning of Hybridomas in Soft Agar

Autoclave 3% and 5% Bacto-Agar in saline. Prepare a 5-ml base layer (0.5% agar) in a 5-cm Petri dish by dissolving 5% agar in boiling water bath, sucking up the viscous solution and diluting 1:10 with HAT-20% FCS in a $44^{\circ}C$ bath. Similarly prepare 0.3% agar, measure out 5-ml aliquots in tubes in the $44^{\circ}C$ bath, and add 1, 2, or 3 drops of 0.3×10^6/ml cells immediately before pouring onto the base layer. Pick off colonies 10-14 days later with a Pasteur pipette and grow up cloned cells in microwells.

H. Plasmacytoma Lines

NSI and other mouse plasmacytomas may be obtained from Flow Laboratories. Until recently plasmacytomas were obtainable from the Salk Institute (Cell Distribution Center, San Diego, California 92112). This service has now been taken over by the American Type Culture Collection. For human plasmacytomas, inquiries should be directed to Dr. H.S. Kaplan, Cancer Biology Research Laboratory, Department of Radiobiology, University School of Medicine, Stanford, California 94305, or to Dr. Paul Edwards, Ludwig Institute for Cancer Research, The Haddow Laboratories, Clifton Avenue, Sutton, Surrey SM2 5PX, United Kingdom.

III. APPLICATIONS OF MONOCLONAL ANTIBODIES

A recent review (Parkinson, 1981) of industrial investment in the production and application of McAb reveals 70-80 companies active in this area, many of them newly established solely to exploit this commercial potential. It has been forecast (Green, 1981; Shoemaker, 1981) that by 1985 McAb will have 25-50% of the market in which polyclonal reagents are currently employed. More dramatically, Dixon (1981), a vice-president of Abbott Laboratories Diagnostic Division, has stated "5 years from now [1981] every [antibody-based] diagnostic product will be based on monoclonal antibodies."

The essential advantages of McAb as analytical reagents lie in their specificity and permanent availability. However, each McAb appears to have unique properties and their reaction characteristics may differ markedly from polyclonal antibody such that simple substitution of one reagent by the other may not be possible. In the murine system nine antibody isotypes are defined (γM, γ1, γ2a, γ2b, γ3, γA, γD, and γE), each having characteristic physicochemical properties and biological activities. The fundamental parameters of reactivity are the avidity (valency, affinity) and epitope specificity of a given McAb. However, other activities, such as binding to staphylococcal protein A (SpA), complement fixation/activation, Fc receptor binding, etc., may also be exploited to develop new assay systems. It is important to emphasize that individual McAb may have properties that differ from those previously considered to be characteristic of the isotype to which they belong. Thus it has been a common experience to produce IgM antibodies that do <u>not</u> fix or activate complement; γ1 antibodies may bind SpA, etc.

It is important that the profile of activities of a given McAb be established, rather than assumed, before application in systems where secondary activities may be relevant. The diversity of properties of McAbs specific for the same antigen or epitope extends their potential as reagents since a McAb may be selected that has the appropriate combination of properties to allow optimization of a given assay procedure.

In the following sections the application of McAb to immunoassay will be considered in relation to the technique or procedure employed (Hunter, 1982).

A. Direct Measurements of Immune Complex Formation

The lattice theory of immune precipitation (Marrack, 1938) assumes multivalent antigen and divalent antibody of varying epitope specificity. Thus McAb might appear not to be applicable to techniques which depend on direct measurement of immune complex formation. However, it has recently been demonstrated that nonspecific, Fc-mediated aggregation of small complexes is also important in the formation of immune precipitates (Jacobsen and Steensgaard, 1979). Also, the addition of polyethylene glycol (PEG) to a medium in which an antigen-antibody reaction is taking place may dramatically enhance the rate and degree of immune complex formation. The ability of PEG to render soluble complexes insoluble is exploited as a technique for quantitating immune complexes present in pathological sera. Therefore, in principle, it might be possible to optimize conditions such that individual or simple mixtures of McAb could be applied within such methodologies.

In our laboratories we have successfully applied McAb to the quantitation of human immunoglobulins. It should be appreciated that immunoglobulins are a special, but not unique (cf. hemoglobin), case since the symmetry and conformation of the molecule results in it being antigenically divalent. A study of the size of complexes formed at varying antigen/McAb ratios in the absence of PEG shows that relatively simple complexes (trimers, tetramers) predominate throughout (Steensgaard et al., 1982). However, in the presence of PEG these complexes may precipitate. Thus, it has been possible to develop a Mancini assay for IgG using a single McAb incorporated into agarose containing 6% PEG (Lowe et al., 1982).

Automated immunoassay procedures depend on the rapid formation of insoluble immune complexes, in the presence of PEG (approximately 4%) and their detection by measurement of turbidity development or scattered light (nephelometry). A series of McAb when used individually failed to yield insoluble complexes under these conditions. However, combinations of two McAb directed against structurally and spatially distinct epitopes resulted in turbidity development comparable to that obtained with polyclonal antibody (Steensgaard et al., 1980). Simple mixtures or combinations of antibody can be readily reproduced and thus standardization of antiimmunoglobulin reagents becomes possible. Such systems can be further exploited to allow the quantitation of antigenically defined subpopulations of immunoglobulin molecules. A combination of McAb specific for the heavy chain of IgG and the kappa light chain allows IgG_{kappa} molecules to be quantitated. IgG_{lambda} molecules may be quantitated similarly. Such systems have been developed and applied to precise quantitation using the centrifugal analyzer (Jefferis et al., 1980; Deverill et al., 1981), the Mancini technique (Prior et al., 1981) and the laser nephelometer (May, personal communication). For immunoglobulins the technique could, in principle, be extended to the quantitation of subclasses, allotypic variants, etc. It may also have application to the quantitation of other polymorphic proteins. Some antihuman IgG McAb cross-react with the immunoglobulins of other species with sufficient avidity to allow these antigens to be quantitated also (Jefferis et al., 1980; May et al., 1982). This suggests new perspectives for standardization of reagents.

While systematic studies of other antigen/McAb systems have not been reported, Lachmann et al. (1980) failed to precipitate C3 using a combination of three antibodies. Similarly in our laboratory human serum albumin was not precipitated by a combination of three McAbs in the presence of 7% PEG. It is not known whether these antibodies are directed against structurally and spatially distinct epitopes.

B. Observation of Secondary Phenomena

Secondary phenomena which may be employed in immunoassay include agglutination, complement fixation/activation, SpA binding, and direct binding of McAb revealed using a labeled polyclonal antibody to mouse Ig. In our studies we have used agglutination of passively sensitized sheep red blood cells as the primary screen for the presence of McAb in culture supernates (Lowe et al., 1981). Initially the same technique was adopted for specificity testing, species cross-reactivity studies (Jefferis et al., 1982), etc. The titer of antibodies within a panel of 40 anti-Fc McAbs varies between $\log_2 4$ and $\log_2 25$, while their concentration varied, maximally, by a factor of five. Since all antibodies are of the γ1 class it might be anticipated that the agglutination titer is, principally, a reflection of affinity. However, affinity measurements demonstrate a much narrower range of values (Jacobsen et al., 1982). In our experience antibodies having a low agglutination titer may be very effective in precipitation systems and vice-versa. Also, some antibodies of high hemagglutination titer have been shown to be poorly active or inactive in ELISA assay (Skvaril; Jaafar; personal communications). The protocol of both Skvaril and Jaafar required the antigen (human IgG) to be immobilized on plastic plates. It is assumed that this procedure either results in denaturation with the destruction of certain epitopes or that binding occurs preferentially through certain functional residues so that some epitopes are not, sterically, available for reaction with antibody.

It must be emphasized that ELISA and radioimmunoassay techniques have been most popularly employed as the initial screens for antibody producing hybridomas. Antibodies selected in this way will obviously be applicable in the same technique for further quantitative assay. Our experience suggests that if McAb to a specific epitope is an essential requirement it would be advisable to screen hybridoma products by at least two assay techniques, each dependent on different structural or physicochemical requirements.

Similar considerations are particularly relevant to the development of McAb for immunohistology. McAbs produced as a result of immunization with native antigen in soluble form or as expressed on the surface of live cells may vary in their applicability to visualization of antigen in fixed tissue. The experience with two McAbs each having specificity for human IgG1 subclass proteins illustrates several of the points made above. Antibodies JL512 and NL16 have hemagglutination titers of $\log_2$ 19 and $\log_2$ 12, respectively (Lowe et al., 1982); JL512 but not NL16 precipitates antigen in the presence of 6% PEG. However, NL16 but not JL512 is applicable to ELISA, indirect immunofluorescence of membrane IgG1 on live cells (Cooper, personal communication), and indirect immunoperoxidase detection of cytoplasmic IgG1 in tissue-fixed in 10% formalin (MacLennan, personal communication).

C. Use of Labeled Antigens

There is the potential for the further development of the sensitivity and specificity of radioimmunoassay (RIA) procedures using McAb. However, RIA does depend upon the use of antibodies of high avidity while the affinity constants of monoclonal antibodies, selected on the basis of specificity, are modest and in the range $Ko(M^{-1})$, 10^6-10^{10}. Concentration on immunization schedules that promote maturation of the immune response, hence the production of high-affinity antibody, and a hybridoma screening assay that selects for specificity and high affinity, should result in McAb with the required properties.

Sufficient data are available to suggest that McAb may allow more specific quantitation of structurally related hormones or drugs and their metabolic products. Ivanyi and Davies (1980) have studied the specificity of McAbs to human growth hormone (HGH) and their cross-reactivity with human chorionic somatomammotropin (HCS) which has 85% sequence homology with HGH. McAbs exhibiting complete or partial cross-reactivity were observed and also McAbs showing complete specificity for HGH. McAbs to T_3 and T_4 (Wang et al., 1981) have been compared with polyclonal reagents. The affinity constants of the McAbs were $Ko(M^{-1})$, 1.5×10^8 and 1.4×10^9, respectively, while the polyclonal reagents used in RIA gave values of 1.6×10^9, respectively. Studies of the specificity and affinity of antibodies to cardiac glycoside digoxin suggest that they offer further refinement in the detection and quantitation of structurally related drugs and their metabolic products (Margolies et al., 1981). A McAb of affinity constant $(Ko(M^{-1}))$ 4.7×10^9 was shown to be applicable to clinical radioimmunoassay of serum digoxin levels and comparable to a sheep antidigoxin antibody of affinity constant 1.2×10^{10}. The first McAb (clone 5004) marketed for RIA of hCG was launched in 1981 (Medix). The affinity of 5004 is 8×10^{10} mol/liter allowing an RIA sensitivity of 3 U/liter. Cross-reactivity with hLH and hTSH is quoted at 0.3% and 0.02%, respectively.

D. Use of Labeled Antibodies

Hunter (1982) has shown that mouse McAb may be reproducibly labeled with ^{125}I, by the chloramine-T method, to provide stable reagents that offer major advantages in immunoradiometric assay (IRMA) procedures. He predicts that the availability of McAb will result in the replacement of many RIA by IRMA, particularly for proteins, because of the improved speed of the latter. The development of IRMA may be expected to benefit from the ability to label McAb biosynthetically by the incorporation of radioactive amino acids. This offers the advantage of simplicity, and therefore cost, control over the level of activity incorporated, and minimized loss of activity due to the labeling procedure.

E. Clinical Applications

The broader implications of the monoclonal revolution can receive only brief mention here. Complex families of molecules such as immunoglobulins and histocompatibility antigens will be much more efficiently analyzed and new subdivisions are already being recognized. Previously undetected subpopulations of cells (e.g., in lymphoid and nervous tissue) will be revealed and the various stages of differentiation pinpointed by detection of minor antigens. The antigenic structure of important pathogens (bacteria, viruses, parasites) will be more fully understood. Microbial antigens purified by affinity-chromatography on McAb columns might be used in vaccines. McAb from human hybridomas may be used in passive immunization (e.g., in the prevention of rhesus sensitization). In the tumor field McAbs to cell surface antigens have been successfully used for tumor imaging and there has been a revival of interest in the "magic bullet" using antitumor McAb conjugated to a toxic agent such as ricin. It is safe to predict that there will be a whole new approach to the analysis and preparation of biologically and clinically important molecules.

REFERENCES

Cobbold, S. P., and Waldmann, H. (1981). A rapid solid-phase enzyme-linked binding assay for screening monoclonal antibodies to cell surface antigens. J. Immunol. Methods 44: 125.

Deverill, I., Jefferis, R., Ling, N. R., and Reeves, W. G. (1981). Monoclonal antibodies to human IgG: Reaction characteristics in the centrifugal fast analyser. Clin. Chem. 27: 2044.

Dixon, H. F. (1981). Biotechnology's new thrust in antibodies. Business Week. May 18th, p. 147.
Fazekas de St Groth, S., and Scheidegger, D. (1980). Production of monoclonal antibodies: Strategies and tactics. J. Immunol. Methods 35: 1.
Galfré, G., Howe, S. C., Milstein, C., Butcher, G. W., and Howard, J. C. (1977). Antibodies to major histocompatbility antigens produced by hybrid cell lines. Nature 266: 550.
Galfré, G., Milstein, C., and Wright, B. (1979). Rat x rat hybrid myelomas and a monoclonal anti-Fd portion of mouse IgG. Nature 277: 131.
Goding, J. W. (1978). Use of staphylococcal protein A as an immunological reagent. J. Immunol. Methods 20: 241.
Green, H. (1981). Bioeng. News 1 (2): 2.
Haas, J. B. and R. H. Kennett. (1980). Characterisation of hybridoma immunoglobulins by sodium dodecylsulphate-polyacrylamide gel electrophoresis. In: Monoclonal Antibodies, R. H. Kennett, J. McKearn, and K. B. Bechtol (eds), Plenum Press, New York, p. 407.
Hunter, W. M. (1982). Monoclonal Antibodies in Clinical Chemistry. Edinburgh Royal Society Proceedings.
Iscove, N. N., and Melchers, F. (1978). Complete replacement of serum by albumin, transferrin and soybean lipid in cultures of lipopolysaccharide-reactive B lymphocytes. J. Exp. Med. 147: 923.
Ivanyi, J., and Davies, P. (1980). Monoclonal antibodies against human growth hormone. Mol. Immunol. 17: 287.
Jacobsen, C., and Steensgaard, J. (1979). Evidence of a two-stage nature of precipitin reaction. Mol. Immunol. 16: 571.
Jacobsen, J., Frich, J. R., and Steensgaard, J. (1982). Determination of affinity of monoclonal antibodies against human IgG. J. Immunol. Methods 50: 77.
Jefferis, R., Deverill, I., Ling, N. R., and Reeves, W. G. (1980). Quantitation of human total IgG, IgG_{kappa} and IgG_{lambda} in serum using monoclonal antibodies. J. Immunol. Methods 39: 355.
Jefferis, R., Lowe, J. A., Ling, N. R., Porter, P., and Senior, S. (1982). Immunogenic and antigenic epitopes of immunoglobulins I. Cross-reactivity of murine monoclonal antibodies to human IgG with immunoglobulins of certain animal species. Immunology 45: 71.
Jonak, Z. L. (1980). In: Monoclonal Antibodies, R. H. Kennett, J. McKearn, and K. B. Bechtol (eds), Plenum Press, New York, p. 405.
Kennett, R. H., Denis, K. A., Tung, A. S., and Klinman, N. R. (1978). Hybrid plasmacytoma production: Fusions with adult spleen cells, monoclonal spleen fragments, neonatal spleen cells and human spleen cells. Curr. Top. Microbiol. Immunol. 81: 77.

Klinman, D. M., and McKearn, T. J. (1981). Dialysable serum components can support growth of hybridoma cells in vitro. J. Immunol. Methods 42: 1.

Kohler, G., and Milstein, C. (1975). Continuous culture of fused cells secreting specific antibody. Nature 256: 495.

Kohler, G., and Milstein, C. (1976). Derivation of specific antibody-producing tissue culture and tumor lines by cell fusion. Eur. J. Immunol. 6: 511.

Lachmann, P. J., Oldroyd, R. G., Milstein, C., and Wright, B. W. (1980). Three rat monoclonal antibodies to human C3. Immunology 41: 503.

Ling, N. R., Bishop, S., and Jefferis, R. (1977). Use of antibody-coated red cells for the sensitive detection of antigen and in rosette tests for cells bearing immunoglobulins. J. Immunol. Methods 15: 279.

Lowe, J. A., Hardie, D., Jefferis, R., Ling, N. R., Drysdale, P., Richardson, P., Raykundalia, C., Catty, D., Appleby, P., Drew, R., and MacLennan, I. C. M. (1981). Properties of antibodies to human immunoglobulin kappa and lambda chains. Immunology 42: 649.

Lowe, J., Brid, P., Hardie, D., Jefferis, R., and Ling, N. (1982). Monoclonal antibodies (McAbs) to determinants on human gamma chains: properties of antibodied showing sub-class restriction or subclass specificity. Immunology 47: 329.

Margolies, M. N., Mudgett-Hunter, M., Smith, T. W., Novotny, J., and Haber, E. (1981). Monoclonal antibodies to the cardiac glycoside digoxin. In Monoclonal Antibodies and T-cell hybridomas. Elsevier/North-Holland Biomedical Press, Amsterdam, p. 367.

Marrack, J. R. (1938). The chemistry of antigens and antibodies. Special Report Series No. 230. p. 1. His Majesty's Stationery Office, London.

May, K., Senior, S., Garni, H. M., Porter, P., Jefferis, R., and Ling, N. R. (1981). Use of monoclonal antibodies in the characterization of immunoglobulin classes and sub-classes in animal species. Protides Biol. Fluids 29: 785.

Parkinson, D. (1981). Monoclonal antibodies; another success for biotechnology? E.E.C. DG12 Biosociety Sub-Programme. Technology Policy Unit, Preliminary Report, University of Aston, Birmingham, U.K.

Posner, M. R., Antoniou, D., Griffin, J., Schlossman, S. F., and Lazarus, H. (1982). An enzyme-linked immunosorbent assay (ELISA) for the detection of monoclonal antibodies to cell surface antigen on viable cells. J. Immunol. Methods 48: 23.

Prior, M., Ling, N. R., Lowe, J., Evans, S., May, K., and Jefferis, R. (1981). Quantitation of human IgG_{kappa} and IgG_{lambda} in normal and pathological sera. Protides Biol. Fluids 29: 789.

Secher, D. (1980). Monoclonal antibodies by cell fusion. Immunol. Today 1: 22.

Secher, D., and Burke, D. C. (1980). Monoclonal antibody for large-scale purification of human leucocyte interferon. Nature 285: 446.

Shulman, M., Wilde, C. D., and Kohler, G. (1978). A better cell line for making hybridomas secreting specific antibodies. Nature 276: 269.

Shoemaker, H. (1981). Bioeng. News 1 (2): 2.

Steensgaard, J., Jacobsen, C., Lowe, J. A., Hardie, D., Ling, N. R., and Jefferis, R. (1980). The development of difference turbidimetric analysis for monoclonal antibodies to human IgG. Mol. Immunol. 17: 1315.

Steensgaard, J., Jacobsen, C., Lowe, J. A., Ling, N. R., and Jefferis, R. (1982). Theoretical and ultracentrifugal analysis of immune complex formation between monoclonal antibodies and human IgG. Immunology 46: 751.

Sunderland, L. A., McMaster, W. R., and Williams, A. F. (1979). Purification with monoclonal antibody of a predominant Leukocyte-common antigen and glycoprotein from rat thymocytes. Eur. J. Immunol. 9: 155.

Wang, L., Hexter, C. S., and Inbar, M. (1981). Monoclonal antibodies to thyroid hormones — anti-thyroxine and anti-triiodothyronine. In Monoclonal Antibodies and T-cell hybridomas, G. J. Hammerling, U. Hammerling, and J. F. Kearney (eds.), Elsevier/North-Holland Biomedical Press, Amsterdam, p. 357.

9
Free Hormones in Blood: Their Physiological Significance and Measurement

ROGER EKINS
Middlesex Hospital Medical School,
London, England

I. INTRODUCTION

The thyroid hormones thyroxine (T_4) and triiodothyronine (T_3) circulate in blood largely bound to a variety of binding proteins: thyroxine-binding globulin (TBG), thyroxine-binding prealbumin (TBPA), and albumin. A minute fraction (0.03% and 0.3%, respectively) of each hormone also circulates in non-protein-bound or "free" form, the free and protein-bound moieties coexisting in a state of dynamic equilibrium governed by the laws of mass action. Analogous noncovalent links are likewise formed between many of the steroid hormones and corresponding steroid-binding proteins present in blood, for example, corticosterone-binding globulin (CBG), sex-hormone-binding globulin (SHBG), progesterone-binding protein (PBP), and albumin. Binding of the polypeptide and protein hormones to specific serum proteins is generally supposed not to take place, albeit the demonstration of such binding — were it to exist — would constitute a more difficult technical problem than is presented by ligands of small molecular size as exemplified by the thyroid and steroid hormones. The possible occurrence of such protein-protein binding phenomena cannot therefore be totally dismissed at the present time. Meanwhile, many other biologically active substances — vitamins, drugs, and others — participate in similar interactions with binding proteins present in blood, and are thus known to circulate in partially bound, partially free forms.

The physiological effects of this widespread phenomenon and, in particular, the exact nature of the physiological role of the specific binding or "transport" proteins which have evolved in many species, raise many interesting questions, not all of which have been satisfactorily resolved. In the case of certain examples, the specific binding protein concerned appears to mediate directly — by its attachment to the target cell surface — the transport of the ligand into the interior of the cell. This is claimed, for example, to represent the mechanism of delivery of vitamin A by retinol-binding protein (RBP) into retinal and intestinal mucosal cells (Rask and Peterson, 1976). In the majority of cases, however, it is generally believed that the specific binding protein does not facilitate cellular entry in this way, and that the molecules of the ligand traverse capillary and cell walls in free form unattached to binding proteins and without their direct intervention in the transport process. Considerable evidence has, for example, accumulated to support this view of the mechanism of delivery of the thyroid hormones to target tissues. Such evidence includes the absence of any physiological consequence in subjects in whom, for genetic reasons, TBG is either depressed or totally absent, and observations relating to the relative rates of diffusion of labeled thyroid hormones and of proteins such as TBG from blood into adjacent tissue spaces.

Further evidence for this widely held view includes the many observations that demonstrate a high degree of correlation between free (thyroid) hormone levels in peripheral blood and such physiological and clinical parameters as thyroid hormone turnover, metabolic status, etc., in circumstances in which the specific serum-binding protein concentration — and, concomitantly, the bound hormone concentration — are abnormally depressed or elevated. For example, both serum TBG and "total" T_4 levels rise significantly in pregnancy; meanwhile the free thyroxine concentration as measured by traditional dialysis techniques is not greatly changed, if at all, in these circumstances, while the pregnant individual generally appears essentially eumetabolic.

Such observations have led to the general belief that the serum-binding proteins possess no specific biological function other than that of acting as serum reservoirs or buffer stores for individual hormones. Such reservoirs are seen as essentially serving only to attenuate short-term fluctuations in the secretion of or peripheral demand for the hormones concerned. As a corollary of this belief, the absence of any overt physiological impairment in individuals in whom a specific binding protein is genetically absent is sometimes suggested as indicating the elimination, under contemporary social conditions, of the kind of environment in which the presence of serum protein-bound hormone reservoirs would be useful or confer a special biological advantage. Nevertheless an element of mystery continues to attach to the evolution of many of the serum-binding proteins and to the changes in the concentration of many of them that, in humans and in many other mammals, occur in pregnancy.

It is not the prime purpose of this review to examine in detail other postulated roles for the serum-binding proteins. Nevertheless, it is clearly pertinent to any discussion of methods of measurement of the free concentrations of hormones and of other ligands present in blood to scrutinize the foundations of the "free hormone" concept and the validity of the general belief that the free concentration of (the majority of) biologically active substances constitutes the parameter that primarily governs their individual biological effects.

It should immediately be emphasized that the demonstration that hormone molecules move across capillary wall and cellular membrane barriers when in "free" form does not necessarily imply that the serum free hormone concentration as measured at equilibrium (for example in serum in a test tube "at rest") represents the fundamental determinant of hormone action and effect. This is because, under the "dynamic" conditions which pertain in vivo, it cannot be assumed that an equilibrium state of the kind existing in serum in a test tube is maintained. Hormone must, in such circumstances, be visualized as flowing from the blood into various metabolizing tissues where it is cleared or otherwise utilized. This in turn implies that the normal equilibrium state existing in vitro is disturbed, and that a variety of rate-limiting processes govern the passage of hormone into various tissue compartments. As a corollary of these effects, the free hormone concentration which exists in capillary blood in vivo may differ significantly from that observed in vitro.

A basic understanding of the effects of the existence in serum of binding proteins in the dynamic conditions which arise in vivo is of such importance in the context of a discussion of "free hormones" and of the methodology of their measurement that a brief overview of this topic will be presented in the following section.

II. CONCEPTS OF HORMONE DELIVERY: THE ROLE OF SERUM-BINDING PROTEINS

A. Basic Physicochemical Effects

The essential effect on the in vivo free hormone concentration of a net outward flow of hormone (or of other comparable ligands) from capillary blood into adjacent tissue may be portrayed by reference to a highly simplified mathematical model. Although it does not purport to represent the in vivo situation exactly, this model nevertheless provides an insight into the kinetic events likely to be involved in hormone transport, and the possible consequences of the existence of binding proteins in serum.

Let us consider the passage along a capillary of serum containing a hormone (the "total" concentration of which is represented by [H]) accompanied by a single specific binding protein (whose total concentration is given by [P]) which reacts with the hormone with an equilibrium constant K. Let us, in particular, examine the events taking place in a small segment or "element" of the capillary (Fig. 1) of length l and radius r. Assuming, for the sake of clarity and simplicity, that the extracellular (free) hormone concentration ($[fH]_t$) is zero, and the capillary wall permeation rate constant is given by k_p (mass of hormone transported/unit wall area/unit concentration gradient/unit time), then the rate at which hormone will pass

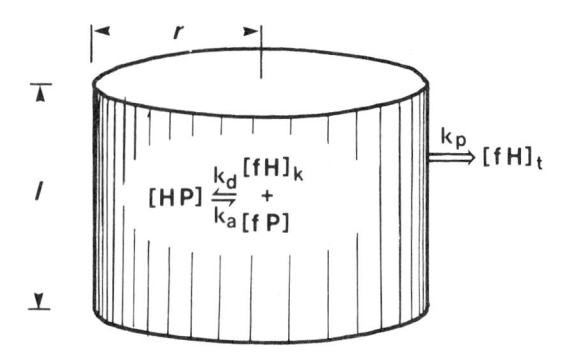

FIGURE 1 Capillary segment of length l and radius r through which serum is in transit.

from the serum contained within the segment into surrounding tissue will be given by:

$$2\pi rl[fH]_k k_p \qquad (1)$$

where $[fH]_k$ = intracapillary free hormone concentration.

Meanwhile within the capillary segment, the normal association and dissociation reactions between hormone and binding protein will be proceeding as conventionally described:

The total rate of hormone release from binding protein = (2)
$$= \pi r^2 l[PH]k_d$$

where [PH] is the concentration of bound hormone and k_d is the dissociation rate constant, while the rate of hormone reassociation to binding protein

$$= \pi r^2 l[fH]_k [fP]k_a \qquad (3)$$

where [fP] = is the free binding protein concentration and
k_a = association rate constant.

We may assume that, within the capillary segment, a new, dynamic, in vivo equilibrium state is established in which the rate of hormone release from binding protein exactly balances the sum of the disappearance rates of free hormone molecules arising in consequence of their reassociation to protein and outward permeation loss. Thus:

$$\pi r^2 l[PH]k_d - \pi r^2 l[fH]_k [fP]k_a - 2\pi rl[fH]_k k_p = 0 \qquad (4)$$

Free Hormones in Blood

From this equation we may derive the intracapillary free hormone concentration:

$$[fH]_k = \frac{[PH]}{K[fP] + \frac{2k_p}{rk_d}} \quad (5)$$

Meanwhile the free hormone concentration ($[fH]_e$) which exists in the same serum when in "undisturbed" equilibrium (for example, at rest in a test tube) is given by the conventional expression:

$$[fH]_e = \frac{[PH]}{K[fP]} \quad (6)$$

Thus the free hormone concentration within the capillary is <u>reduced</u> with respect to the free concentration which exists under undisturbed equilibrium conditions <u>in consequence of the net outward flow from the capillary into adjacent tissue</u>. The relative magnitude of this depression in the intracapillary free hormone concentration is determined by, among other factors, the relationships between the ratio of the permeation and dissociation rate constants k_p/k_d, the radius of the capillary r, and the term $K[fP]$. Meanwhile the rate of outward flow of hormone into surrounding tissues is given by:

$$\text{hormone flux/unit capillary length} = 2\pi r [fH]_k k_p$$

$$= \frac{2\pi r [PH] k_d}{\frac{k_a}{k_p}[fP] + \frac{2}{r}} \quad (7)$$

Notwithstanding the highly simplified nature of this account, and of the assumptions upon which it is based, certain very important conclusions emerge from the analysis. The first is that any depression in the intracapillary free hormone concentration is likely to vary from tissue to tissue since it is dependent on the values of parameters k_p, r, etc., which characterize the blood supply to individual tissues. (Although implicitly disregarded in the above analysis, the rate at which hormone is metabolized or otherwise cleared from target tissue is also an important factor in governing the depression in the local intracapillary free hormone concentration.) Secondly, the rate at which hormone is conveyed to target cells is governed by a complex expression [Eq. (7)] which includes, among its constituent parameters, the concentration of bound hormone present within the capillary.

Clearly, the value for this expression can lie anywhere within the range:

$$\frac{2\pi r[PH]k_p}{K[fP]} \qquad \qquad \pi r^2[PH]k_d$$

depending on the relative magnitudes of the terms k_a/k_p [fP] and $2/r$.

This implies that at one extremity of this range, the hormone delivery rate to target cells is essentially determined by the free hormone concentration as observed under "equilibrium" or "at rest" conditions; at the other, the delivery rate is governed by the product of the bound concentration and the dissociation rate of the hormone/binding-protein complex. Indeed, in a capillary of a diameter approaching zero, it is evident that the dissociation rate of the bound hormone <u>inevitably</u> controls the rate of transport of hormone across capillary walls (assuming that $k_p \neq 0$). The rate of hormone delivery to individual target organs clearly lies, in principle, anywhere within the above limits. This implies, in short, that the level of protein-bound hormone may, in the case of certain target organs in the body (though not necessarily in all) significantly influence the rate at which hormone is conveyed to them.

B. Current Concepts

Before we discuss further the implications of these concepts, it is illuminating and relevant to examine currently accepted beliefs relating to the kinetics of hormone transport to target tissues, concerning which there appear to exist marked differences in opinion among endocrinologists working in different fields. More particularly it is notable that the views entertained by the majority of thyroidologists on this issue — based largely on hypotheses originally advanced by Robbins and Rall (1979) — totally contradict those of many steroidologists, whose ideas appear to have been essentially derived from concepts enunciated originally by Tait and Burstein in the mid 1960s (1964). For these reasons it is important to clarify the present perceptions of the relationship between the serum free hormone concentration and hormone delivery which are, respectively, entertained by investigators in these two areas.

In considering the kinetics of hormone delivery, both Robbins and Rall and Tait and Burstein expressly addressed their respective attentions to the hepatic extraction of the thyroid hormones on the one hand and to the steroid hormones on the other. The liver, though not perhaps a target organ in the conventional sense, is characterized by a high rate of tissue clearance of certain steroid and thyroid hormones, and is therefore one in which the kinetic consequences of the passage of hormone from blood to cells are likely to be especially marked.

The essence of Robbins and Rall's interpretation of experimental data relating to the hepatic clearance of T_4 and T_3 is depicted in Fig. 2. Their central conclusion was that, even in the face of the high rate of hor-

Free Hormones in Blood 223

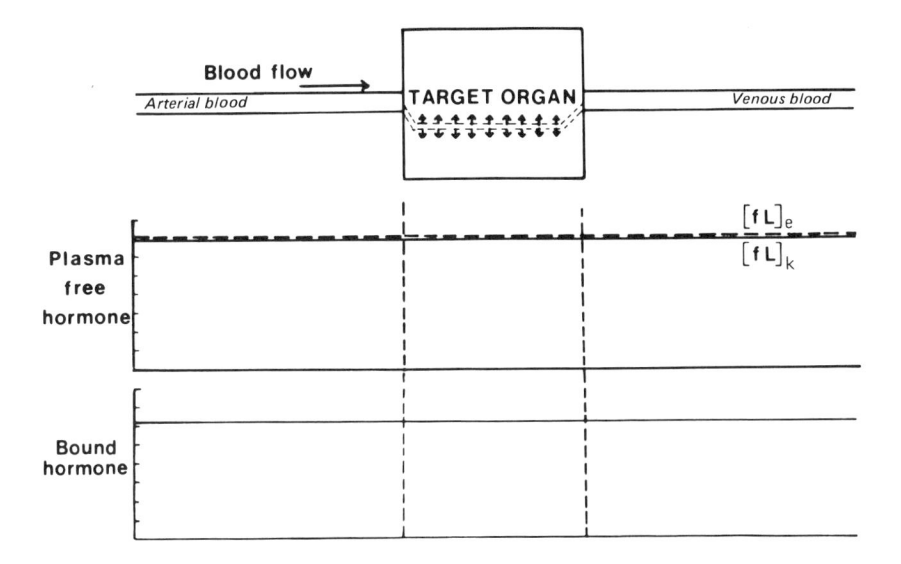

FIGURE 2 Robbins and Rall concept of hormone delivery. The intracapillary free hormone concentration ($[fH]_k$) is maintained at its equilibrium value ($[fH]_e$) in the face of permeation of free hormones to target tissue.

mone clearance by the liver, the intracapillary free thyroid hormone concentrations are maintained at their "equilibrium values" by virtue of the extreme rapidity of dissociation of the corresponding bound hormone moieties present in blood (implying that the term $2k_p/rk_d$ in Eq. (5) is negligible). Based on this fundamental precept, Robbins and Rall concluded that the free hormone concentration in blood supplying all other tissues in the body must likewise be maintained at the same, uniform, "equilibrium" value (i.e., that which exists in serum "at rest"); thus the maintenance of the intracapillary free hormone level by local "instantaneous" dissociation of the bound complex constitutes the essential basis for experimental observations correlating serum free thyroid hormone concentrations (as measured in vitro) with hormonal status and physiological effect. Among the logical consequences of this fundamental hypothesis it can be deduced that — assuming the bound hormone concentration in blood is not significantly depleted during capillary transit — hormone permeation across capillary walls into adjacent tissues will be uniform along the entire length of individual capillaries, while both the rate of blood flow along the capillary and the capillary transit time are of essentially no consequence in relation to hormone supply to individual target organs.

The description of analogous events occurring in the liver in regard to cortisol uptake offered by Tait and Burstein is, as suggested earlier, diametrically opposed to the views of Robbins and Rall. The essence of the Tait/Burstein hypothesis is represented in Fig. 3. According to this interpretation, free hormone is entirely removed from the blood during its time of transit through the liver while the specifically bound hormone passes through essentially undissociated and unmetabolized (i.e., the term $2k_p/rk_d$ in Eq. (5) is implicitly regarded as very large). Thus the amount of hormone cleared is essentially given by the product of the free hormone concentration preexisting within the blood as it enters the liver, and the hepatic blood flow rate. The experimental observations relied on by Tait and Burstein in their analysis were not, however, totally consistent with this prediction: in particular they noted that the hepatic clearance rate considerably exceeded the product of the free hormone concentration and the blood flow rate. They therefore suggested that some of the albumin-bound hormone moiety, being "loosely bound," must also be extracted during the transit of blood through the organ. (It should be noted in this context that mass action considerations indicate that free and albumin-bound hormone concentrations in serum are normally closely correlated. Hence the establishment of a correlation between, for example, hormonal status and serum free hormone concentration implies a corresponding correlation between hormonal status and the albumin-bound level.)

The essential contradiction between the Tait and Burstein and Robbins and Rall views can thus be summarized as follows. The Robbins and Rall hypothesis is based on the supposition that the dissociation rate of the TBG-thyroid complex is so rapid that the free hormone concentrations throughout the body are maintained at a uniform "equilibrium" level in the face of tissue delivery. Conversely the Tait and Burstein view implies that the dissociation rate of the CBG-cortisol complex is so slow that the free (and albumin-bound) steroid concentrations in serum fall to values close to zero during capillary transit, while the contribution to hormone delivery to individual target tissues — including the liver — deriving from the specifically bound moiety is regarded as negligible. The contradiction implicit in these two views is heightened by the observation that the value of the dissociation rate constant (k_d) of the CBG-cortisol complex (0.087/sec) (Dixon, 1968) is some five times greater than that of TBG-T_4 (0.018/sec) (Hillier, 1971).

Both the Robbins and Rall and Tait and Burstein hypotheses are, nevertheless, consistent with the notion that the free hormone concentration constitutes the essential determinant of hormone delivery and effect, and that variations in specifically bound hormone levels are therefore not of physiological consequence. They clearly differ in that the rate of blood flow is seen as essentially irrelevant to the rate of hormone delivery in the Robbins and Rall view, but is regarded by Tait and Burstein as being of equal importance to the free hormone concentration per se in regulating hormone supply to individual tissues.

Free Hormones in Blood

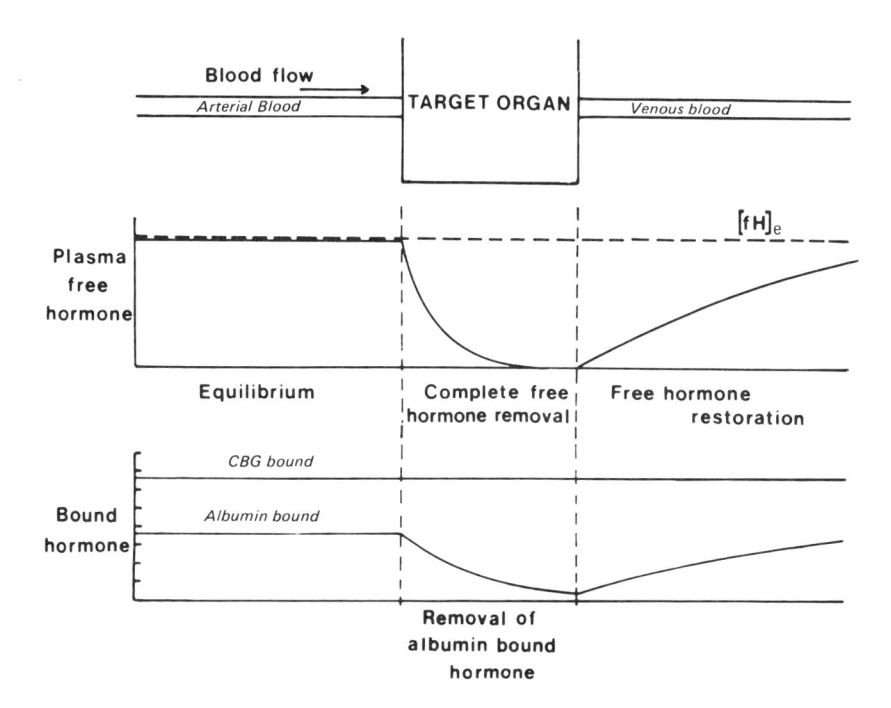

FIGURE 3 Tait and Burstein model of hormone delivery. The free (steroid) hormone level is visualized as falling essentially to zero in consequence of target organ clearance: albumin-bound hormone is also partially or totally cleared. Specifically bound hormone does not dissociate during capillary transit, implying that delivery of hormone to the target tissue is proportional to the free hormone concentration.

Unfortunately, Tait and Burstein did not extend their consideration of hormone delivery to encompass organs other than the liver. However, their views appear to have formed the basis of the belief (which appears to be commonplace among steroidologists) that albumin-bound steroid hormones are to be regarded as essentially "free" within the context of hormone transport, a postulate the latest expression of which is to be found in a recent extensive review by Pardridge (1981). Clearly such a concept finds no place within the overall perceptions of those thyroidologists who implicitly subscribe to the views of Robbins and Rall on hormone transport.

The distinction between these two viewpoints has been discussed in some detail, first in an attempt to highlight the conflicting conceptual bases on which the "free hormone concept" currently rests, and also to provide added justification for the notion that a more valid representation of the molecular events governing hormone transport to target tissues may lie somewhere between the contradictory physicochemical extremes postulated respectively by Tait and Burstein and by Robbins and Rall. The essence of the intermediate view is that the rates of dissociation of hormones from specific binding proteins are neither so rapid nor so slow as to be disregarded in any consideration of the molecular events occurring in the microcirculation, but may be such as to influence the delivery of hormone to different target organs in the body. However, this suggestion clearly implies an abandonment of the free hormone concept in its commonly accepted form in that the hormone supply to different target organs is postulated as being, in varying degree, dependent on the bound hormone concentration in the blood. Among the further implications of this view is the possibility that a major function of the specific binding proteins relates to the distribution of hormones to different target organs, and that <u>changes</u> in the levels of specific binding proteins serve to modify the pattern of hormone delivery within the body (see Fig. 4.). An experimental basis for this hypothesis is provided by the studies of Keller and co-workers (1969), who have likewise hypothesized that specific binding proteins exist to increase the specificity of endocrine systems and "to determine the distribution of [hormonal] signals" to different target organs in the body. (The molecular basis for this suggestion is, however, entirely different from

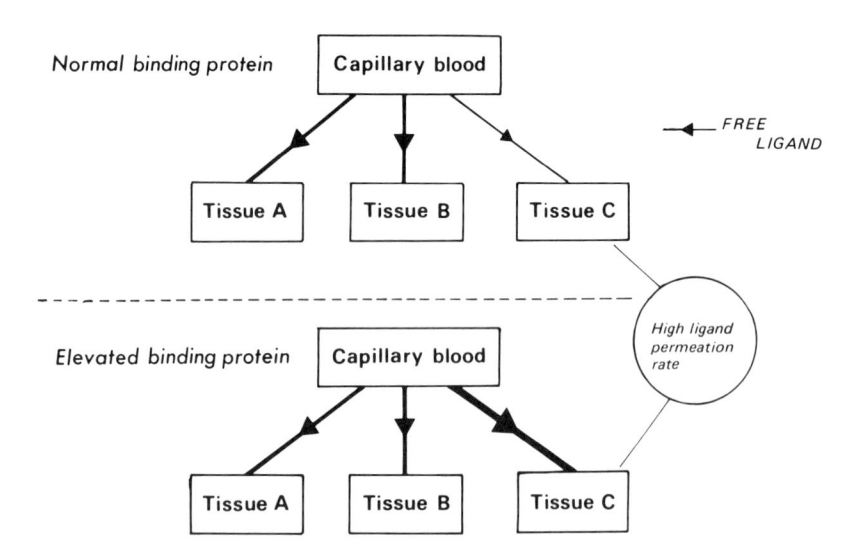

FIGURE 4 Physicochemical consequences stemming from the author's model of intracapillary protein-binding kinetics. An increased binding-protein concentration enhances hormone delivery to rapidly metabolizing target tissues possessing high capillary-permeation rate constants.

that discussed in this chapter.) A similar concept has also been advanced recently by Pardridge (1981) who — on the basis of theoretical concepts and experimental data which, it must be admitted, have attracted considerable criticism (Ekins, et al., 1982; Tait, 1982) — has suggested that a rise in bound hormone concentration serves to redirect hormone to target tissues characterized by relatively extended capillary transit times such as the liver.

Were the redistribution of hormones to particular target tissues within the body to represent the essential biological function of certain of the serum-binding proteins, then it is evident that free hormone levels <u>in isolation</u> would not adequately reflect hormonal status, and that the bound hormone concentration in blood would also constitute a relevant parameter in relation to hormone supply to <u>certain target organs</u>. The particular relevance of this conclusion clearly lies in the context of mammalian reproduction, since it is characteristically in the event of pregnancy that binding proteins such as TBG, CBG, PBP, and others, undergo major changes in their serum concentrations. Such changes may be of prime importance in relation to the redirection, in this physiological situation, of the hormones they respectively transport to particular target tissues. Thus an exclusive reliance on measurements of serum free hormone concentrations might well conceal changes in other endocrine parameters which are of prime importance either in pregnancy or in other physiological situations characterized by analogous alterations in the levels of the binding proteins.

In summary, this discussion has endeavored to illuminate some of the uncertainties that underlie the commonly held view that the free concentrations in serum of, for example, the thyroid and steroid hormones, govern their respective physiological activities. A simple physicochemical consideration of the effects stemming from the presence of serum-binding proteins indicates that these may specifically serve to modulate the pattern of hormone delivery in certain physiological circumstances and that the serum free hormone concentration as measured in vitro may not represent the prime determinant of hormone transport to <u>all</u> target tissues. Such a hypothesis does not, it must be emphasized, conflict with experimental data, established in many clinical studies suggesting that the serum free (thyroid and steroid) hormone concentrations normally reflect hormonal status, that is, that <u>for most of the time</u> they govern hormonal delivery to <u>most tissues</u> in the body. That they govern hormonal supply to <u>all</u> tissues under <u>all</u> circumstances — implying that the binding proteins constitute a biological irrelevance — is a physicochemically questionable and possibly dangerous assumption.

III. MEASUREMENT OF FREE HORMONES IN BLOOD

A. Introduction

Notwithstanding the doubts expressed above regarding the unqualified validity of the "free ligand" or "free hormone" concept, there is little doubt that serum free thyroid hormone levels correlate more closely with overall thyroid status than the corresponding "total" serum hormone concentrations. Evidence accumulated over many years supporting this assertion is too extensive to warrant further detailed discussion here. Acceptance of the validity of this belief forms the basis of the current reliance by clinical chemists and endocrinologists on a variety of indirect measures of free thyroid hormone concentrations such as the free thyroxine index (Clark and Horn, 1965), the T_4/TBG ratio, and other similar parameters which variously contrive to reflect both the T_4 (or T_3) and iodothyronine-binding protein concentrations present in serum.

Although analogous techniques have not been so extensively employed in the case of the steroid hormones, it is generally accepted that their free concentrations are likewise frequently of greater diagnostic value than the total or protein-bound levels. Consequently techniques are increasingly employed to estimate the concentration in blood either of the free steroid of interest or of its specific carrier protein.

It is not proposed, in this presentation, to review in great methodological detail the variety of traditional techniques whereby some insight — if not necessarily an absolute measure — of the free hormone concentrations in blood may be gained. Nevertheless it is appropriate to discuss their underlying principles, and to attempt a broad classification of the more commonly used methods in order to place the new generation of immunoassay techniques in perspective.

The traditional techniques may usefully be subdivided into "indirect" and "direct" methods (Ekins, 1979). The indirect methods have relied essentially on the combination of two unrelated measurements, for example, that of the total serum hormone concentration, and of the free hormone "fraction." The latter has been customarily estimated by the addition of a tracer quantity of radioactively labeled hormone to serum and the subsequent separation, either by equilibrium dialysis, ultrafiltration, or some other analogous physicochemical technique, of the free hormone moiety existing within the sample. The distribution of the radiolabeled hormone between free and bound compartments indicates the fraction of the total hormone which is "free"; this value, combined with the estimate of the total concentration of hormone present in the sample, yields an estimate of the absolute free hormone concentration. This approach is typified by the classic technique for the measurement of serum free thyroxine developed by Sterling and Hegedus (1962) which relies on equilibrium dialysis to isolate free and protein-bound hormone moieties. Similar techniques have likewise been developed for the measurement of free steroid levels.

An alternative "indirect" approach relies on the combination of estimates of the total hormone concentration and of the corresponding serum binding protein(s) present in serum. Generally one such specific serum-binding protein will be responsible for transporting the larger part of the total hormone present in blood under normal physiological conditions, for example, TBG binds approximately 75% of serum thyroxine, the remaining fractions being distributed between thyroxine-binding prealbumin (~8%) and albumin (~16%). Thus a combination of estimates of total hormone concentration and of the concentration of the dominant serum binding protein (in the form, typically, of the hormone/binding protein ratio) yields a relatively simple and reliable measure of the free hormone concentration present in the test sample. (More complex algebraic equations may nevertheless be employed to derive a more accurate estimate of free hormone concentration based on a knowledge of the concentrations of each of the binding proteins and relevant hormones in serum.)

Another group of indirect techniques is exemplified by the combination of measurements which yield various so-called "free thyroxine indices." A free thyroxine index (FTI) generally comprises an empirically defined variable whose value essentially correlates with the free thyroxine concentration in serum: such an index is typically derived by combining the total serum T_4 concentration (ascertained by, for example, a conventional radioimmunoassay) together with a T_3 or T_4 "resin uptake" measurement. The latter is a determination of the fractional uptake onto resin (or other suitable adsorbent) of radioactive hormone initially added to the test sample. The latter measurement is closely analogous to the measurement of the free T_4 fraction appearing in a serum dialysate as described earlier, albeit (for a variety of reasons that need not be discussed in detail here) the substitution of powerful solid adsorbent for a small volume of dialysate generally implies, among other things, a major disturbance to the equilibrium between free and bound moieties in the test sample. This may, in turn, lead to poor correlation between the FTI and fT_4 in samples in which the binding protein concentrations are grossly abnormal (see below).

The common feature shared by all the indirect methods described above is, as emphasized earlier, their reliance on the estimation of two separate and independent variables, one of which generally includes the total serum hormone concentration, the other revealing, in one way or another, the extent to which the total hormone is bound. Within this group of techniques, however, a further subdivision may be made, which it is perhaps appropriate to discuss briefly at this point. However, a similar distinction also exists within the class of "direct" free hormone assays described later in this chapter. In particular certain methods may be described as "absolute" in that — in principle if not perhaps in practice — they yield an estimate of the free hormone concentration present in test samples without reference to serum standards of known free hormone content. In contrast, other techniques provide only an "index" or "response parameter" whose value — though correlating with the free hormone concentration existing within the test sample — does not provide a measure of the free hormone concentration per se unless the system is "calibrated" using serum samples whose free hormone content has been previously established: this group of methods may be described as "comparative."

The distinction between absolute and comparative free hormone assay methods is of considerable importance. Only a few of the methods (either indirect or direct) currently available are capable of yielding estimates of absolute free hormone concentrations in serum. The majority, including all the newer direct immunoassay techniques discussed later in this chapter, must be classed as "comparative." This creates major problems in relation to free hormone assay standardization and quality control which are discussed in greater detail below.

A reconsideration of the indirect methods described above reveals that the traditional dialysis (and ultrafiltration) techniques — typified by the method of Sterling and Brenner for the measurement of free T_4 — may be categorized as "absolute" methods. The free T_4 index and T_4/TBG approaches may, in contrast, be described as "comparative." In practice, for a variety of reasons that are neither entirely apparent nor totally persuasive, it has not been conventional in the past to calibrate these "comparative" assay systems using standards of known free hormone content, and results have therefore usually been expressed in terms of the arbitrary "index" or response variable yielded by the particular system used. A notable exception to this rule is provided by the fT_4 assay kit recently developed and marketed by Corning (Odstrchel et al., 1978). This essentially comprises a technically refined fT_4 index-based method which has departed from convention by including a series of standards of known fT_4 content within the package. This has enabled the manufacturers to proclaim, entirely legitimately, that the kit constitutes an assay method for fT_4. The numerical values of free hormone which the kit yields are, of course, entirely dependent on, and hence reflect, the absolute method used to calibrate the standards provided. In this respect, however, the kit does not differ from almost all the recently developed commercial kits that "directly" measure free hormone concentrations in serum. (Further discussion of the Corning technique — which incorporates some interesting and novel features — will be deferred to the section dealing with direct radioimmunoassay of free hormone in serum.)

Thus, in summary, the traditional "indirect" free hormone assays can be further subdivided into the "absolute" methods, which yield estimates of the free hormone concentration in measured samples without reference to standards of known free hormone content, and the "comparative" methods which only provide values for the free hormone concentrations in unknown samples when calibrated using standards whose free hormone content has been established using one of the absolute methods.

B. "Direct" Methods

This group of techniques embraces the newer free hormone radioimmunoassays which represent one of the major conceptual innovations in immunoassay methodology in the last few years. Nevertheless, other free ligand assay techniques also fall within the category of "direct" methods, and merit a brief description here.

1. Dialysis/Ria

Probably the first example of a "direct" method of free hormone measurement is provided by the technique developed by the author (Ellis and Ekins, 1973) for the assay of free thyroid hormones in serum. This relies on a dialysis system for the sequestration of free hormone, and involves the incubation of the sample in small equilibrium dialysis cell (see Fig. 5) at 37°C for sufficient time to achieve reequilibration of free and bound hormone concentrations throughout the entire contents of the cell. Following the reestablishment of equilibrium (typically after overnight incubation), the free hormone concentration in the dialysate may be measured by a sensitive RIA technique.

The assay system is calibrated using gravimetrically (or spectrophotometrically) defined dilutions of the relevant hormone in buffer; it therefore represents an absolute method which can be employed to establish the free hormone content in the serum standards utilized in "comparative" techniques. Although the method necessarily depends both upon the availability of immunoassays of very high sensitivity and careful experimental technique, it avoids reliance on the purity of the radiolabeled hormones which form the basis of the traditional indirect dialysis methods of free

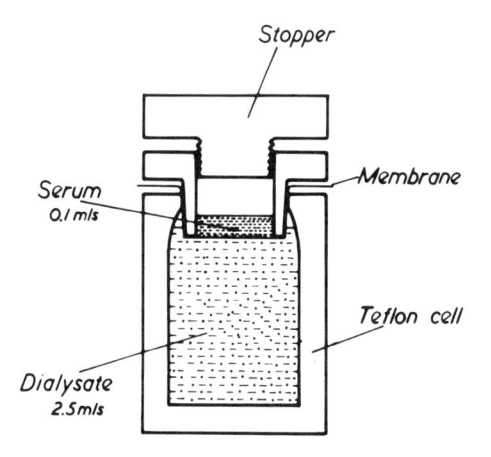

FIGURE 5 Simple dialysis cell used in the author's laboratory to estimate serum free thyroxine concentrations. The concentrations of T_4 and T_3 in the dialysate (determined by RIA) are essentially identical to the free hormone concentrations existing in undialyzed, undiluted serum.

hormone measurement. It is probably in consequence of the avoidance of this source of error that free thyroid hormone values obtained with this technique have generally been somewhat lower than those obtained using the older methods (Ekins, 1979).

The method raises an important conceptual point which sometimes generates misunderstanding, moreover, the issue concerned is of general significance and therefore merits brief discussion. It is evident that the diffusion of free hormone into the dialysate compartment results in a disturbance of the thermodynamic equilibrium between free and bound moieties initially existing within the serum compartment; reestablishment of equilibrium throughout the cell can only be achieved by a net dissociation of protein-bound hormone. This in turn results in a slight reduction in the bound hormone concentration and, implicitly, a concomitant reduction in the free hormone concentration with which the former remains in equilibrium. In short, the net effect of subjecting a serum sample to equilibrium dialysis in the manner described above is a <u>reduction</u> in the free hormone concentration present within the dialysate as compared with that existing in the sample prior to dialysis. However, the magnitude of this reduction is clearly dependent both on the relative volumes of dialysate and dialysand (i.e., the serum sample under test), and the relative concentrations of free and bound hormone initially present in the test serum. Provided that an "insignificant" (i.e., less than, for example, 5%) fraction of the <u>total</u> hormone originally present in the test sample is ultimately located within the dialysate compartment, then the free hormone concentration in the dialysate, at equilibrium, closely approximates that originally present in the serum sample under test.

This point is of general importance since virtually all techniques of free hormone measurement necessarily involve <u>some</u> disturbance to the preexisting equilibrium, and a consequent dissociation of the bound hormone complex. If such dissociation is small, then its effects on the measured free hormone concentration are generally negligible and can either be ignored or a small correction applied. (Such a correction is implicit, in "comparative" techniques, in the use of serum samples of known free hormone content to calibrate the procedure.) However, if the extent of dissociation of the bound complex is large (for example 10% and upwards), then significant errors will arise in the assay of test samples. In particular, variations in the concentrations of binding proteins (and concomitantly of bound hormone) will result in severe biasing effects, particularly, for example, in samples from TBG-deficient subjects, in which the quantitative consequences of dissociation of protein-bound hormone will inevitably be magnified. Distortions of this kind are characteristically observed in certain of the recently launched commercial kits for the measurement of free T_4; they also arise particularly in the majority of conventional FTI procedures.

The biasing effects on assay results arising from this source include generally the artifactual raising of measured fT_4 values in samples in which TBG concentrations are high (as seen, for example, in pregnancy serum) and the depression of values when TBG concentrations are abnormally low. Recognition of the existence of such effects is essential to an interpretation of assay results yielded by certain free hormone methodologies in conditions, such as pregnancy, in which binding protein concentrations depart significantly from normal. Meanwhile it must be emphasized that relatively large amounts of hormone may be dissociated from serum-binding proteins in the course of a free hormone measurement without transgression of the basic rule that the bound hormone should not undergo significant fractional depletion for the free hormone measurement to be reliable.

2. Resin Adsorption/RIA

A method bearing close conceptual similarities to the dialysis/RIA techniques discussed above has been described by Romelli and colleagues (1979) and forms the basis of the methods for the measurement of fT_4 and fT_3 currently marketed by Lepetit. This method essentially relies on the exposure of serum to Sephadex LH-20 resin beads (usually contained in a short column): these can be regarded as microdialysis sacs permitting the entry therein of free hormone while excluding the protein-bound moiety. Equilibrium is rapidly established throughout such a system, as a result of which the amount of hormone finally residing within the resin matrix is proportional to the serum free hormone concentration. Following complete removal of the serum, the hormone remaining adsorbed to the resin beads is eluted from the column and measured by RIA.

Although this technique is slightly cumbersome compared with some of the more recently developed direct RIAs discussed in the next section, it possesses a number of advantages. Among these is the fact that it constitutes an "absolute" method, since the Sephadex columns employed may be calibrated using buffer solutions of thyroxine and triiodothyronine, and not serum standards of known free hormone content. A second advantage is that the technique closely resembles, in principle and practice, the dialysis/RIA methods originally developed in the author's laboratory which are now widely accepted as yielding the most reliable estimates of serum free hormone concentrations. The Sephadex adsorption method is concomitantly less vulnerable to the presence in serum of a variety of disturbing factors which, as discussed below, afflict some of the newer direct RIA techniques, and which may distort free hormone measurements in certain circumstances.

3. Direct RIA Methods

Several recently developed free hormone measurement techniques may be grouped together under this heading since, although they differ in methodological detail, they all rely on a common basic proposition. This is that the introduction into serum of a "small" amount of antibody — conveniently, but not necessarily, coupled to a solid support such as cellulose particles, or the interior surface of a plastic tube — results in a fractional antibody occupancy (by hormone) which is dependent on the free hormone concentration in the final equilibrium mixture (see Fig. 6). Moreover, provided that the amount of antibody introduced into the system is sufficiently low to ensure that the net amount of hormone transferred from serum protein-binding sites to antibody is negligible (e.g., 1% or less), then the final free hormone concentration existing in the mixture very closely approximates that existing in the "undisturbed" serum sample prior to the introduction of antibody. Under these circumstances the final antibody occupancy closely reflects the initial serum-free hormone concentration.

Many newcomers to the free hormone assay field are undoubtedly perplexed by this simple basic concept, and, in particular, evidently fail to grasp the notion that although a substantial redistribution of hormone from binding proteins to antibody may take place during the incubation period (implying that much greater amounts of hormone than initially existed in the free moiety may ultimately be bound to antibody) the final antibody occupancy will nevertheless reflect the ambient free hormone concentration in the system (see, for example Stockigt et al., 1982; Midgley and Wilkins, 1982). Another common misconception is that — assuming the introduction into serum of only a "trace" amount of exogenous antibody — all antibody-binding sites will become occupied by hormone. This is not the case, even though the factorial excess of total hormone over antibody binding sites in the system may be 1000-fold or more.

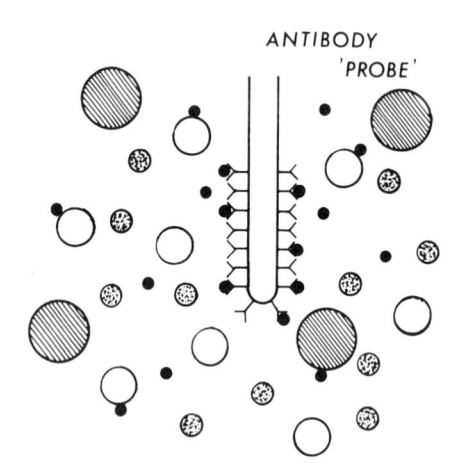

FIGURE 6 Fundamental principle of free hormone immunoassay. Antibody-binding sites on an antibody "occupied" to an extent governed by the ambient free hormone concentrations in the surrounding fluid milieu.

Free Hormones in Blood 235

A close parallel to the principles involved is provided by the introduction of a cold thermometer into a warm glass of water. This will result in a net transfer of heat from the water to the thermometer until the later has reached the temperature of its immediate surroundings and is in thermal equilibrium with them. A consequent fall (albeit small) will inevitably occur in the temperature of the water resulting from the passage of heat to the thermometer; nevertheless, when equilibrium is reached, the latter will, usually with little error, reflect the initial temperature of the water in the glass.

More formally the basic theoretical principles underlying free hormone RIA may be expressed as follows. The free hormone concentration present in undisturbed equilibrium conditions is given by:

$$[fH]_e = [H] - [fH]_e \sum_i^n \frac{K_i[P_i]}{1 + K_i[fH]_e} \qquad (8)$$

where $[fH]_e$ = [free hormone] at equilibrium,

$[H]$ = [total hormone],

K_i = equilibrium constant of ith protein-binding site, and

$[P_i]$ = concentration of i-th protein-binding site.

When specific antibody is introduced directly into the serum sample, a new equilibrium is established wherein the (reduced) free concentration $([fH]_{Ab})$ is given by:

$$[fH]_{Ab} = [H] - [fH]_{Ab} \sum_i^n \frac{K_i[P_i]}{1+K_i[fH]_{Ab}} - \frac{[Ab]_o[fH]_{Ab}K_{Ab}}{1+K_{Ab}[fH]_{Ab}} \qquad (9)$$

where $[Ab]_o$ = [total antibody]

K_{Ab} = equilibrium constant of antibody-binding sites.

Assuming that the amount of hormone sequestered onto antibody sites is small (i.e., approximately 1% or less) compared with the total hormone present in the system ($[H]$) then:

$$[fH]_e = [fH]_{Ab} \quad \text{approximately.} \qquad (10)$$

Meanwhile the fractional "occupancy" of the antibody-binding sites by hormone following reestablishment of equilibrium throughout the system is given by:

$$\frac{[fH]_{Ab} K_{Ab}}{1 + K_{Ab}[fH]_{Ab}} \qquad (11)$$

$$= \frac{[fH]_e K_{Ab}}{1 + K_{Ab}[fH]_e} \quad \text{approximately} \qquad (12)$$

Thus, the bringing into contact of a trace amount of antibody with serum ultimately results in a fractional occupancy by hormone of the antibody-binding sites which, to a close approximation, is a function of the free hormone concentration which exists in the serum at equilibrium.

The extent of antibody occupancy consequent upon its exposure to an ambient free hormone concentration can be assessed in a number of ways. The different possible approaches have been exploited by different immunoassay kit manufacturers, and currently form the basis of the most commonly used commercial kits for the measurement of free thyroid hormones.

Labeled Hormone, Back Titration Technique ("Two-Step" or "Sequential" Free Hormone RIA): The basic principles relating to this technique are illustrated in Fig. 7. Following incubation of solid-phased antibody with the test serum, the serum is removed, the antibody preparation is washed, and subsequently exposed to labeled hormone for a further period of incubation. During the second incubation stage, labeled material will be bound to residual, unoccupied, antibody-binding sites to an extent essentially proportional to their concentration. Thus the greater the free hormone concentration in the test sample, the lower the concentration of unoccupied sites at the termination of the first incubation period, and therefore the greater the amount of labeled hormone ultimately bound to the solid-phased antibody.

It is inappropriate to provide here a complete theoretical account of this approach or to describe in detail the experimental procedures whereby it may be implemented in practice. Nevertheless it should be noted that the procedure rests on a somewhat different theoretical basis from that which underlies the conventional RIA of total hormone concentrations in blood, and different criteria govern the design of such a system. For example, it may be shown that the form of the dose-response curve is (within certain limits) independent of the concentration of antibody employed. On the other hand, the equilibrium constant of the antibody must be carefully selected since the use of antibodies either of an inappropriately high or inappropriately low equilibrium constant prohibits the successful measurement of the free hormone concentration.

As a good working rule, it is generally advisable to select antibody possessing an equilibrium constant roughly equal to the reciprocal of the free hormone concentration one wishes to measure, and, as emphasized earlier, to employ it at a concentration such that no more than approximately 1-2% of the total hormone present in the sample is sequestered onto antibody-binding sites at the termination of the first incubation stage. This implies, in turn, that antibodies possessing very high equilibrium constant are required to measure fT_4 and fT_3 concentrations in serum, but that a much lower affinity antibody is obligatory for the measurement of, for example, serum free cortisol.

This technique was originally developed in the author's laboratory (Ekins et al., 1980) and has formed that basis of an "in-house" method of measurement of free thyroid hormones for the past 3-4 years. A similar approach was independently developed by Clinical Assays Inc. and forms the basis of the Gammacoat two-step fT_4 method marketed by this company.

The principal experimental problem encountered with methods in this category stems from the sequential addition of the reagents and the "disequilibrium" nature of the assay system, which renders such systems particularly liable to assay drift. Such a phenomenon can nevertheless be minimized by the choice of an antibody possessing relatively slow association and disassociation rate constants and careful attention to the timing of the successive steps in the procedure when performed in the laboratory.

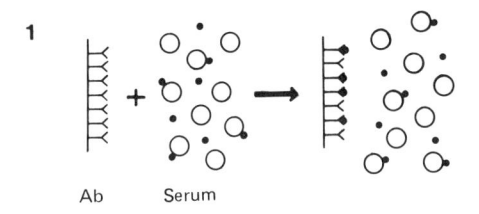

1   Ab   Serum

2   **Remove serum; wash**

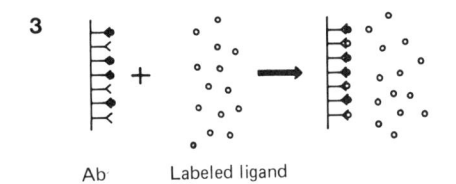

3   Ab   Labeled ligand

4   **Remove labeled ligand; wash**

5   **Count Ab-bound activity**

FIGURE 7 Principle of "two-step" labeled antigen back-titration technique of free hormone measurement.

Another difficulty experienced by some users of commercial kits relying on this approach is the relatively low precision they are prone to yield. This almost certainly arises as a result of poor assay design rather than from a basic deficiency in assay principle. When selected antibodies and a carefully optimized assay protocol are used, results based upon this approach in the author's laboratory have shown completely acceptable reproducibility (Fig. 8).

Nevertheless, although the two-step methods may suffer from problems relating to assay drift, and poor precision occasionally and are somewhat less convenient than the "single-step" techniques described below, they are undoubtedly potentially less susceptible to artifactual errors arising from abnormal constituents in serum than are the more recently developed single-step methods. For example, the existence of abnormal binding proteins, endogenous antibodies, etc., in test samples is unlikely to influence assay measurements significantly. Indeed, perhaps the sole major source of bias to which the two-step techniques are vulnerable (one which they share with all RIA methods) is the presence in test samples of substances that directly affect the antibody-hormone interaction, either by altering the effective amount of antibody present, or by changing the equilibrium constant of the reaction. For these reasons, the two-step methods, despite their evident disadvantages, remain the preferred approach in the author's laboratory.

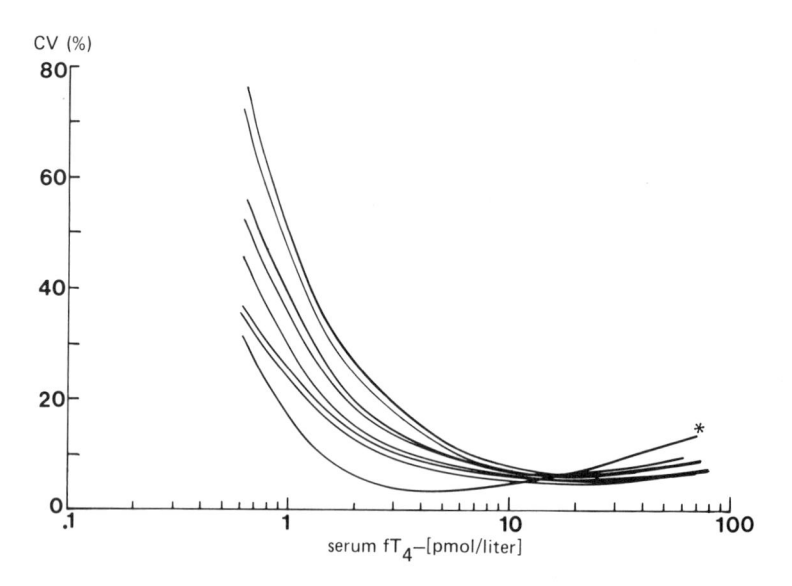

FIGURE 8 "Precision profiles" for routine in-house fT_4 assay conducted in the author's laboratory during a representative period.

Labeled Hormone Analog, Simultaneous Addition ("Single-Step" RIA): The small practical disadvantages of the sequential two-step, free hormone assays may be overcome by a further methodological refinement: the concurrent incubation of serum and antibody together with a labeled analogue or derivative of the hormone which, while retaining the ability to react with exogenous antibody, is chemically restricted from reaction with serum-binding proteins. This variant (which can be regarded as relying on a simultaneous rather than sequential back-titration of antibody binding sites) is portrayed in Fig. 9. In order for this methodological stratagem to succeed, however, it is clearly imperative that binding of the labeled hormone analog by serum hormone-binding proteins be either totally absent or essentially insignificant. Assuming that this condition is fulfilled, and assuming also that the amount of antibody introduced into the assay system is small (as defined earlier in this chapter) it may readily be shown that the distribution of the labeled analog following the attainment of equilibrium between antibody bound and free states is given by the following equation:

$$b^2 - b\left(\frac{K[fH]}{K^*[T^*]} + \frac{1}{K^*[T^*]} + S + 1\right) + S = 0 \qquad (13)$$

where b   = fraction of labeled analog bound to antibody,
       $[T^*]$ = concentration of labeled analog,
       K^* = equilibrium constant of antibody binding of labeled analog,
       K   = equilibrium constant of antibody binding of hormone,
       S   = $\frac{[Ab]}{[T^*]}$

In initial experimental studies carried out in the author's laboratory using this procedure, T_4 was coupled to albumin and other proteins in attempts to identify a form of label which, while binding to T_4 antibody, was totally unreactive with endogenous serum proteins. (A similar approach was adopted by Schall et al. (1979) in the development of an enzyme-labeled T_4, fT_4-uptake method.) During the course of these experimental studies, Amersham International revealed their independent development of a similar technique which has subsequently been launched as a commercial fT_4 assay kit (Midgley and Wilkins, 1980; Smith et al., personal communication, 1980). The nature of the labeled T_4 analogue used in the Amersham procedure has not, however, been disclosed. Other commercial kits based on similar principles have also been subsequently developed by Clinical Assays Inc. (Gammacoat one-step) and Diagnostic Products Inc. (Coat-a-Count).

Although, clearly, the single-step methods described above are experimentally more convenient than the original two-step techniques, and, since the antibody-hormone reactions are generally permitted to proceed essentially to equilibrium, are less prone to assay drift, such techniques are clearly liable to artifactual error arising from the existence of binding, or abnormal binding, of the T_4-analog by endogenous binding proteins present in test samples. Most particularly, the presence of endogenous T_4 anti-

1

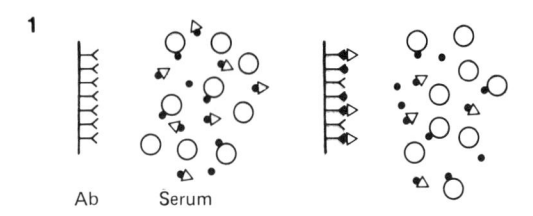

Ab   Serum

○ Binding protein
● Free ligand
◆ Labeled ligand analogue
  nonreactive with binding-protein

2 Remove serum; wash

3 Count antibody-bound activity

FIGURE 9 Principle of single-step, labeled ligand analogue technique for free hormone measurement.

bodies or of one of the abnormal thyroid hormone binding proteins — for example, an abnormal T_4-binding albumin whose existence has recently been revealed (Henneman et al., 1979) and on which much attention has been focused — may give rise to grossly distorted free thyroxine measurements (Stockigt et al., 1981).

The existence of such abnormal proteins characteristically gives rise to apparently elevated free hormone values which, in general, do not reflect the endocrine state of the subject under investigation. Although anomalies of this kind should usually not escape notice nor ultimately lead to serious misdiagnosis, they are undoubtedly a somewhat disturbing feature of the present generation (at least) of single-step free-hormone RIAs. Fortunately the frequency of occurrence of aberrant serum T_4-binding proteins in the population at large appears to be relatively low. However, the recognition of the existence of such abnormalities is likely to lead to a spate of related observations in the immediate future, and it is impossible at the present time to be certain of the true incidence of binding protein abnormalities of this kind.

Meanwhile more subtle effects arising from the binding to normally occurring proteins (such as albumin) of the thyroxine analog employed in current, commercial, single-step procedures has also been suspected of leading to a distortion of free hormone measurements. At this stage it is perhaps premature to speculate too extensively on the consequences of such putative effects, particularly since the manufacturers of commercial kits have generally been reticent regarding the nature of the reagents on which they rely, and relevant data are therefore difficult to acquire. Nevertheless, observation of the effects on, for example, measured serum fT_4 values deriving from single-step methodologies in heparinized patients displaying high blood levels of nonesterified fatty acids (Wilkins et al., 1982) suggests the possibility of interaction of the T_4 analog presently used in the commercial kits with serum albumin (Ekins et al., 1983), these interactions being variously affected by the levels of other serum constituents. Other unconfirmed reports have indicated that patients with albumin levels much below normal display artifactually low fT_4 levels when assessed using current labeled thyroxine analogue, single-step, free hormone assay techniques.

Labeled Antibody, Simultaneous Addition ("Single-Step") Immunoradiometric Assay (IRMA): An alternative "simultaneous" single-step assay technique relying on the use of labeled antibodies is depicted in Fig. 10. This approach, which should strictly be termed an immunoradiometric assay, depends on the use of labeled antibody rather than a labeled hormone analog, and relies on observation of the distribution of the labeled antibody between exogenous solid-phased hormone (or hormone analog) and endogenous free hormone. This distribution is given by the following expression

$$b^2 - b \left(\frac{[fH]}{[Ab^*]} + \frac{1}{K[Ab^*]} + s + 1 \right) + s = 0 \qquad (14)$$

where K is equilibrium constant of reaction between antibody and hormone and s is $\frac{[immunosorbent]}{[Ab^*]}$.

This simplified equation rests on the assumption that solid-phased hormone or hormone conjugate is bound with the same avidity as the native hormone.

This approach is currently under development in the author's laboratory. Similarly to the labeled hormone analog method described above, it is potentially liable to artifactual error arising from the existence in serum of abnormal binding proteins or of other factors capable of significantly influencing antibody binding to the solid-phased hormone. The avoidance of such effects constitutes one of the prime objectives of present studies on the method. There are certain general methodological and logistic advantages deriving from the use of labeled antibodies in immunoassay procedures which, nevertheless, make the development of such a technique attractive.

4. Other Methods

Two further commercially available methods of free hormone measurement are described in this section. One, the Corning Immophase method referred to earlier cannot, however, strictly be portrayed as a "direct" method, but essentially constitutes a calibrated free thyroxine index measurement. Nevertheless, it closely resembles some of the newer direct RIA methods

1

Coupled   Serum
ligand

○ Binding-protein
● Free ligand
⊸ Labeled antibody

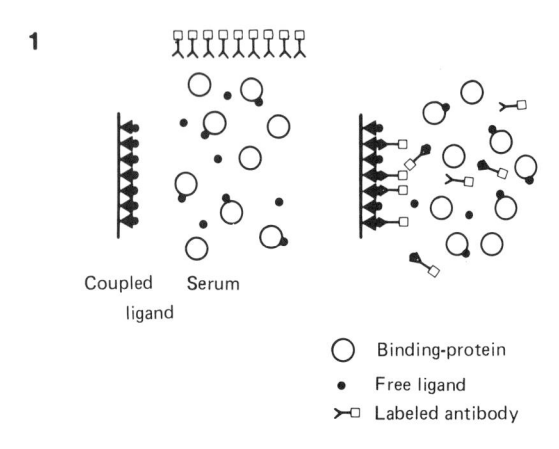

2 Remove serum; wash

3 Count coupled-ligand-bound activity

FIGURE 10 Principle of single-step labeled antibody technique for free hormone measurement.

in basic principle, and its mode of operation is probably more readily understood if the details of the method are discussed at this point.

<u>The Corning Immophase Method:</u> Figure 11 illustrates the basic protocol for this technique. Although frequently described as a "kinetic" method by the manufacturers (Odstrchel, 1982) (who have advanced a theoretical basis for it which relies heavily on considerations of the rate of uptake of serum free hormone by exogenous antibody binding sites), the Corning Immophase technique does not (in principle at least) especially depend upon differences in antibody uptake rates resulting from differences in ambient free hormone concentration in test sera. The method can, indeed, be viewed more simply as another example of the approach depicted in Fig. 6, in which the degree of antibody occupancy by hormone (either at equilibrium or before equilibrium is attained) is assessed, not by either of the "two-step" or "single-step" back titration techniques discussed in the previous section, but by the labeling of endogenous hormone with radioactive hormone prior to the exposure of the serum sample to antibody. Under these circumstances, a measure of the extent of antibody occupancy is derived by observation of the amount of labeled hormone remaining bound to the solid-phased antibody following its separation from the system.

It should, of course, be noted that the calculation involved in estimating antibody occupancy from this observation requires a knowledge

Free Hormones in Blood

1   **Estimate 'total' serum ligand**

2   **Add labelled ligand to serum**

3   **Calculate specific activity of 'diluted' ligand**

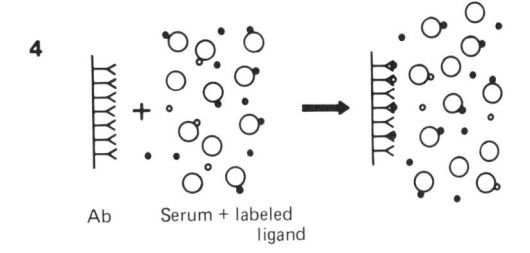

4

Ab   Serum + labeled ligand

5   **Remove serum; wash**

6   **Count Ab-bound activity**

FIGURE 11 Principle of Corning Immophase technique for free hormone measurement.

of the (diluted) specific activity of the labeled hormone in each individual serum sample under test; the method therefore necessitates an associated measurement of the total hormone concentration present in the test sample. Thus, the Corning technique, despite its close conceptual similarity in many respects to the direct free hormone RIAs, requires an associated assay of the total serum hormone level in each sample. This value, combined with the observation of the amount of labeled hormone bound to exogenous antibody in which may be termed the "uptake test" component of the overall protocol (as depicted in Fig. 11) yields a response variable that reflects the serum free hormone concentration.

    In practice, the Corning procedure relies on the observation of labeled thyroxine uptake onto solid-phased antibody from separate aliquots of the test serum contained in two tubes. In one tube is included an inhibitor (merthiolate) of serum protein thyroxine binding (as is customary in most conventional total thyroxine radioimmunoassays). The amount of labeled hormone bound to antibody therefore indicates the total hormone concentration in the sample in the usual way. The second tube contains no such inhibitor, and the antibody uptake reflects the free hormone level as discussed above. It is evident, in summary, that the procedure is closely comparable to other conventional fT_4 index measurement techniques. Its principal methodological novelty lies in the use of antibody rather than, for example, resin or charcoal as labeled-hormone adsorbent as is usual in most uptake tests. The other distinguishing feature of the Corning kit is that serum standards of known free hormone content are included, so that a calibration curve relating the calculated "$fT_4 I$" to fT_4 may be constructed, and values of fT_4 in unknowns expressed directly in concentration units rather than in terms of an arbitrary free thyroxine index.

One important criticism that may perhaps be leveled against the Corning fT_4 kit per se, though not against the principle on which it relies, is that, as currently marketed, it relies on an assay design which results in a substantial proportion (10-20%) of the total hormone in the sample being transferred from binding proteins to antibody in the course of the "uptake" measurement. This contravenes the fundamental principles of free hormone assay as enunciated earlier, since the large associated reduction in the free hormone level which occurs during the incubation will inevitably differ between different samples and the standards included in the kit (depending on the levels of binding proteins and bound hormone they individually contain), and thus give rise to variable bias in the fT_4 measurement from sample to sample. Thus the TBG concentration per se is likely to influence assay results, with samples containing elevated levels of TBG displaying artifactually raised levels of fT_4 and vice versa.

The Damon Microencapsulated Antibody "Liquisol" Method: This novel technique relies on the use of T_4 antibody prereacted with labeled T_4 contained in small "microcapsules" (Fig. 12). The microcapsules are bounded by semipermeable membranes which, while permitting passage of labeled and unlabeled T_4 molecules, are totally impermeable to molecules of antibody or binding proteins. When the microencapsulated labeled T_4/anti-T_4 preparation is placed into serum, unbound labeled T_4 present in the capsules is free to diffuse out into the surrounding medium (being replaced by unlabeled T_4 diffusing into the capsules and combining with vacant antibody-binding sites). The rate of loss of labeled T_4 from the microcapsules is claimed to be essentially dependent solely on the ambient free hormone concentration in the medium (Buehler, 1982).

The underlying physicochemical basis for this assertion has not been extensively expounded in the literature, and it is therefore somewhat impertinent, perhaps, to attempt to describe the detailed molecular events occurring within the microcapsules. These provide the justification for the manufacturer's claim that the Liquisol kit constitutes a free hormone assay method. If we bear this reservation in mind, the system may be visualized as comprising molecules of labeled hormone and antibody undergoing repeated association and dissociation reactions within the microenvironment of each individual capsule. The labeled hormone molecules are free to "escape" into the surrounding medium only when temporarily unbound to

Free Hormones in Blood 245

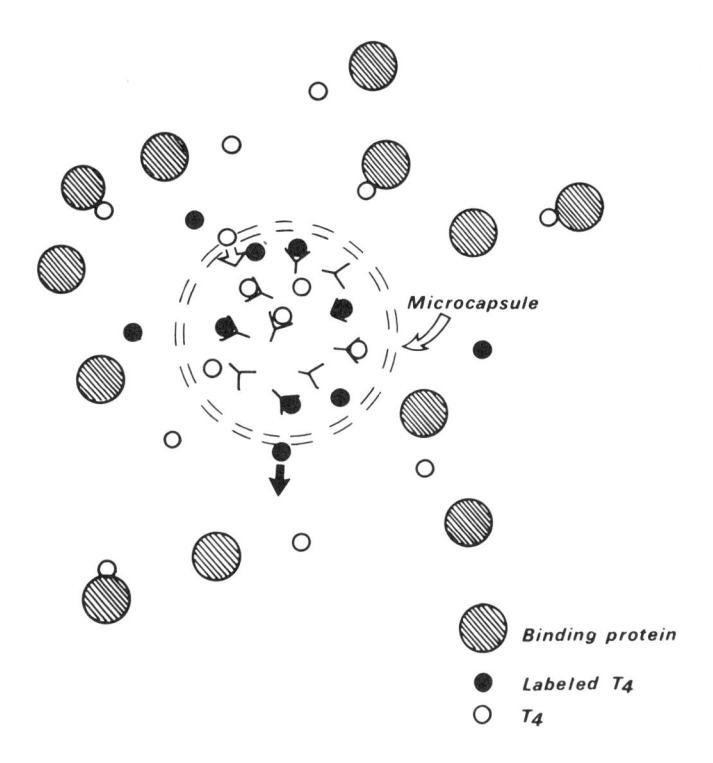

FIGURE 12 Principle of Damon microencapsulated antibody Liquisol method.

antibody. Meanwhile as labeled hormone molecules diffuse from capsules into medium, a concurrent inflow of unlabeled T_4 molecules into the individual microcapsules will take place at a rate essentially determined by the ambient free hormone concentration in the medium. The existence of unlabeled hormone molecules within the microcapsules — because of their competition for the restricted number of antibody binding sites present — essentially decreases the probability of recombination of labeled hormone molecules with antibody, while conversely increasing the probability of their escape into the surrounding medium. This implies, in turn, a greater labeled-molecule escape rate in the presence of a greater ambient free hormone concentration in the medium.

These simple basic concepts are complicated, however, by a second phenomenon: the influence of the total (free and bound) concentration of unlabeled hormone in the medium. A labeled hormone molecule, having escaped from a microcapsule but nevertheless in close proximity to it, clearly possesses a finite possibility of reentry. Moreover, the probability of this is dependent on the binding protein (and bound hormone) concentrations in the medium. For this reason, the net rate of efflux of labeled hormone molecules from the microcapsules is likely to be dependent, not only on the free hormone concentration in the medium, but also on the bound hormone concentration present. The relative effects of the bound and free hormone concentrations in serum respectively upon the rate of labeled hormone depletion from the capsules thus ultimately depend upon the magnitude of a number of rate constants (including the association and dissociation rate constants of the antibody-hormone reaction, the permeation rate constant governing hormone transport through the microcapsule envelopes, etc.) as well as on the size of the microcapsules, the concentration of antibody-binding sites within them, and other similar physical parameters. The system is, in short, one of considerable theoretical complexity, and the method obviously depends upon the particular reagents employed and the way in which they are combined. Meanwhile, in a recent publication, Buehler (1982) has confirmed a theoretical dependence on the Liquisol technique on the level of specific binding protein (TBG) in the serum which (although the mathematical basis for this finding is not elaborated) must be presumed to derive from effects of the kind described above. Though not so great as to vitiate the diagnostic usefulness of the method, this particular effect of variations in TBG concentration upon assay results undermines the reliable use of the Liquisol technique to reveal <u>subtle</u> changes in free hormone level which may occur, for example, in pregnancy, or in situations in which binding protein concentrations are grossly abnormal. Thus the existence of a high concentration of TBG is likely to bias fT_4 results upwards and vice versa. Paradoxically, such an effect, though concealing the real reduction in fT_4 concentration which is now claimed by many authors as occurring in pregnancy, may nevertheless enhance the diagnostic efficiency of the method by artifactually restoring observed pregnancy fT_4 levels to normal values.

The manufacturers have recently claimed to have applied the same methodology to the measurement of free steroids, including serum free cortisol and serum free testosterone (Buehler, 1982).

IV. STANDARDIZATION AND QUALITY CONTROL

As discussed earlier in this chapter, the majority of the new direct free thyroxine assay methods which have become commercially available in the past 2-3 years are "comparative" methods which rely on serum standards containing known concentrations of free hormone for calibration purposes. This implies a prior determination by the manufacturer of the free hormone concentration in the serum standards using some form of "absolute" method. Unfortunately, conventional absolute free hormone measurement techniques are somewhat unreliable, and a survey of the literature reveals considerable heterogeneity in results using methods often differing only in relatively minor detail. In the thyroid hormone field in particular, the absolute method traditionally relied on has been the Sterling-type dialysis procedure, and, although manufacturers have generally not revealed the methods whereby they ascribe fT_4 concentrations values to the serum standards they provide, it must be presumed that the majority rely on this particular technique. Unfortunately, many methodological variants of this procedure exist, involving different purification techniques for the labeled T_4 employed, different compositions of the buffer used in the dialysate, different dialysate-dialysand ratios, etc., each of which is liable to influence assay results. Not unexpectedly, absolute values deriving from the modifications of the Sterling method vary widely, and such variations can be expected to be transmitted in turn to free hormone values yielded by the different commercial RIA kits.

In the author's laboratory, the direct dialysis/RIA method described earlier is employed to calibrate the serum standards used in the "in-house" free thyroid hormone immunoassays which have been developed. Amersham International likewise use this technique to calibrate the serum standards they supply (Giles, 1982). This approach avoids the biasing effects deriving from labeled hormone impurities which undoubtedly contribute significantly to the heterogeneity of results deriving from variants of the original Sterling dialysis method. The normal and pathological ranges based on direct dialysis/RIA procedures are generally lower and more consistent than those obtained in the past using the traditional techniques.

Thus, until manufacturers either agree on the use of a single, common, absolute method for the establishment of the free hormone concentrations assigned to the standards which they distribute or, alternatively, eliminate the biasing effects arising in the absolute methods on which they individually rely, the results of free hormone measurements deriving from different commercial free hormone kits are, unfortunately, likely to vary widely.

Another major problem currently afflicting the fT_4 assay field is the variable susceptibility of the different methods to a variety of biasing effects arising from variations in the composition of test serum samples. Earlier in this chapter, particular attention was drawn to the effects of TBG (and implicitly of bound T_4) upon fT_4 measurements in certain assay methods. As also previously emphasized, other perturbations arise from the existence of abnormal binding proteins, drugs, heparin, etc., each of which can potentially influence individual methods in different ways (see, for example (Kaptein et al., 1981; Bayer and McDougall, 1982; Melmed et al., 1982). The confusion thereby created is compounded by the fundamental difficulty in establishing the "true" concentration of free hormone in an individual serum sample, notwithstanding a widespread tendency to place ultimate trust in the validity of methods based on dialysis. These problems will inevitably generate difficulties in the areas of interlaboratory standardization, internal and external quality assurance, etc., as free hormone measurements progressively supplant assays of total serum hormone levels in the clinical laboratory.

The fundamental difficulties involved in establishing the analytical validity of any particular method of free hormone measurement are also bound to impede resolution of the questions raised earlier relating to the validity of the free hormone hypothesis per se, and to the physiological significance of the existence of binding proteins in serum. For example, does the finding that fT_4 falls throughout pregnancy as suggested (Kurtz et al., 1979) by the present author (and, subsequently, by other workers using commercial kits ([Midgley and Wilkins, 1981]) represent a methodological artifact, a mild degree of hypothyroidism among women throughout pregnancy, or a breaching of conventional views relating to thyroid hormone transport to target tissues, and the mode of operation of the pituitary/hypothalamic feedback system?

V. SUMMARY AND CONCLUSIONS

In this chapter, an attempt has been made initially to focus attention on the concept of free hormone per se, and on the widely accepted conventional view that it is the free hormone concentration which determines hormone delivery to target tissue. Particular emphasis has been placed on the hitherto unperceived contradictions that underlie this view, and the insecure theoretical basis on which it therefore rests. Moreover, in recent years, at least three groups of workers (including the author's) have seriously questioned these conventional concepts. Each group has advanced different molecular models as the basis of their alternative hypotheses regarding the role of serum transport proteins in hormone delivery. These hypotheses concur, nevertheless, in ascribing an important distributional role to the specific serum hormone-binding proteins, and implicitly place particular importance on the protein-bound hormone moiety in regard to hormone delivery to certain target tissues in the body.

An attempt has also been made to clarify the basic principles underlying a number of new methods of free hormone measurement which have come into existence in the past 2-3 years. Some have undoubtedly proved conceptually perplexing to many workers. Indeed, in some cases commercial manufacturers have themselves misunderstood the mode of operation of their own methodologies and thereby contributed to a measure of scepticism among endocrinologists, clinical chemists, and others regarding the validity of the assays they offer.

Nevertheless it is virtually certain that the initial difficulties, confusions, and doubts that have inevitably surrounded the introduction of these entirely new methodologies will be resolved, and that free hormone measurements will ultimately figure prominently in the diagnostic tests employed by the endocrinologist and clinical chemist. Already a great deal of relevant clinical data has been acquired using the new techniques, and is being reported in the literature. Although some diagnosticians might nevertheless feel that the direct free hormone radioimmunoassays do little more than provide a simple replacement of older technologies on which they have traditionally relied (such as fT$_4$ index measurement), it is the author's belief that these techniques, because of their simplicity, precision, and widespread availability will ultimately provide fresh insights into the functioning of the endocrine system as a whole, and reveal subtleties in mechanisms of hormone delivery which have hitherto remained concealed behind the veils of unrefined methodology.

REFERENCES

Bayer, M. F., and McDougall, R. (1982). The value of free thyroxine in hospitalized patients (using a two-step, antibody coated tube, radioimmunoassay). In Free Hormones in Blood, A. Albertini and R. P. Ekins (eds.), Elsevier Biomedical, Amsterdam, p. 275.

Buehler, R. J. (1982). Applications of microencapsulated antibody in free hormone radioimmunoassays. In Free Hormones in Blood, A. Albertini and R. P. Ekins (eds.), Elsevier Biomedical, Amsterdam, p. 121.

Clark, F., and Horn, D. B. (1965). Assessment of thyroid function by the combined use of the serum protein-bound iodine and resin uptake of ^{131}I-triiodothyronine. J. Clin. Endocrinol. Metab. 25: 39.

Ekins, R. P. (1979). Methods for the measurement of free thyroid hormones. In Free Thyroid Hormones, R. Ekins, G. Faglia, F. Pennisi, and A. Pinchera (eds.), Excerpta Medica, Amsterdam, p. 72.

Ekins, R. P., Filetti, S., Kurtz, A. B., and Dwyer, K. (1980). A simple general method for the assay of free hormones (and drugs); its application to the measurement of serum free thyroxine levels and the bearing of assay results on the "free thyroxine" concept. J. Endocrinol. 85: 29.

Ekins, R., Edwards, P., and Newman, B. (1982). The role of binding-proteins in hormone delivery. In Free Hormones in Blood, A. Albertini and R. P. Ekins (eds.), Elsevier Biomedical, Amsterdam, p. 3.

Ekins, R.P., Jackson, T., Edwards, P., Salter, C., and Ogier, I. (1983). Euthyroid sick syndrome and free thyroxine assay. Lancet 2:402.

Ellis, S., and Ekins, R. (1973). Direct measurement by radioimmunoassay of the free thyroid hormone concentration in serum. Acta Endocrinol. (Kbh) Suppl. 177: 106.

Dixon, P. F. (1968). The kinetics of the exchange between transcortin-bound and unbound cortisol in plasma. J. Endocrinol. 40: 457.

Giles, A. F. (1982). An improved method for the radioimmunoassay of free thyroxine in serum dialysate. Clin. Endocrinol. 16: 101.

Henneman, G., Docter, R., Krenning, E. P., Boz, G., Otten, M., and Visser, T. J. (1979). Raised total and free thyroxine index but normal free thyroxine. A serum abnormality due to inherited increased affinity of iodothyronines for serum binding protein. Lancet 1: 639.

Hillier, A. P. (1971). Human thyroxine-binding globulin and thyroxine binding pre-albumin: Dissociation rates. J. Physiol. 217: 625.

Kaptein, E. M., MacIntyre, S. S., Weiner, J. M., Spencer, C. A., and Nicoloff, J. T. (1981). Free thyroxine estimates in nonthyroidal illness: comparison of eight methods. J. Clin. Endocrinol. Metab. 52: 1073.

Keller, N., Richardson, U. I., and Yates, F. E. (1969). Protein binding and the biological activity of corticosteroids: In vivo induction of hepatic and pancreatic alanine aminotransferases by corticosteroids in normal and estrogen-treated rats. Endocrinology 84: 49.

Kurtz, A., Dwyer, K., and Ekins, R. (1979). Serum free thyroxine in pregnancy. Br. Med. J. 2: 550.

Melmed, S., Geola, F. L., Reed, A. W., Perkary, A. E., Park, J., and Hershman, J. M. (1982). A comparison of methods for assessing thyroid function in nonthyroidal illness. J. Clin Endocrinol. Metab. 54: 300.

Midgley, J. E. M., and Wilkins, T. A. (1980). European Patent Application No. 0026103, September, 1980.

Midgley, J. E. M., and Wilkins, T. A. (1981). The direct estimation of free hormones by a simple equilibrium radioimmunoassay. Amersham International, Amersham, p. 1.

Midgley, J. E. M., and Wilkins, T. A. (1982). What do radioimmunoassay methods for free thyroxine using "unbound analogues" actually measure? (A reply). Lancet 2: 712.

Odstrchel, G. (1982). A kinetic method for the measurement of free thyroxine and its application in the diagnosis of various thyroid diseases states. In Free Hormone in Blood, A. Albertini and R. P. Ekins (eds.), Elsevier Biomedical Press, Amsterdam, p. 91.

Odstrchel, G., Herte, W., Ward, F. B., Travis, K., Lindner R. E., and Mason, R. D. (1978). New concepts for the assay of unbound thyroxine (FT4) and thyroxine binding globulin (TBG). In Radioimmunoassay and Related Procedures in Medicine, Vol. II, IAEA, Vienna, p. 369.

Pardridge, W. M. (1981). Transport of protein-bound hormones into tissues in vivo. Endocr. Rev. 2: 103.

Rask, L., and Peterson, P. A. (1976). In vitro uptake of vitamin A from the retinol binding plasma protein to mucosal epithelial cells from the monkey's small intestine. J. Biol. Chem. 251: 6360.

Robbins, J. and Rall, J. E. (1979). The iodine-containing hormones. In <u>Hormones in Blood</u>, Vol. 1, 3rd ed., C. H. Gray, and V. H. T. James (eds.), Academic Press, London, p. 575.

Romelli, P. B., Pennisi, F., and Vancheri, L. (1979). Measurement of free thyroid hormones in serum by column adsorption chromatography and radioimmunoassay. <u>J. Endocrinol. Invest.</u> <u>2</u>: 25.

Schall, R. F., Crutchfield, T., Kern, C. W., and Tenoso, H. J. (1979). An enzyme labeled immunoassay for the measurement of unsaturated thyroid hormone binding capacity in serum and plasma. <u>Clin. Chem.</u> <u>25</u>: 1078(076).

Sterling, K., and Hegedus, A. (1962). Measurement of free thyroxine concentration in human serum. <u>J. Clin. Invest.</u> 41: 1031.

Stockigt, J. R., Barlow, J. W., White, E. L., and Csicsmann, J. (1981). The plasma binding abnormality of familial euthyroid T_4-excess. <u>Annales d'Endocrinologie</u>, 11th Annual Meeting of the European Thyroid Association, Pisa, September 1981, p. 34A.

Stockigt, J. R., White, E. L., and Barlow, J. W. (1982). What do radioimmunoassay methods for free thyroxine using "unbound analogues" actually measure? <u>Lancet</u> <u>2</u>: 712.

Tait, J. F. (1982). The biological availability of free and protein-bound steroid. In <u>Free Hormones in Blood</u>, A. Albertini and R. P. Ekins (eds.), Elsevier Biomedical, Amsterdam, p. 65.

Tait, J. F., and Burstein, S. (1964). In vivo studies of steroid dynamics in man. In <u>The Hormones</u>, Vol. V., G. Pincus, K. V. Thimann, and E. B. Astwood (eds.), Academic Press, New York, p. 441.

Wilkins, T. A., Midgley, J. E. M., and Giles, A. R. (1982). Treatment with heparin and results for free thyroxin: An in vivo or an in vitro effect? <u>Clin. Chem.</u> <u>28</u>: 2441.

10
Data Analysis and Quality Control of Assays: A Practical Primer

R. P. CHANNING RODGERS*
International Institute of Cellular and Molecular Pathology
Université Catholique de Louvain, Brussels, Belgium

I. INTRODUCTION

This chapter sets challenges for both author and reader. Most conferences or books dealing with binder-ligand assay feel obliged to include a session or chapter on "the mathematics of" or "the theory of" but this is usually regarded by the intended audience as something more to be endured than enjoyed — if not evaded altogether. Quite properly, most assay users are more concerned with how to put assay results to work than they are with the internal workings of the assay itself. Assayists form a diverse group which includes clinicians with patients to care for as well as basic and applied scientists who have pressing scientific and technical problems to solve. To the extent that these people are interested in the assay at all, it is usually in aspects of the procedure which are closest to their own discipline—often the chemical aspects of assay.

The challenges to the author are:

1. To convince you that an understanding of the basic mathematical tools of assay is just as important as knowing how to obtain a good antibody or handle a pipette properly
2. To demonstrate that the basic concepts of proper data processing and quality control are not arcane mathematics, but reasonable and consistent with common sense
3. To leave you with sufficient distrust of the accepted widsom that comprises laboratory common sense that you will know when to consult an assay biostatistician

---
*Present Affiliation: University of California School of Medicine, San Francisco, California.

The challenge to the reader is to throw aside any inhibitions or antagonisms he or she may have regarding assay mathematics and to plunge ahead, trying to look for ways in which the practical content of this chapter can improve the application of assay results.

Overlooking the proper mathematical analysis of assay data can introduce just as much error or inefficiency into an assay as poor reagents or inadequate skill in the laboratory. Properly approached, assay data processing and quality control are not dry mechanics, but a vital way in which to improve your own work.

To keep this process lively for both of us, I depart from the conventional didactic format of such chapters. Each of the main sections will be separated by interludes, in which I will pose a practical question to the reader. The enthusiastic reader may want to scan these sections before reading the rest of the text. These interludes are meant to point out the direct application of the basic mathematical principles of assay, with an emphasis on the principles, not the mere manipulation of numbers. They are also meant to be cathartic for the author, and may be provocative to some assayists, particularly those who have assimilated some of the misleading folk wisdom in which the assay discipline abounds.

The emphasis throughout will be on basic definitions and concepts: detailed step-by-step instructions for the manual computation of assay results require much more space and are provided elsewhere (Rodgers, 1981). Although assayists must understand how to calculate results properly by hand, it is hoped that by the end of this chapter, they will be so thoroughly convinced of the advantages of automated calculation that they will not wish to obtain answers by the much more difficult and hazard-prone manual techniques.

I will assume that the reader already has basic statistical knowledge, understands the distinction between parametric and nonparametric statistics, and is comfortable with the basic terms of parametric statistics, such as mean, variance, standard deviation, and standard error or standard deviation of the mean. There are many good introductory statistical texts; a particularly appropriate one for readers interested in biological applications and assays is that of Colquhoun (1971).

Finally, remember that the methods described here are completely general, and apply equally well to all forms of continuous-response assays in use today, including radioimmunoassay, immunoradiometric assay, chemiluminescent assay, enzyme-linked immunoadsorbent assay (ELISA), particle-counting assay (PACIA), or any of the myriad other related forms of assay. To denote the generality of the data processing techniques, I have in the past referred to all of these methods collectively as binder-ligand assay. Older generic names such as saturation assay and competitive protein binding assay (CPBA) are not appropriate for some of the newer types of binder-ligand assay.

II. THE LANGUAGE AND GOALS OF ASSAY DATA PROCESSING AND QUALITY CONTROL

A. What is Quality Control?

Quality control has gained renewed prominence of late due to the spectacular rise of Japanese industry. The Japanese themselves are quick to attribute much of their success to the assiduous application of the principles of industrial quality control. Japanese industry has commemorated the role of one of its primary teachers, the American statistician W. E. Deming, by naming two annual quality control awards in his honor (industries in the United States and France have subsequently inaugurated Deming medals as well).

The spirit, aims, and methods of assay quality control are largely borrowed from industrial quality control (Bennett and Franklin, 1954; Duncan, 1974), and it is vital to bring to assay the same broad view taken by the industrial statisticians. To rephrase Deming (1967, 1975) slightly, we could state that assay quality control consists of all efforts made toward assuring the maximal utility of the product of an assay (which is generally a clinical decision or scientific inference), given the constraints imposed by economic, social, and technical considerations. This is too broad a definition to treat in a single chapter, so the reader is referred to the growing literature on medical decision theory to see how assay results relate to clinical actions (Griner et al., 1981; Weinstein et al., 1980). We will confine ourselves to a narrower definition of assay quality control as those efforts taken to assure the smallest possible error in the analyte concentration estimate obtained from an assay, consistent with the imposed constraints.

During its birth, an assay result exists in three different milieux: the psychosocial, the biophysical, and the mathematical. The psychosocial millieu is that in which the assayist and the assay end-user operate. The biophysical encompasses the physical and biochemical manipulations that comprise the analytical method. Finally, the numbers obtained from the assay detection device must enter the mathematical milieu to produce a usable result. Both disastrous and subtle errors can arise in any of these three milieux. For example, consider the psychosocial milieu. Clerical mistakes can cause loss of samples or mistaken exchange of two specimens from different sources, or an unpleasant working environment can degrade the performance of an assayist. Good management ability and psychological insight are valuable skills in optimizing conditions in the psychosocial milieu.

The biophysical milieu also requires attention. Instruments must be properly calibrated, reagents must be of adequate and consistent quality, and the analytical method must be sound. This requires technical skill, good planning, and an understanding of the physical principles of assay.

A complete quality control scheme must consider each of the three milieux, although we will concentrate here on just one: the mathematical. Helpful advice concerning the other two can be found in numerous sources (Büttner, 1968; Broughton and Raine, 1969; Copeland, 1973; Davidsohn and Henry, 1972; Hainline, 1982; Henry and Geigel, 1977; Inhorn, 1978; Jeffcoate, 1981; Rappoport, 1971; Tonks, 1970; Whitehead, 1977). Of notable value is the prototype laboratory check list presented by Ottaviano and DiSalvo (1977).

B. The Language of Assay Quality Control

A binder-ligand assay attempts to quantify the concentration or mass of some substance of interest, the analyte, which will also be referred to as the ligand of interest. The assays in this large family of methods generally exploit the specific chemical affinity of another molecule, the binder, for the ligand of interest. The binder is often an antibody molecule.

To accomplish this task, an assayist submits a set of known analyte concentrations, the standards, to a series of chemical and physical manipulations known as the analytical method. In radioimmunoassay, for instance, this consists of combining the standards with a constant concentration of binder and labeled ligand, allowing the binder-ligand reaction to proceed, and then separating the labeled ligand which is bound to binder from that which is still free in solution. The final measurement device of a radioassay is either a gamma counter or scintillation counter, which detects the radioactivity in the bound fraction, the free fraction, or both. The final detection device will vary according to the particular type of assay in use, and could equally well be a colorimeter, particle-counting device, photometer, or other device.

The results obtained from the end detection device may be preprocessed, as for the decay and background corrections required in radioassay. These numbers may then be used to calculate the assay response metameter. The response will be some monotonic (continuously increasing or decreasing) function of the analyte concentration, and a plot of the response vs. the analyte concentration constitutes a calibration curve. This curve is a graphic representation of a quantitative relationship between what we actually measure (the numbers used to calculate the response) and what we desire to measure (the concentration of analyte). This process constitutes the calibration of the assay method. Now, the assayist may attempt to determine the analyte content of some test specimen, often a physiological fluid such as blood or urine. Sometimes a preliminary purification step is required before the test sample may be submitted to the assay. The added complications of such a step will not be discussed in this chapter. A small volume of the sample is subjected to the same analytical method used in calibration, and a response is obtained. The calibration curve is now used to determine the analyte concentration that corresponds to this response: this process is known as analyte concentration interpolation, or simply interpolation.

Calibration and interpolation are the core procedures of assay, and it is tempting to stop at this point and consider the assay process complete. Too many assayists and too many publications dealing with assay do. Unfortunately, the assay analyte concentration estimates are flawed; they are subject to error, and this error must be quantified and controlled. This is the task of assay quality control.

It is useful to divide the error inherent in an assay result into two components: random error and bias (or systematic) error. Random error is concerned with the reproducibility of a result, and can be estimated (and to a limited extent controlled) by replication of a given measurement. Bias error presents a more formidable problem, as there are potentially an infinite number of sources. All of the likely sources must be guessed at and studied separately: an understanding of physiology and assay chemistry is vital here. One especially important tool in this procedure is the quality control (QC) sample. There are many types of QC samples, the most commonly employed being the spot sample. This is a sample of known analyte concentration, in a medium identical to the unknowns (often a pooled urine or blood collection). It is processed just as an unknown, in every assay. Various statistical tests can be carried out on the results to ascertain if the assay is behaving normally.

Although information about reproducibility can be garnered from the standards, the unknowns, and the spot quality control specimens, the spot samples are the major source of information about bias error in an assay.

III. INTERLUDE 1: COUNTING STATISTICS AND COMMON SENSE

Albert Einstein is said to have remarked that common sense is the collection of prejudices acquired by age 18. All too often, assayists are content to conduct their work with the aid of a few basic principles, acquired early in their training, which are sometimes antithetical to the quality and efficiency of their assays.

These principles have become part of the established "common sense" of assay work, yet if these same principles are scrutinized critically, they are found to be misleading, and more powerful principles emerge. Once they have been used for some time, these new ideas can be seen to conform even better to true common sense. A remark of R. L. Ackoff (1968) is pertinent: "Common sense. . .has the very curious property of being more correct retrospectively than prospectively."

A simple example is provided by the counting statistics of radioimmunoassay. The raw numbers for calculations come in the form of counts of radioactive decays. As events dispersed randomly in time, these counts are subject to poissonian statistical variations. The statistical quantity, variance (or var), which corresponds to a given number of counts is equal to the number of counts itself. The standard deviation (s.d.) is simply the square root of the variance. The coefficient of variation (c.v.) of a measurement is equal to the standard deviation divided by the measurement value. If the number of counts (including background) in the bound fraction of a given sample of a given assay is B, then these error relationships can be expressed as follows:

$$\text{var}(B) = B \tag{1a}$$

$$\text{s.d.}(B) = \sqrt{B} \tag{1b}$$

$$\text{c.v.}(B) = \sqrt{B/B} = 1/\sqrt{B} \tag{1c}$$

Many assayists will follow a counting strategy in which each assay tube is counted until a preset number of counts or preset counting time has been reached, whichever goal is achieved first. The preset number of counts is often 10,000. The standard deviation in such a count is 100, yielding a c.v. of 100/10,000, or 1%. Following this naive strategy is generally safe, but as we shall see below, it can lead to a drastic waste of counter time. This strategy is indicative of a problematic attitude which appears repeatedly in daily assay practice: the focusing of attention on intermediate assay quantities instead of the final assay measurement itself. We wish to control error in the final assay measurement, not in the raw counts coming from the counting device. In the following discussion, we will present a more rational approach to counting, developed by Ekins and co-workers (1978), which can result in a two- to 17-fold improvement in the efficiency of counting device utilization.

As we shall see in Sect. VI., under certain statistical assumptions the error in an assay concentration estimate is proportional to the error in the corresponding response. Therefore we can confine our attention to controlling error in the assay response. Suppose that the response metameter in use is the bound over total ratio, which we represent as b. The value of total counts (treated as an errorless scaling constant in the following discussion) is represented by T, and the counting time used for obtaining B counts in the bound fraction by t. The response may be obtained as b = B / (tT), neglecting background corrections (which does not influence the thrust of the statistical arguments to follow). The formulas describing the counting error in b are:

$$\text{var}(b) = b / (t \cdot T) \tag{2a}$$

$$\text{s.d.}(b) = \sqrt{b / (t \cdot T)} = \sqrt{B} / (t \cdot T) \tag{2b}$$

$$\text{c.v.}(b) = \sqrt{b / (t \cdot T)} / b = 1/\sqrt{b \cdot t \cdot T} = \sqrt{B}/B = 1\sqrt{B} \tag{2c}$$

Note that although the standard deviation of b is generally different from that of B, their coefficients of variation are identical, as would be expected, as b is merely a rescaling of B.

This reexpression of counting error in terms of the response, although ultimately helpful, does not directly improve our counting strategy. First we must realize that counting error is only one component of the total error influencing a given sample. A host of other random errors (in pipetting and timing, for example) have already done their mischief by the time the sample arrives at the counter. Even if the sample is counted forever, the resulting value of b will reflect at least this minimum amount of error. These noncounting errors are often referred to as experimental error. The final total amount of error in the value of b will reflect both counting error and experimental error. Such errors are said to add in quadrature. If we represent the variance in b due to counting error as s_c^2, and the variance due to experimental error as s_e^2, then this addition can be expressed as:

$$s^2 = s_e^2 + s_c^2 \tag{3}$$

Note that the variances, and not the standard deviations, are summed together. The final standard deviation in b will be equal to $\sqrt{s^2}$.

A further word about experimental error is in order. When an assay is in control, this error tends to be a clearly defined function of b, and this relationship can actually be plotted following the methods described in Sec. VI. The relationship between b and s_e^2 or s_e can then be described by some appropriate mathematical formula. Several data processing packages (Ekins et al., 1978; Dudley, 1981) utilize linear relationships such as:

$$s_e^2 = \alpha + \beta b$$

here, s_e^2 varies as a straight line with respect to b, the location of the line being determined by the constants α and β, the values of which must be determined for the particular assay being run.

We are now prepared to derive the formula which will be the keystone of our new counting strategy. If we define f as a number which, when multiplied by the experimental error present in a given sample, yields the total error we desire to allow in the value of b, the appropriate counting time, t, may be defined as follows:

$$t = \frac{1}{T(f^2 - 1)((\alpha/b) + \beta)} \tag{5}$$

This formula is obtained by combining Eqs. (2a), (3), and (4).

The practical impact of this equation can be better appreciated, if we first examine some results from Eq. (3). If we divide both sides of this formula by b^2, we end up with a sum in quadrature of coefficients of variation rather than standard deviations. Suppose that the experimental error in a given sample is 5%; counting this sample until 10,000 counts

TABLE 1 Calibration Curve-Fitting Methods[a]

Empirical
  Manual curve-fitting:[b] 63, 70, 73, 85, 144, 167
  Point-to-point linear segments:[b, c] 70 (23, 161a)
  Polynomials:[b, c] 2
    First-order (straight line): 161
    Second-order (parabolic): 85 (13)
    Third-order (cubic): 81, 154 (24)
    Fourth-order (quartic): 157
    Adjustable-order: (108)
    Other:[b, c]
      Ratio of polynomials: 85 (86)
      Inverted polynomial (fitted on y rather than x): 63
  Polygonal interpolation:[b, c] 142, 143 (84)
  Splines:[b, c] 149
    Interpolating: 85, 132
    Smoothing: 70, 87a, 132, 142, 143, 144 (88, 104a, 158)
  Other:
    Logistic-like (but empirically obtained):[b, c] (63)
    Power series on y (resembles IC6b above):[b, c] (160)
    Exponential function of x:[b, c] 81, 154
    Log(log y) vs. inverse second order polynomial on x:[b, c] 153
    Log x:[d, c] (89, 97, 161a)
    Log y:[d, c] (89)
    Rectangular hyperbola:[d, c] 9, 101, 125
    Sigmoidal transformations resembling logit:[d, c]
      Arcsine: 2, 148, 149, 159
      Angular transformation (79c)
      Arctangent: 148, 149 (76)
      Hyperbolic sine: 148, 149
      Hyperbolic tangent: 148
      Probit or Gaussian fit: 79c, 98, 111[f], 148 (35a, 76)
Semi empirical
  Log-log:[d, c] 5, 55, 146, 171 (161a)
  Reciprocals:
    1/B or T/B:[d, c] 8, 85, 146
    F/B:[d, c] 41, 146
    Bo/B:[d, c] 18, 62, 146
    Modified 1/B:[b, c] 81, 154
  Logistic:
    Two parameter (logit-log):
      Standard:[d, c] 2, 70, 73, 74, 79c, 80, 85, 103, 111[f], 127, 142, 143, 144, 145, 146, 148, 149 (19, 20, 24, 36, 38, 54a, 66, 76, 89, 93, 104, 125, 130, 139, 140, 161a, 167a)

TABLE 1 (Continued)

Modified:[d,c] 31f, 73, 74, 111[f], 142, 143, 145, 146 (64, 82, 83)
Three Parameter:[b,c] 2, 63, 81, 154 (14, 110)
Four parameter:
  Standard:[b,c] 17, 51, 73, 74, 85, 107[g], 113, 142, 155g (32, 51, 56, 59, 60, 95, 108, 109, 112, 115, 127, 128)
  "Two plus two" method:[d,c] (29, 37, 173)
Five-, six-, and seven-parameter:[b,e] 43, 112, 132 (95, 109)

Model-Based[h]
  Two-parameter (K, q):[b,e] 71, 79 (165)
  Three-parameter
    (K, p*, q):[b,e] 43, 96 (87)
    (K, q, B_n): 150

  Four-parameters:
    (K, q, p*, B_n):[b,e] (166)

    (K, K*, p*, q): (103)
    (K, p*, p, q): 47, 47a
  Five-parameter:[b,e] 113
  Other:[b,e]
    Two- to five-parameter adjustable: 132 (135)
    N-parameter multiple binders: 54, 85
    Other: 45, 168

---

[a]Where a program for assay data processing is mentioned as available, or where sufficient detail is offered to allow easy implementation, the reference number appears in parentheses.
[b]Method that fits a relationship directly to nonlinear data, without prior transformation.
[c]Method for which linear regression methods may be employed.
[d]Method that attempts to linearize data by transformation, and then fits a straight line.
[e]Method in which nonlinear regression techniques must be used.
[f]Method applied to quantal response assay.
[g]Robust regression method used.
[h]For definitions of K, K*, p, p*, q, and B_n see section IV.A.2.

have been obtained will produce a counting error of 1%. However, these two errors summed in quadrature yield a final c.v. of only 5.1%. If the sample is only counted for 1000 counts, the counting c.v. is 3.16%, but the total error is 5.9%. We have gained only 0.8% in precision at a cost of 10 times as much counting time!

Let us work one example using Eq. 5. First, note that the term $((\alpha/b) + \beta)$ in the denominator is equivalent to (s_e^2 / b). Suppose that T=10,000, b=0.5, (s_e / b)=0.05, and f=1.2. The values of f and c x v x (b) imply that we are beginning with a 5% experimental error in b, and are willing to allow a total error of (1.2)·(0.05), or 0.06 (that is, 6%). The value of t obtained from Eq. (5) is 0.18 min. Since T=10,000 cpm and b=0.05, this corresponds to only about 909 counts. This achieves, for this tube, a 10-fold increase in efficiency of counter use over the strategy of counting until 10,000 counts have been obtained.

This formula allows the counting error to be adjusted at will over the entire range of the assay. The appropriate application of this method of counting requires careful consideration of the application requirements of the assay: there may be analyte concentration levels at which it is desirable to control error carefully, and other levels at which the random error may not be very important. It also requires counting equipment with appropriate microprocessor or computer programs controlling counter operation; such devices already exist (see Ekins et al., 1978, and Dudley, 1981); regrettably, the additional data processing done by the first of these devices is less than acceptable (see Challand, 1978).

There are other ways in which the application of this counting method seems to contradict current laboratory common sense. For example, given that a sample is represented by both bound and free fractions, and that it is desired that only one fraction be counted, it may in certain circumstances be preferable to count the fraction with the least counts; I refer the reader to the original sources for a fuller discussion of this seeming paradox.

This is but one example of how the careful application of assay mathematics and statistics can have a direct and potent practical impact upon assay performance.

IV. CALIBRATION CURVE-FITTING

A. Formulas and Transformations for Fitting Calibration Curves

The extensive literature dealing with automated calibration curve-fitting is summarized in Table 1. Rodbard (1978a) has reviewed it in detail. The sheer number of articles dealing with this topic has had a deleterious impact. Assayists have been led to believe that calibration curve-fitting is the sole essential task of assay data processing. They have been bewildered by the seemingly huge variety of methods employed for fitting the curve.

So much attention has been spent on presenting new formulas for curve-fitting that the role of good statistical and computing procedure has been slighted. The importance of curve-fitting per se was perhaps exaggerated in the minds of some readers of early assay quality control papers which suggested the use of curve location parameters as indices of quality control (Rodbard et al., 1968; Rodbard, 1971). A shift in the location of a calibration curve certainly indicates that the physical properties of an assay reaction system are changing, but does not indicate the impact of such a change on assay quality. The performance of the assay as judged by a statistically valid criterion of performance such as bias or imprecision at a given analyte concentration could actually be improving, or not shifting significantly. Thus such indices are only indicators of assay stability, and do not yield direct information about assay error or quality.

There are two basic problems posed by calibration: (1) to find an appropriate function to fit to the calibration curve, and (2) to fit it correctly. We will examine both of these issues in this and the following section.

The reader should not be intimidated by the large number of calibration formulas which have been proposed. They tend to fall into just a few basic groups, and many "new" methods turn out, on closer inspection, to be equivalent to older methods. Thus the "new mathematical model" of Harding and colleagues (1973) turns out to be identical to a three-parameter logistic, and the "application of radioimmunoassay theory" of Fernandez and Loeb (1975) in practice turns out to be identical to a simple technique proposed much earlier (see Fernandez et al. 1983; Hales and Randle, 1963). I propose an arbitrary classification of these methods which divides them into three categories based upon the theoretical justification for their use in radioimmunoassay: the empirical, the semiempirical, and the model-based.

1. Empirical

These methods are labeled empirical because their use is not based upon some physicochemical model for the assay process, but rests entirely upon success in practice. Of manual plotting and point-to-point line segments, we will say more in the following interlude.

Polynomials: Polynomial curve-fitting methods are employed in many branches of science and technology. If we refer to the horizontal (analyte concentration) axis as the x axis, and the vertical (response) axis as the y axis, the formula for a polynomial is as follows:

$$y = a_0 + a_1 \cdot x + a_2 \cdot x^2 + a_3 \cdot x^3 + \ldots a_n \cdot x^n \quad (6)$$

This is said to be a polynomial of order n. The values $a_1 \ldots a_n$ are known as the parameters of the equation, and completely determine its location for a given x value. A straight line is simply a polynomial of order one. Commonly encountered are equations of order two (parabolic), three (cubic), and four (quartic). It is a general property of polynomials that, given (n+1) data points, a polynomial of order n can be fitted so as to pass exactly through each of the (n+1) points. At first thought, this might seem to make the polynomial an ideal candidate for calibration. The catch is that although the polynomial will pass through each of the data points, its behavior between these points can be highly erratic, with huge oscillations. This tendency worsens as the degree of the polynomial increases. Generally, the order selected should be much lower than the available number of points to be fitted (generally of order two or three). Such a curve will tend to be smooth but may not approximate the data very well.

Splines: Another approach is to divide the data into small groups, and fit low-order polynomials to each of the separate groups, constraining the fitting process so as to create one smooth curve. This is the spline technique, which has been very heavily developed (DeBoor, 1978). This approach has been likened to a mathematical equivalent of the draftsman's French curve, and is in frequent use in automotive and aeronautical industrial design. Interpolating splines pass through each data point, while smoothing splines, sometimes known as tensioned splines, allow the specification of a smoothing factor which will bring the final curve closer to a straight line. Therefore such a curve will not pass directly through the points being fitted.

Polygonal Interpolation: This method has been touted for its computational simplicity and appropriateness for use on the smallest of computing devices. It resembles a simplified spline, but consists of a number of short line segments rather than a smooth curve. The algorithm divides the horizontal distance between neighboring data points into several sections of equal length, and draws a straight line segment within each of these sections in such a way that the final result approximates a single smooth curve drawn through each of the data points.

Other: Numerous other methods have been proposed. They often rely on transformations of the assay response variable which result in the calibration curve approximating a straight line. Some of these make particular use of the fact that assay data often can be plotted in such a way that the calibration curve is a symmetrical sigmoid (s-shaped) curve.

2. Semiempirical

These techniques can be regarded as semiempirical for assay methods closely related to radioimmunoassay, because, under certain very rigid simplifying assumptions, there are theoretical justifications for why they should fit calibration data. Easily the most important of these methods is the logistic technique. This comprises a large family of different equations, the most commonly encountered at present being the four-parameter logistic.

Data Analysis and Quality Control 265

Logistic equations have been used for many years, and appear in studies of populations, tumor growth, and economic models (Reed and Berkson, 1929). They are also important in studies of medical survival and in diagnosis (Hosmer et al., 1978). The four-parameter logistic was introduced to the assay community by Healy (1972) and subsequently heavily developed by Rodbard (1974, 1978a) and others (DeLean et al., 1978; Rodbard and Hutt, 1974). It is described by the following equation:

$$y = \frac{a - d}{1 + (x/c)^b} + d \qquad \left(\frac{x}{c}\right)^b = \frac{a-y}{y-d} \qquad (7)$$

This equation is fitted to experimental assay data in Fig. 1. The four parameters a, b, c, and d can be readily interpreted in terms of the shape of the resulting symmetrical sigmoid. The value of a and d correspond to the upper and lower asymptotes of the curve, respectively. The value of c is the value of x (analyte concentration) which corresponds to the center, or inflection point, of the sigmoid. The value of b is related to the slope of the center of the sigmoid in a way that will become clearer after a discussion of the two-parameter logistic equation.

how does the error transform?

Note that Eq. (7) is equivalent to:

$$\log \frac{y - d}{a - y} = (-b) * \log(x) + b * \log(c) \qquad (8)$$

which suggests that it might be useful to define the following transformation:

$$Y = \log \frac{y - d}{a - y} \qquad (9)$$

This is the logit transformation, and Eq. (8) now becomes:

$$Y = (-b) * \log(x) + (b * \log(c)) \qquad (10)$$

which is simply the equation for a straight line for Y vs. log(x) with a slope of [-b] and y intercept of [b*log(c)]. The earliest application of a logistic function to binder-ligand assay was known as the logit-log technique (Rodbard et al., 1969). This procedure is equivalent to using Eqs. (8) and (9) after setting the values of a and d to 1 and 0, respectively. The value used for y is the bound counts for a given analyte concentration divided by the bound counts at zero analyte concentration (after correcting all bound count values for the effects of nonspecific binding — see Secs. VIII. and IX.). The y values are subjected to the logit transformation. Then a straight line is fitted to the results. This is equivalent to fixing the end points of the four-parameter model, reducing it to a two-parameter logistic. Occasionally, this arbitrary fixing of the ends of the curve results in logit-transformed data which clearly deviate from a straight line. With the advent of the four-parameter method, the logit-log technique has become less often used.

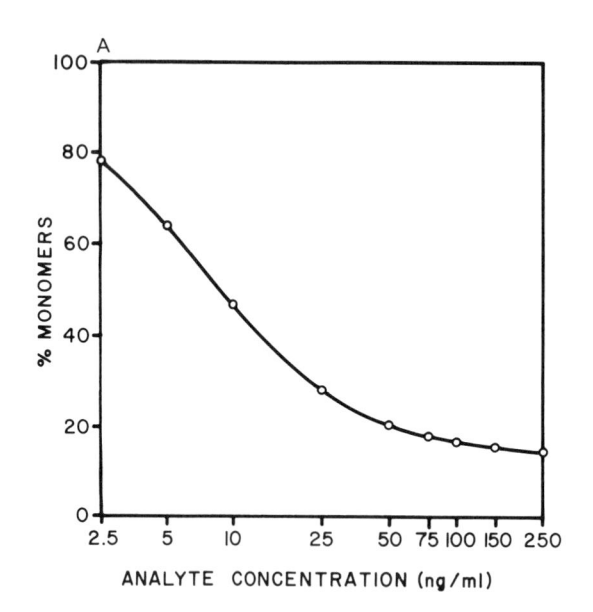

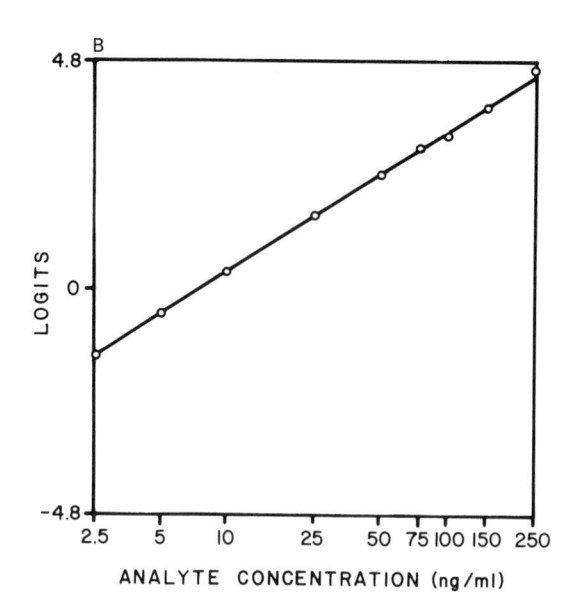

C

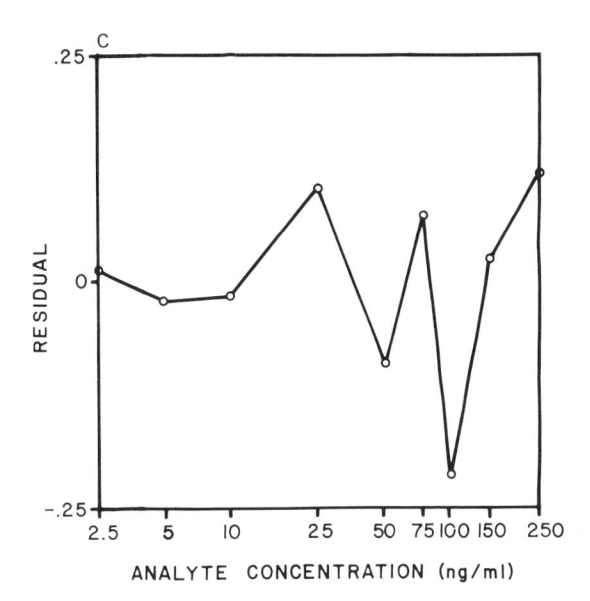

[Graph: Residual vs Analyte Concentration (ng/ml), y-axis from -.25 to .25, x-axis values 2.5, 5, 10, 25, 50, 75, 100, 150, 250]

FIGURE 1 (A) The four-parameter logistic equation (Eq. 7) fit to assay data from a PACIA assay for human IgA. The chemistry of PACIA is radically different from most other binder-ligand assays, in that the analyte is added to a system of antibody-coated latex spheres, provoking an agglutination reaction. The number of monomers (unbound latex spheres) is detected by light scattering, and the response is the percentage of spheres remaining monomers (Masson et al., 1981). (B) The logit-log transform (Eq. 9, 10) of the data of (A), using the values of a and d obtained from the four-parameter fit. The straightness of this line merely reflects the accuracy of the four-parameter method. Were values for a and d arbitrarily fixed, as in the older two-parameter logit-log method, the resulting curve would not be as straight. (C) A plot of the residuals for the fit of (A). They fluctuate randomly about zero, consistent with a good fit.

The practical difficulty with the four-parameter logistic method is that it is nonlinear, in an algebraic sense, with respect to its parameters b and c (note that b appears as an exponent and c as a denominator).* This requires that nonlinear regression techniques be employed. However, Zivitz and Hidalgo (1977) pointed out a strategy, subsequently dubbed the 2+2 logistic technique by which linear regression may still be applied to the four-parameter logistic (Davis et al., 1980). If one starts with good estimates for a and d based on previous experience, the logit transformation can be performed on the calibration curve data, and a straight line fitted to the results, using weighted linear least squares. This yields estimates for b and c. One can then calculate the value of

$$x' = \frac{1}{1 + (x/c)^b} \qquad (11)$$

for each datum, and fit a straight line to y vs. x'. This yields new estimates for a and d. The repeated alternation of these two linear curve-fitting procedures can converge to acceptable estimates for the values of a, b, c, and d. The most difficult component of this otherwise simple method is the proper determination of weights to be used for the two fitting procedures.

A limitation of the four-parameter logistic model is that it is of necessity symmetrical through a central inflection point. If this does not suit the data, then the five-parameter logistics of Richards or Nelder (Raab and McKenzie, 1982) or the five-, six-, or seven-parameter models described by Rodbard and co-workers (1978) may be employed: these allow for asymmetries of various degrees.

Rodbard has also described a generalization of the logistic for any number of parameters. Equation 7 can be reexpressed in the form:

$$y = \frac{a - d}{1 + \exp(f(\ln(x)))} + d \qquad (12)$$

where $f(\ln(x)) = b(\ln(x) - \ln(c))$.

---

*This algebraic linearity must not be confused with geometric linearity. Although a straight line is linear in both the geometric and algebraic senses of the word, a polynomial is considered to be linear with respect to its parameters but is certainly not geometrically linear if its order is greater than one. A function can be said to be linear in its parameters if none of the parameters appear as exponents or denominators, or inside functions such as trigonometric or logarithmic functions. These requirements are clearly violated in the case of the logistic equation.

Note that the function f(ln(x)) should describe the shape of the data when plotted as a logit-log graph as in Fig. 1B. If the transformed data seem to follow a straight line, then the four-parameter model probably applies. Otherwise, any other function that describes the data may be used. An obvious approach would be to use a polynomial for f(x). Because of the linearity of the function chosen for f(x), a 2+n logistic method could be followed, in analogy to the 2+2 technique, using only linear regression methods. The term n here refers to the order of the polynomial plus one, or the number of parameters to be fitted to describe the polynomial chosen. One risk with this technique is that the resulting curve may not be monotonic; however, for low-order polynomials simple tests can be applied at the end of the fit to make sure that this is not the case. For well-behaved data and low-order polynomials (two to four) it is not likely to be a problem. Such methods should become steadily less useful, as the power of inexpensive computing devices is just beginning to allow almost universal access to nonlinear curve-fitting programs which can fit equations like the four-parameter logistic rapidly and in a statistically valid way.

3. Model-Based

The final group of calibration curve-fitting methods is based upon simple physical chemical models for the assay reactions in use. Although there is a wide variety of chemical schemes in use for the various forms of binder-ligand assay, most of the models apply only to radioimmunoassay and closely related methods. A simple scheme for the central reactions of such an assay is:

$$P + Q \underset{k_r}{\overset{k_f}{\rightleftarrows}} PQ \tag{13}$$

$$P^* + Q \underset{k_r^*}{\overset{k_f^*}{\rightleftarrows}} P^*Q \tag{14}$$

where P^* represents the labeled ligand, P is the unlabeled ligand (the analyte), and Q represents the binder, often an antibody. These two competing reactions are generally assumed to be governed by the simple first-order mass action law:

$$K = \frac{k_f}{k_r} = \frac{[PQ]}{[P][Q]} \quad \text{and} \quad K^* = \frac{k_f^*}{k_r^*} = \frac{[P^*Q]}{[P^*][Q]} \tag{15}$$

where the square brackets indicate the concentrations of the contained species at chemical equilibrium, and the K values are the equilibrium constants describing these two reactions. For purposes of further discussion and for use in Table 1, the total concentration of binder will be represented by q, and the total concentrations of labeled and unlabeled ligand by p* and p, respectively.

An example of such a model is that of Naus and colleagues (1977), which attempts to adjust the values of K, K*, p*, and q until the curve so obtained fits the calibration data as well as possible. The potential number of such models is unlimited. Some of those presently available assume that K=K*. Others add a parameter to describe nonspecific binding, B_n, which is often assumed to be a set fraction of the ligand present (an assumption which is rarely validated). Several models allow for a multiplicity of binding sites (binder heterogeneity).

B. Principles of Curve-Fitting

In the early years of immunoassay, most results were obtained by manual techniques. Calibration results were plotted and a curve drawn by hand through the results. If an assay is stable, the location of the plotted points is often reproducible, and with some experience it becomes easy to sketch a curve rapidly through each new set of points. The appealing simplicity of this approach can lull an assayist into a false sense of security. For compelling statistical reasons, a properly implemented automated fit is preferable.

The most commonly encountered technique for automated curve fitting is the method of least squares. An example of a least squares fit of a straight line appears in Fig. 2A. Least-squares techniques make three major assumptions: (1) that the values along the horizontal axis (the x values) are known exactly; (2) that the formula selected for fitting truly describes the data, and; (3) that the data are typical; no extraordinary errors exist in any of the points (that is, there are no outliers). Additionally, it is helpful if the errors in the vertical (y) direction may be described by the normal distribution: this allows the application of parametric statistical tests connected with least-squares methods and permits an interpretation of confidence limits for assay results based on probability, to be discussed in a subsequent section.

The vertical distance between a point and the fitted curve is known as the residual at that point. The least-squares method fits a curve by minimizing the sum of the squares of all of the residuals. If the equation to be fitted is linear in its parameters, as is a polynomial, then the location of the curve can be quickly found by solving a set of simultaneous algebraic equations. This procedure is known as linear least-squares analysis. If the formula is nonlinear in its parameters, such as the four-parameter logistic, then an iterative method must be used, which is slower and technically more complex; this is known as nonlinear least-squares analysis.

Data Analysis and Quality Control   271

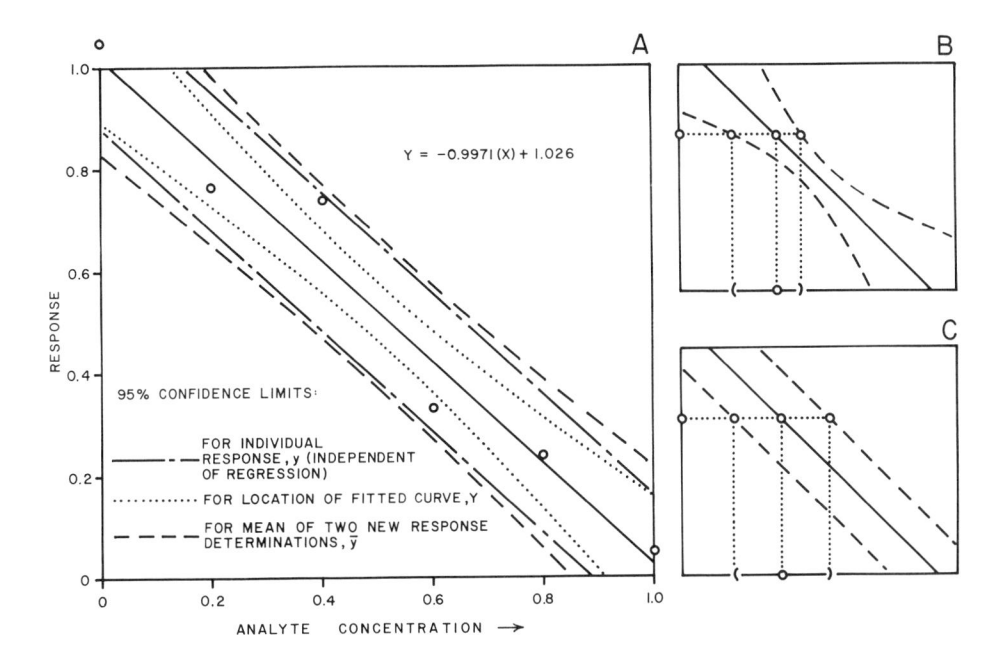

FIGURE 2 (A) A graphic demonstration of the various important errors associated with a calibration line. The magnitude of the random error in the y position of points is purposely exaggerated for purposes of clarity. A rigorous method for determining analyte concentration confidence intervals must account both for uncertainty in the location of the calibration curve and the added uncertainty in the measured response of an unknown. (B) The limits to be used for rigorous computation of the confidence interval are curved inward toward the center of the calibration line. Except at the center of the line, the resulting limits will be asymmetrical about the analyte concentration estimate. (C) In the frequently used slope-error approximation method, the error in the response y is assumed to be constant in the region of the curve where the unknown response value lies, and errors in the uncertainty of the calibration line are ignored. The resulting confidence limits are symmetrical and the interval is slightly smaller than the actual interval.

The number of parameters is important in determining how flexible a given equation will be. Thus a four-parameter polynomial will be able to assume a wider variety of shapes than a two-parameter polynomial (a straight line). It is often said jokingly that a four-parameter model can be used to describe an elephant, and that with a fifth parameter the elephant can be made to wag its tail. Excessive generalization on this point is misleading, as not all n-parameter equations are equivalent. Just as important as the number of parameters is the way in which the parameters appear in the formula. Thus a four-parameter polynomial, which can show wild oscillations, is not equivalent to a four-parameter logistic, which always assumes a well-defined symmetrical sigmoid shape.

There are two basic problems in fitting an equation to points containing error. The equation itself may not accurately describe the data, producing a bias in the fit. Equally, the location of the fitted curve may be inaccurate due to the random error present in the particular set of points being used. How can the bias and imprecision of the fit be determined?

The problem is most easily approached if the data can be fitted by linear least squares. Then the powerful method known as analysis of variance, or ANOVA, can be brought to bear on the results. Analysis of variance allows the variations in a set of results to be partitioned into separate components.

Consider the straight line fitted in Fig. 2A, the data for which appear in Table 2. These synthetic data were produced using the equation for a straight line of the form y = (-1) x + 1.0; this line has a slope of -1 and y intercept of 1. Six x values were selected at which to create simulated experimental y values. The effects of random error were simulated by means of a pseudorandom number generator. It was assumed that each of the y values was drawn randomly from a normal population with a standard deviation of 0.1 (this level of irreproducibility is rather exaggerated for the purpose of greater clarity in the figure). Each plotted point of Fig.

TABLE 2 Calibration Data Used for Figure 2[a]

Standard no.	Analyte conc. (x)	y_1	y_2	$\bar{y}$	y_{true}	$y_{predicted}$
1	0.0	1.049	1.048	1.048	1.0	1.026
2	0.2	0.7057	0.8248	0.7652	0.8	0.8265
3	0.4	0.8308	0.6486	0.7397	0.6	0.6271
4	0.6	0.3559	0.2855	0.3207	0.4	0.4277
5	0.8	0.3347	0.1436	0.2392	0.2	0.2283
6	1.0	0.03734	0.06567	0.05151	0.0	0.02885

[a]These data are plotted as a calibration curve in Fig. 2. The individual response readings (y_1 and y_2) are shown, together with their mean ($\bar{y}$), which is plotted. The value of y_{true} corresponds to the equation used to generate these simulated data, and the value of y_{pred} is that obtained from the equation for the fitted regression line. If there were no random error present in the y values, the values of y_{pred} would equal those of y_{true}.

2A is the result of averaging two different simulated experimental readings. Linear least-squares analysis yielded a fitted line of the form y = (-0.9971) x + (1.026). ANOVA was also applied to the data, the results appearing in Table 3. The actual calculations involved in this one-way analysis of variance can be found in Colquhoun (1971), Natrella (1963), or Rodgers (1981) and will not be repeated here. What is important to understand is that any linear regression problem can be regarded as an ANOVA problem, and that this method can help evaluate the quality of the fit of the equation.

The grand mean for all the y values of Table 2 is about 0.5. The deviation of all the individual y values from this mean value can be attributed to three sources: (1) y varies as a function of x (that is, the slope of the line is nonzero); 2) there is random error present in all of the y values, and; 3) the underlying structure of the data, were it free from random error, is perhaps not a straight line. Under certain assumptions about the nature of these various deviations, ANOVA allows estimation of their relative importance.

The SSD (sum of squared deviations) values of Table 3 are used to derive MS (mean square values) which are equivalent (in this instance, although not generally) to variance values. The total variance in the system can be broken into two main components: within-replicate error and between-replicate error. If we assume that the random error in y is constant for all values of x, the within-replicate MS is an estimate of the variance in the y values. The square root of this value is 0.086, which is a fairly good approximation to the known standard deviation of 0.1. (It would tend to be even closer as we increased the amount of data.)

TABLE 3 ANOVA Table for Regression of Figure 2

Source	df	SSD	MS	F	P	Interpretation
Due to linear regression	1	1.393	1.393	186.3	<<0.001	Slope of line > zero
Deviations from linearity	4	0.05804	0.01451	1.942	>0.10	Data may not follow line
Total between replicates	5	1.451	0.2902	38.84	<<0.001	
Within replicates	6	0.04483	0.007472			Replicate s.d. approximately 0.086
Total	11	1.496				

The second major component of error, that due to between-replicate variations, can be further subdivided into components due to linear regression* and due to deviations from the fitted line. The portion due to linear regression is large and merely reflects that y really does change with x. The component due to deviations from linearity serves as a form of check on the validity of the formula selected, in this instance, a straight line.

Further statistical interpretation of the results is made by taking the ratios of the MS values for deviations from linearity and for linear regression to the MS value for within-replicate error. Then F tests may be performed to evaluate the significance of the ratios. This procedure yielded a p value of 0.001 for the error due to linear regression. This indicates that if the assumptions underlying ANOVA are correct, such a result would be expected from random chance only one in a thousand times. This strongly suggests that the value of y actually changes as x changes (that is to say, the slope of the line is not zero). The probability value for the component of error due to deviations from linearity (0.10) is quite large and such a result might be expected to occur by chance in 1:10 experiments. This single set of data does not allow us to draw strong conclusions regarding the appropriateness of the straight line model for the system under study. This is due to the large replication error present. Refinements in the technique to reduce this error, or a large increase in the number of replicates used would remedy this problem. The replication errors in working assay systems are not as bad as in this example, so that the appropriateness of a model for calibration curve-fitting is not generally in such doubt; this example is offered simply to demonstrate that ANOVA does afford one method of assessing the appropriateness of a given calibration formula in a statistically formal manner. In routine assays, there are rarely sufficient data available to allow ANOVA to be useful for calibration. However, the method is often applied to the analysis of the results obtained from quality control samples, so it is important to understand its basic principles.

Analysis of variance techniques have been developed for linear models. If the function to be fitted to the calibration data is nonlinear, as is often the case, ANOVA is no longer strictly applicable, although it is possible and on occasion useful to perform an approximate analysis of variance for a nonlinear model. Nonparametric tests of quality of fit are more appropriate but are not very powerful in this context due to the limited number of calibration standards and low level of replication of most assays. The quality of fit may be studied by plotting the residuals as a function of x.

---

*Note that the sum of squared deviations and MS due to linear regression do not measure a true error, but rather the change in y as a function of x. The deviations of all y values from the grand mean y value are due in part to this association and in part to deviations from the straight line model (a sort of bias error) and random deviations about the straight line.

These should vary randomly about zero, and if a pattern is detected in repeated fits to numerous independent sets of calibration data, the equation is probably inappropriate. A number of nonparametric statistical tests may also be useful in analysis of the residuals. One rather easy test examines the number of times the sign of the residual changes.

The fitting process is complicated by several additional factors. The random error in the y values tends to vary as a function of y (and hence as a function of x) rather than being constant. This is known as heteroscedasticity, and requires the use of weighted least-squares regression methods (Fischer, 1983; Rodbard, 1983).

The weights used are inversely proportional to the variance in the y values, so that points with less precision play a smaller role in determining the location of the fitted curve. The problem of determining these weights is discussed in Sec. VI.B. Residuals should also be adjusted by these weights.

Another common problem is the outlier. This term is used in several ways. For instance, suppose that a calibration standard is analyzed in triplicate, and that one of the three responses is far removed from the remaining two. This single value is possibly affected by some rogue error. Suppose that the three values agree well, but that the average, once plotted with the rest of the calibration data, stands well away from where the calibration curve seems to run. This set of replicates might also be an outlier. If there is some obvious explanation for the extra error, it is justifiable to reject the outlying data. More commonly, a single value or average will not be clearly erroneous, but the assayist may be tempted to reject it all the same. Cavalier censoring of the data is dangerous and may mask underlying problems in the assay method.

The rejection of outliers is an old problem for statisticians, but nevertheless has never been satisfactorily resolved (Barnett and Lewis, 1978). One of the early attempts to develop an outlier rejection procedure has not proved useful in assay practice (Dixon, 1953). The subdiscipline known as robust statistics may hold some of the answers. These methods assume that although the data may generally behave according to some well-defined distribution, such as the normal distribution, they may be contaminated by a certain number of very erroneous values. Various methods have been evolved to detect and reject such contaminants (Healy, 1979), and one such method has been developed for use in an automated data processing program (Healy, 1972). A comparison of automated outlier rejection methods found Healy's approach to be the best of those tested (Wilkins and Chadney, 1981; Hawker and Challand, 1981). This method, requiring the use of duplicates throughout the assay, was subsequently generalized for higher levels of replication (Healy and Kimber, 1982). These results must be regarded with caution, and further development of robust methods encouraged. Robust regression methods have also been put forward. These fit a calibration curve which is relatively unaffected by outlying points. One such method is that of Tiede and Pagano (1979), which has been criticized on several counts by Raab (1981b).

The final implementation of any outlier rejection scheme must not bypass the judgment of the assayist. Outliers should be flagged and the assayist's own judgment brought into play. The presence of a larger than usual number of outliers may be a signal of a loss of quality control. Although robust statistical methods will undoubtedly find a place in assay data processing, their implementation must not mask quality control problems in the assay.

V. INTERLUDE 2: WHICH OF THESE CURVE-FITTING METHODS IS THE BEST?

Figure 3 presents a set of points to which three different curves have been fitted: a linear least-squares line, a nonlinear least-squares fit of a four-parameter logistic, and a manually drawn curve. Which is the best curve fit?

I hope that the reader will refuse to answer this question without additional information. It is hopelessly vague, but it serves to point out a number of misconceptions associated with calibration curve-fitting.

In the lack of any information about the random error inherent in the plotted points, no meaningful distinction can be drawn between the three methods of Fig. 3. If the points are known to be extremely precise, then clearly method A is inadequate: the data clearly do not conform to a straight line. Using a straight line will produce a bias in assay results due to lack of fit.

Very little can be learned from a single set of data. A large number of sets is required to tell if deviations between the plotted points are due to random error inherent in the data (in which case the selected curve-fitting equation may be perfectly adequate) or is due to bias inherent in the

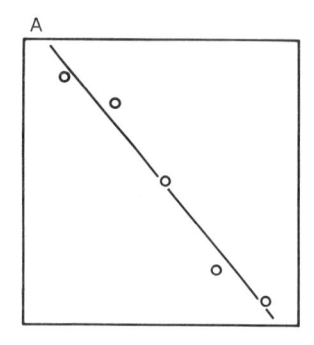

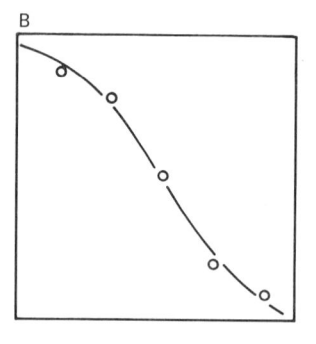

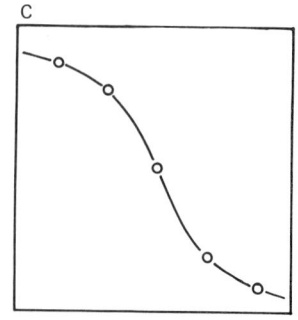

FIGURE 3 (A) A straight line is fit by linear least squares to data from a hypothetical assay. (B) A four-parameter logistic curve is fit by nonlinear least squares to the same points. (C) A manual curve is drawn through the same points, so as to pass through each of the points. Which of the methods is preferable?

model (in which case another formula should be tested). Even if a formula fits routine assay data very well, an occasional set of data may not appear to fit due to the presence of outliers.

Some assayists endow the straight line with almost magical properties. If the central portion of the data looks linear, they will prefer to fit a straight line to this portion and reject all of the data outside and central region. Alternatively, a linearizing transform such as the logit-log is applied and a straight line drawn through the results. Another possible strategy (less often encountered) is to use some unusual transform of the x values to yield a straight line. Each of these approaches poses problems: the first involves an unnecessary loss of information, the second requires the use of complex weighting schemes to overcome the severe heteroscedasticity of response produced by a logit transformation, and the third introduces formidable statistical problems into the fitting process and subsequent use of the calibration curve. With the ready availability of good nonlinear regression techniques for assay, there would seem to be no justification for following some other procedure simply because it yields a straight-line calibration curve.

Perhaps the ultimate tribute to the psychological potency of the straight line appears in a paper (Shaw et al., 1977) which advises assayists to modify the chemical design of their assays so as to yield results that can be fitted to a straight line transform. (An identical conceptual error is promulgated by Halfman (1979), who recommends modifying assay chemistry so as to obtain calibration data that follow a hyperbolic relationship.) Such suggestions stand good practice on its head: the goal of a good data-processing method should be to maximize the quality of analysis regardless of calibration curve shape. It should adapt to the optimal analytical design of the assay. Modifying the assay chemistry to suit the data processing technique could entail a great sacrifice in assay quality.

It is commonly believed that the definitive method for fitting a calibration curve is the manual method. This is a dangerously false notion. Although an experienced assayist will be able to produce results roughly comparable to those from a good automated package, manual calibration and interpolation are by no means definitive procedures. Different assayists will draw slightly different calibration curves, and even the same assayist will produce a slightly different result every time he or she tries. There are no clear rules for how to draw a curve, and it is almost impossible to provide properly for weighting. Often assayists tend to favor the end points of the curve more than they should (more information is contained in the center of the data), and excessive reliance on the actual location of plotted points tends to turn one form of assay error (imprecision) into another (bias in fit). The high variability in results attributable to manual data reduction has been demonstrated by Jeffcoate and Das (1977) and Pegg and Miner (1982).

This problem of the confounding of random and bias errors in calibration poses a dilemma in the comparison of different curve-fitting methods. It is difficult for some to understand why a curve that passes through no points directly may be superior to one that passes through all the data points exactly. If one examines the residuals of the plotted points, methods such as manual curve-fits, splines, and polygonal interpolation will always appear to be superior (Herndl and Marschner, 1975; Schöneshöfer, 1975, 1977; Sinterhauf et al., 1975, 1976). This is an innately biased technique for comparing regression equations. It is essential to overcome this problem by using intermediate standards for such method comparisons. The equations to be compared are fitted using the conventional standards, but residuals are also plotted for additional standard concentrations falling in between these ordinary standards. When such a procedure is followed, the manual and related methods do not fare very well (Sandel and Vogt, 1978; Vogt et al., 1978; Wilkins et al., 1978). They may pass through the usual standards exactly, but poorly predict the location of the intermediate standards. In general, a properly implemented automated fit will always be superior in a statistical sense.

Some workers seem to assume that attention is better paid to the chemistry of an assay than to its data processing. However, the studies of Jeffcoate and Das (1977) and Pegg and Miner (1982) show that differences in data processing techniques account for a significant portion of between-assay variability. Pegg and Miner also demonstrated that different implementations of the same calibration formulas (the logit-log technique) gave significantly different results. This suggests that greater attention should be paid to the standardization of assay software.

VI. ERRORS AND ASSAYS

A. Imprecision

If an unknown is analyzed in the same assay many times, each time a slightly different result will be obtained, due to random error. Generally the values will tend to cluster about some central value. The statistical estimation of this tendency to scatter estimates the degree of reproducibility of the assay technique. The best estimate of the actual answer is taken to be the average of all the replicate determinations, and as the number of replicates used is increased, the average can be made to approach the true value (in the absence of systematic errors) arbitrarily closely. In practice, it is uneconomical to process more than a few replicates, so it is essential to estimate the extent to which the result may be influenced by random error.

The degree of irreproducibility in a result is most easily represented by a set of confidence limits. These fall to either side of the assay measurement and define a range known as the confidence interval. Such an interval is always associated with some stated degree of confidence. For instance, 95% limits define a range within which the sought after true value would be contained on 95 out of 100 occasions, were the measurement and the associated computation of confidence limits to be repeated many times. Note that the confidence limits only apply to random errors. The confidence interval could be quite small yet the result still inaccurate due to bias.

The ultimate goal of an assay is to obtain an analyte concentration estimate with its associated confidence limits. To understand how these confidence limits are calculated, it is necessary to understand the various sorts of uncertainty associated with calibration and interpolation. There are three quite distinct types of error which are important. In discussing them, and in discussing analysis of variance methods in Sec. IV. I am heavily indebted to the treatment given these subjects by Colquhoun (1971). To illustrate these points we shall resort to the simple example of Fig. 2.

First, there is an uncertainty associated with each of the response determinations made for the calibration standards, which we shall call the random error in y. We assume here that the random errors follow a gaussian distribution, and that the uncertainty in y can be represented by the standard deviation of y. Further, we will assume that the random error in y is constant for all values of x. The plotted points in Fig. 2 were the result of taking the mean, $\bar{y}$, of two independent determinations. Therefore, in fact, the vertical position of the points has an uncertainty described by the standard deviation of the mean, which will be equal to the standard deviation in y divided by the square root of the number of replicates (two in this instance).

A second important type of error is that associated with uncertainty in the location of the calibration line. In Fig. 2A, this zone of uncertainty is represented by two dotted curves that pinch in toward the center of the fitted line. These curves lie inside the confidence intervals for the $\bar{y}$ values because they reflect the added information gained by use of all of the points as opposed to just a single replicate pair. They narrow toward the center of the line because this is the zone of maximum information content about the line's location (points in the center of the line have neighbors that add information about their location, whereas points at the extremes of the data do not). This zone of uncertainty takes into account the error in y, the level of replication used to obtain means used for the plotted points, and the number and concentrations of calibration standards.

A third type of random error associated with the assay process expresses the uncertainty in the location of the point on the calibration curve corresponding to a new mean response obtained from a given number of replicated response determinations. For the example in Fig. 2A, it is assumed that duplicates are averaged to obtain a new point, and the appropriate confidence limits are presented as dotted and dashed curves. These limits reflect the uncertainty in both the location of the calibration line and the location of the new mean response, and are wider than the two previous error zones we have discussed. These limits should be employed for estimating confidence intervals for analyte concentration estimates.

The process of interpolation and confidence interval estimation is represented graphically in Fig. 2B, where a horizontal line is drawn from the mean response obtained for an unknown to the calibration curve and the appropriate confidence limits. From these three points of intersection, vertical lines are extended to the horizontal axis, defining the analyte concentration estimate and its upper and lower confidence limits. Note that because the confidence limits about the calibration line are curved, the confidence limits for the analyte concentration are not symmetrically disposed, a tendency which worsens at either extreme of the analyte concentration range.

It is impractical to perform this procedure graphically, but the exact mathematical equivalent is well known in statistics as Fieller's theorem. Although Fieller's theorem presents a rigorous approach to the calculation of confidence limits, most assay data processing packages use a simpler approximate method. Instead of the confidence limits for the location of a new mean about the calibration curve, the limits for individual measurements are used (properly adjusted for the level of replication by use of the t-distribution). These limits run parallel to the calibration line, and the interpolation process in this case is as in Fig. 3C. The confidence limits so obtained are symmetrical about the analyte concentration estimate. These limits may sometimes be optimistically small, as they neglect the error in the location of the calibration line: its position is assumed to be known exactly. The width of the confidence interval in this method is directly proportional to the standard deviation of y and inversely proportional to the slope of the calibration curve. We will call this the slope-error method of computing confidence limits.

This simple example of two basic approaches to the computation of confidence limits has made many assumptions which are not met by most assays. The calibration relationship is rarely a straight line, and the error in y usually varies as a function of x. The x axis is often plotted on a logarithmic scale, further contributing to confidence limit assymetry.

The only data processing package using Fieller's theorem is that of Finney (1976, 1979), which he presents as more of an educational tool than a program designed for routine use. Because of the nonlinearity of the four-parameter logistic equation it uses for calibration, Finney's application of Fieller's theorem is only an approximation. Its use assumes that, locally at least, the calibration curve and its associated confidence limits behave as for the case of a linear calibration line.

The vast majority of other data processing packages use the slope-error technique, which will generally be a poorer approximation than Finney's approach, neglecting as it does the uncertainty in the location of the fitted calibration relationship.

A third approach is employed in the package of Raab, McKenzie, and Thompson (Raab, 1978; Raab and McKenzie, 1982; McKenzie and Thompson, 1982). Details of this method have yet to be published, but it is apparently more general than Finney's approach. It is easily extended to the calculation of limits when an unknown is analyzed at several different dilutions, and easily applied to various calibration equations (with Fieller's theorem, formulas for the confidence limits must be rederived for each type of calibration equation used).

It is important to remember that all of the published methods are only approximate. Even the exact methods would still be dependent upon the assumption that the calibration equation selected is truly representative of the shape of the calibration data. All methods have difficulty near the extremes of the calibration curve where confidence limits expand explosively. This is rarely a problem as results from these portions of the curve generally lie outside the valid analytical range and must be regarded with caution in any event. Also note that the foregoing calculations are concerned solely with within-assay confidence limits. These are useful for comparison of results obtained within the same analytical run, but if results from different assay runs or different laboratories are to be compared, the (wider) between-assay or between-laboratory confidence intervals must be used. They are calculated in a similar fashion, using error information garnered from quality control specimens analyzed in different assay runs.

Two of the important tools for studying random error are the response-error relationship (RER) or response-variance relationship (RVR), and the imprecision profile (IP). The response-error relationship is a plot of the standard deviation of the response vs. the response. Due to the low level of replication used in most assays, RERs and IPs constructed from a single assay run will not be as useful as one constructed by pooling the results of several assay runs. Some workers pool the results of as many as 30 consecutive runs. This assumes that the assay is stable with respect to random error. Appropriate tests must be done before adding data from the current assay to the pool to ascertain that there is no evidence of the contrary. Baxter (1980) recommends using just spot quality control results, but this seems to entail an unnecessary loss of information. If appropriate tests are done to determine that no differences exist between the imprecision of standards, unknowns, and quality control results, all of these values may be used (Dudley, 1981). Although prepared for differences between these groups, Raab and McKenzie (1982) found that in practice it was always possible to pool the results: Raab (1982) subsequently found exceptions to this pattern. Rodbard and co-workers (1976) offer sound advice on the proper methods for pooling results into bins.

A straight-line or second-order polynomial may generally be fitted through the results to describe how the error in response is related to the response. Finney prefers to plot the variance in the response rather than the standard deviation, and uses a power function to describe the results. The features of his method are discussed by Feldman (1983) and the method was modified by Raab et al. (1980). Rodbard made a comparative study of five different models for fitting variance vs. response (Rodbard et al., 1976). Robust statistics may also be of use in estimating the variance function properly (Healy and Kimber, 1982). A relatively recent technique, developed for use in pharmacokinetic research and christened "extended least squares" by its developers, may eventually come into use in assay. It employs a maximum likelihood approach to curve-fitting, a method more general than least squares, and its use would be tantamount to attempting to fit the imprecision profile and the calibration curve simultaneously using the same data (Sheiner, 1981).

All authors using imprecision profile methods are quick to stress that the results are ultimately rather insensitive to the form of the curve used to fit the response-error or response-variance relationship; all that seems necessary is a roughly accurate description of the functional relationship between error and response (Duddleson et al., 1973, Finney and Phillips, 1977). The response-error relationship may be used in three ways: (1) to weight the automated fit of the calibration curve; (2) to monitor the stability of the assay with respect to random error, and (3) to calculate the confidence intervals using the slope-error approximation. In this latter instance, an accurate determination of the correct form of the RER is presumably more important. (For examples, see Ekins et al., 1978; Ichihara et al., 1977; McDonagh et al., 1977; Pilo and Zucchelli, 1975; Rodbard, 1974, 1978b; Rodbard et al., 1976, 1978; Rodgers, 1981; Wilson et al., 1971). When the RER is to be used as a quality control indicator for monitoring assay random error, it is helpful to have a statistical test which indicates when a shift in an RER is statistically significant; that is to say, when the shift is unlikely to be due to random chance alone, and may represent a genuine change in assay performance. Such tests have not been extensively developed and the only published program offering such a test (based in this instance upon Chi-square statistics) is the calculator program of Dudley (1981).

The imprecision profile is a plot of the width of the confidence intervals about an analyte concentration estimate, plotted as a function of analyte concentration, at a stated level of statistical confidence. These may be plotted to reflect within-assay, between-assay, or between-laboratory errors. Often these are plotted using the slope-error approximation. Ekins (1978) prefers the term "precision profile" for these plots, although this is less consistent with the more formal terminology used by the International Federation of Clinical Chemists (IFCC) (Büttner et al., 1975, 1976) and workers wishing to be consistent with that terminology (Rodgers, 1981; Dudley, 1981).

Imprecision profiles are a compact graphic summary of the impact of random error on an assay, and facilitate quality control monitoring and the comparison of assays. When methods are to be compared, the profiles should be drawn at the same level of confidence. If only the analytical designs (see below) are to be compared, corrections must be made if different levels of replication were used.

Gross changes in a profile may be detected visually, but formal tests to compare imprecision profiles have not been developed. Note that comparison of profiles may be hazardous if different computation methods have been employed. Use of the slope-error technique, for example, disregards the statistical design of the assay (the number and placement of calibrators) whereas the Feiller's theorem approach does not.

It is also simple to estimate the lower detection limit of an assay from its imprecision profile (it is the largest analyte concentration containing zero within its confidence limits). An analogous quantity, the upper detection limit, may be defined as the smallest concentration that contains the highest calibration standard concentration within its confidence limits. The range between these values is known as the valid analytical range of the assay. Rodbard (1978c) has offered a simple formula for determining the lower detection limit (LDL) (which he refers to as the MDC, or minimum detectable concentration) which is equivalent to using the imprecision profile generated by use of the slope-error approximation. This calculation is of general value in analytical chemistry and has been considered in some detail by Oppenheimer and co-workers (1983). They consider various forms of the RER and various aspects of assay design in computing what they refer to as the lowest limit of reliable assay measurement (LLORAM).

B. Bias (Systematic) Error

Even if all random error could be eliminated from an assay, a given result might still not be accurate (that is, close to its corresponding true value) due to bias. Systematic errors are in principle, but almost never in practice, correctable errors. For instance, if the calibration standards were inadvertently made up at one-half their specified concentration, the resulting analyte concentration estimates would be low by a factor of two. The difficulty with bias is that there are very many potential sources, and each likely source must be studied separately.

A central assumption of binder-ligand assay is that the standards and unknowns behave identically, and that the only entity affecting the response is the analyte concentration. Many errors arise because of invalidity of this assumption. A classic method for assessing non-identity of standards and unknowns is known as parellelism testing (also: validity testing, similarity testing). Parallelism testing was initially developed in the context of biological assay (Finney, 1977, 1978). It requires making several different dilutions of an unknown and analyzing them all in the assay. A single analyte concentration estimate with corresponding confidence limits can be derived from the results, but also a curve can be created using the individual values obtained at different dilutions. This curve should be superimposable upon the calibration curve by means of a simple multiplicative rescaling of the x axis; if it is not, then there is some sort of difference between standards and unknowns (Raab, 1982).

Major sources of bias include: (1) cross-reaction, in which a ligand other than the analyte reacts with the assay binder, and (2) interference. This latter term is used in several ways. In binder-ligand assay, it can refer to the influence of a substance other than the primary binder that reacts with ligand. It can also refer to the influence of substances that can modify the final assay measurement without participating in the binder-ligand reaction, as for example a colored substance might interfere with use of a colorimeter. Generally, the magnitude of these bias errors varies as a function of both analyte concentration and the concentration of the offending agent and it is misleading to cite a single figure to describe the impact of such errors. Often it is helpful to create bias profiles which plot the magnitude of bias as a function of the appropriate concentrations.

Assay bias can also change temporally, an event known as drift. This can be ascertained by placing several identical groups of spot quality control samples at different places in the assay stream, or (more powerfully) by placing several sets of calibration standards at several points in the assay (Raab, 1982).

There seems to be a general consensus about rough guidelines for the use of spot quality control samples, which is well-summarized in the review by Ayers et al. (1981). They recommend that spot quality control samples be available in at least three concentrations, spanning the valid range of the assay, and should be placed at multiple points in the assay. The results obtained for these specimens can be charted to assess the stability of the assay (Westgard et al., 1977; Woodward and Goldsmith, 1964; Faure et al., 1980a). Control charts require the development of decision rules to indicate when a result should alarm the assayist (Groth et al., 1981). Edwards (1983) comments on the danger of using too few quality control pool concentrations.

Quality control samples also provide helpful information concerning imprecision, by contributing to the pooled RER. In addition, analysis of variance techniques can use quality pool results to determine between-assay or between-laboratory imprecision (Munson and Rodbard, 1978; Russell et al., 1974). Youden and Steiner (1975) developed an attractively simple method for external quality control based upon analysis of paired quality control samples.

The foregoing remarks pertain primarily to monitoring the performance of established assay procedures. When validating a new assay, rather exhaustive studies may be required to establish the precision and accuracy of the method prior to its being adapted as a routine procedure (Doetsch, 1983). Some of these additional studies may be required later as well to establish the cause of decayed assay performance.

VII. INTERLUDE 3: WHICH OF THESE ASSAYS IS SUPERIOR?

Figure 4 displays the calibration curves from two different assay systems which use the same response metameter and are designed to detect the same analyte. The question is: which of these assays is better? Try to formulate a response before reading any further.

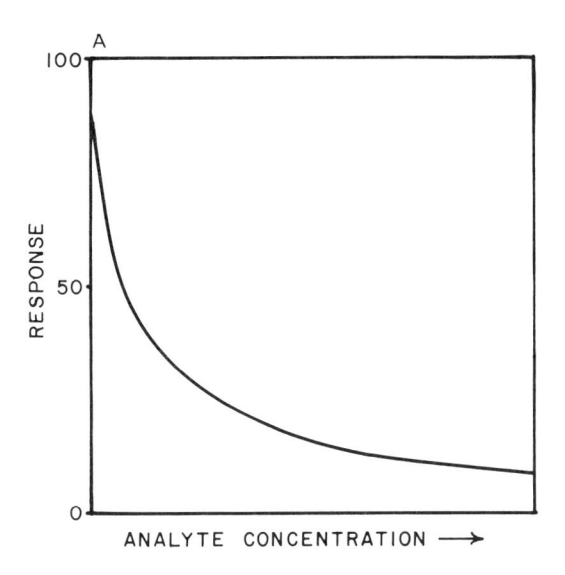

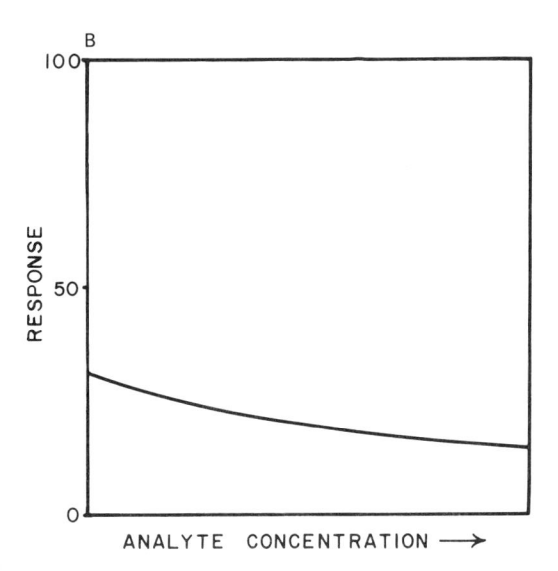

FIGURE 4 Two calibration curves, each from a different assay. The scales of the plots are identical, and the assays are designed to detect the same analyte. Which assay is better?

The appropriate response is neither "A" nor "B" but a low grumbling noise, for this is a very poorly phrased question, and should arouse anger in anyone with a serious interest in assay analytical design. Nothing whatsoever can be said about comparative assay quality by the mere examination of two calibration curves. All too often, the response given to our initial question is: "Assay A is the better, because the calibration curve slope is higher, and therefore the sensitivity is better." The confusion between assay sensitivity and calibration curve slope is the grandfather of all assay design misconceptions. It is another example of the confusion of means and ends in assay work. The calibration curve is only the means to the end; the end is the analyte concentration measurement. It is an easy exercise to replot the curves of Fig. 4 using a different response metameter, and to obtain results in which the assay curve slope for method B is higher. The explanation for this apparent paradox is that the error in the response is also transformed according to the response metameter used. The precision of the final analyte concentration estimate, properly calculated, should be the same regardless of the metameter employed for the calibration curve. This is not to say that the metameter selection is not important, for the difficulty and appropriateness of the calculations (particularly the statistical computations) will vary according to the selection made. Calculations are greatly facilitated if the error in the metameter selected follows a gaussian distribution.

In accordance with the recommended terminology of the IFCC, we should avoid abuse of the vague qualitative terms sensitivity and precision (Büttner et al., 1975, 1976). Assays should be compared only according to concretely defined mathematical indices. For example, the assays A and B could be evaluated by comparison of their lower detection limits: the lower detection limit (LDL) is a clearly defined mathematical quantity which corresponds to the qualitative term sensitivity (Rodgers, 1981). An assay can be said to be more sensitive than another if its LDL is lower than the LDL of the assay it is being compared to. It is entirely possible that the LDL could be lower (better) in assay B than in A, as calibration curve slope is only one component of assay precision and sensitivity. The other is the response error and the statistical level of confidence chosen to express the LDL.

Even if assay A were more sensitive than B, it still might not be a better assay. This brings up the second objection to our initial question of which method is the better: it is only appropriate to ask whether assay A is superior to assay B with respect to a specifically designated criterion. For certain applications, method A may be better while, in other instances, method B might be preferable. Even though assay A might have a small MDC, the precision of the determination at some higher analyte concentration might be poorer than in method B. Assay B might be less precise for all concentrations of analyte, yet be less prone to common sources of bias error, such as cross-reaction with a ligand which is similar to the analyte.

It might also be faster, less expensive, or more rugged (resistant to changing conditions) than method A.

VIII. ASSAY DESIGN

Assay design is concerned with developing rules for the performance of assays that optimize some set of desired performance criteria, such as imprecision at one or more analyte concentrations. The ultimate goal is to be able to design an assay to suit a particular application. The problem may be divided into two largely independent parts: analytical design and statistical design (Rodgers et al., 1983).

Statistical design is concerned with determining the optimal number of standards, concentration of standards, and the level of replication of standards and unknowns. From the point of view of precision, a large number of standards and replicates is desirable, but this is clearly not reasonable. Optimization of statistical design must account for economic and temporal constraints as well as performance requirements, and none of the published work on this topic has addressed the problem squarely.

A few elementary guidelines may be set out: standards and unknowns should be analyzed in at least duplicate to allow for at least some chance of outlier detection. One standard must be at zero analyte concentration, and another at the upper limit of the concentration range which is to be studied. The spacing of the other standards will be influenced by the imprecision profile, being more closely spaced where it is desirable to minimize imprecision for the application at hand. That is, placement of the calibration standards allows the assayist to tailor the imprecision profile to his or her particular analytical problem. Formal statistical tests which would allow the imprecision profile to be used as a guide in this context remain undeveloped. Also, some types of computational approximations used for generating the profiles, such as the slope-error method, ignore the impact of statistical design on final assay imprecision and are useless in this regard. The attempts of Dudley (1981) to apply chi-square statistical techniques to the RER are an attempt to determine when an observed change in the RER (but not necessarily the IP) is genuine, and not merely due to the fluctuations of random error.

Meinert and McHugh (1968) proposed, but apparently never published, a Monte Carlo study of the theory of statistical design of assays. In experimental practice, the process of adjusting the statistical design of an assay can be viewed as an optimization procedure. Optimization requires three steps. First, the assayist must define clear optimization goals (such as the minimization of imprecision at a given analyte concentration level). Second, the optimization variables must be listed; these are the aspects of the assay to be modified in the process of optimization (in the case of statistical design, the number, concentrations, and level of replication of the calibration standards). Finally, an optimization strategy must be defined. This is a set of rules which specifies how the assay design parameters are to be varied. For each set of parameter values, one or several assays are run, and the value of the index corresponding to the optimization goal is determined. This value is fed back into the optimization strategy and another set of parameter values obtained. This process continues until the optimization rules indicate that the index of optimization has achieved its minimum (or maximum) value. Numerous optimization strategies have been developed for use in clinical chemistry (Massart et al., 1978).

The problem of analytical design may be subdivided. First, it is sometimes of interest to design an assay so as to minimize the impact of some potential form of bias. This problem has not been adequately studied.

Second, it is also of interest to know how to design the assay so as to minimize the imprecision at one or more analyte concentrations. This requires an accurate mathematical model for the chemistry of the assay as a function of all the important variables such as concentrations of reagents and reaction times. Also required is a model for the way in which the numerous sources of random error, such as pipetting errors, timing errors, and end-device measurement errors (such as counting error in radioassays) combine to form the final random error in the response (Rodbard and Cooper, 1970).

Most of the energy devoted to assay design problems has been directed at chemical models for the assay process. For some of the simplest of models, see Rodgers (1976); several of the more complex models are reviewed by Feldman (1983). Here the interests of the assayist converge with those of several other exciting disciplines, including the study of the interaction of hormones and drugs with cell surface receptors (Boeynaems and Dumont, 1980; Rodbard et al., 1980).

Several radioassay data processing packages have included "optimizers" which attempt to advise the assayist on how to change the concentrations of label and binder so as to minimize imprecision at some given analyte concentration (Rodbard et al., 1978). These have all been based upon extremely simple chemical models, and make very crude assumptions about assay error (often assuming that the response-error relationship remains fixed). None of these packages has been verified, and their use must be regarded as strictly experimental. Also note that many papers make loose use of the term "optimization" in their titles when in fact they are concerned with tangential issues such as curve-fitting or maximizing calibration curve slope, a fallacious approach to analytical design optimization (Schöneshöfer, 1977; Smigel and Lazar, 1981). "Optimization" is just another in the long list of "key words" which are often used misleadingly in the literature (Morrison, 1983).

A number of experimental strategies for optimization of assay design have been successfully applied to assays, notably those based on Simplex or response-surface methods (Massart et al., 1978, Rautela et al.

1979). When optimizing imprecision, the imprecision profile may be used as a guide (Ekins et al., 1978).

Some remarks about the modeling of assay chemistry are in order. Some experimentalists approach it naively, do it poorly, and then reject it entirely when the results are not helpful. Skillful modeling is as difficult as doing good experimental work, and should not be approached lightly. Even a very simple chemical system will present sources of artifact which must be accounted for if a model is to fit the data and predict assay behavior. Experience, physical insight, and patience are required, as well as a continuous dialogue between the modeler and the experimentalist. For a good example of the application of a multiligand, multibinder model to actual experimental systems, see the article by Munson and Rodbard (1980).

It is vital to remember that the model, even if it fits the data well, can not be used to prove anything: the underlying chemical assumptions it makes might be false. Several models with very different underlying assumptions might fit the data equally well. At its best, a model provides a framework for stimulating give-and-take between theory and experiment. Only when the chemistry of a system is thoroughly understood can the parameters of a model be imbued with physical significance, and even then these numbers must be interpreted with caution due to their random errors.

The methods and pitfalls of modeling are very well summarized by McIntosh and McIntosh (1980), who capture the tentative and exploratory spirit of modeling very well when they write:

> A model is a structure of play with — and in so doing to discover new behavioral possibilities of the system. A model is not definitive. It works out the consequences of proposed solutions to the perceived problem, in order to determine how these results compare with experimental observations. A model should be no more than a guide to constructive and efficient experimentation. It should lead to better models. It should grow, transform, or become obsolete when tested, and yet have served its purpose well. It should never be finished or ready for publication in the "quod erat demonstrandum" form. Models are necessarily simplifications, concerned with one or a very few limited aspects of the system studied.

There are three appropriate applications for theoretical modeling in assay work. First, a good model may achieve insights into unusual assay design strategies which would not be appreciated by experimental means without extraordinary effort. Second, it can serve a didactic role and enhance the intuition of the assayist in physics, chemistry, and statistics. Finally, it may provide an equation which is useful for calibration curve-fitting. However, a model could provide powerful insights into the qualitative aspects of assay chemistry and yet not be useful in calibration; similarly, a model with little or no correspondence to physical reality could provide a useful curve-fitting method. Any formula so obtained must satisfy the same statistical goodness-of-fit criteria that an empirical or semiempirical curve-fitting formula must satisfy, and thus, in the final analysis, all curve-fitting techniques must be regarded as empirical in practice. Assayists should not be misled by papers which confuse these distinct goals (Fernandez et al., 1983) and imply that a formula based on some highly simplified chemical model has advantages in routine assay data processing. Raab (1983) and Finney (1983) have demonstrated the practical defects in model-based calibration curve-fitting.

IX. INTERLUDE 4: WHAT DOES THIS SCATCHARD PLOT MEAN?

Scatchard analysis is a very simple graphic form of theoretical modeling for binder-ligand reactions. Suppose for example that radioassay data is used. The concentration of labeled ligand is known, enabling us to compute, for each standard, the total ligand bound (B) and the bound to free ratio (B/F). If the reaction behaves according to the simple first-order mass action law, as described by Eq. (13-15) (with $K^* = K$), then a plot of B/F vs. B would be expected to be a straight line with slope of -K and x intercept of q, the total concentration of binder.

It is never so simple in practice. The values of B and B/F, even in a simple reaction, must generally be corrected for nonspecific binding. This is usually attributed to binding of the ligand by glassware and substances other than the binder, and is often assumed to be a simple fraction of the total ligand present (although this assumption is almost never examined for validity). Under this assumption, it may be estimated by plotting bound over total (B/T) vs. total (T) ligand, and finding the level of B/T that corresponds to the horizontal asymptote which this curve tends to approach at high ligand concentrations.

The shape of the Scatchard plot is highly sensitive to the value of the nonspecific correction, and the results will be biased if it is not accurate. Often, an accurate value will yield negative B and B/F values for the Scatchard plot, and many assayists either reject these points (a loss of information) or fiddle with the nonspecific value (an introduction of bias).

The results are often, as in Figure 5, a widely scattered set of points which suggests a concave upward curve. This clearly does not accord with the simple model for Scatchard analysis, which should yield a straight line, and this curvature can be caused by at least six different factors: (1) the nonspecific correction may be too low; (2) the affinity of labeled and unlabeled ligand may not be equal, with K^* greater than K; (3) the binder may be heterogeneous; (4) the binder may exhibit negative cooperative behavior; (5) a fraction of free ligand may inadvertently be misclassified as bound; and (6) the addition of labeled ligand may have been delayed, and the reaction interrupted prior to the establishment of equilibrium. There are doubtless many other possible explanations for such a curve which have not yet been explored. The Scatchard plot may assume

Data Analysis and Quality Control   291

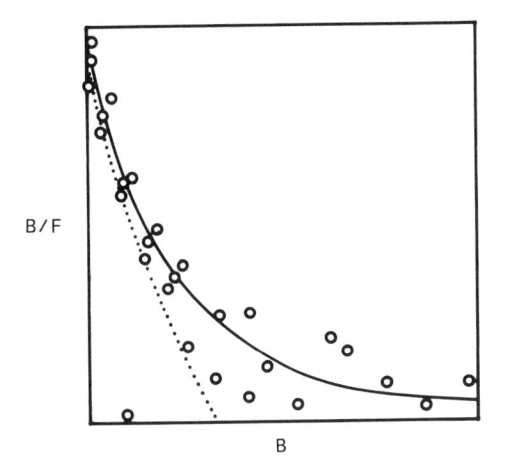

[Scatchard plot figure with B/F on y-axis and B on x-axis]

FIGURE 5 Data from a hypothetical radioassay, plotted as a Scatchard plot. Actual assay data frequently take this form, and it is difficult to draw any sort of meaningful curve through the points. One curve which might have been drawn manually is indicated by the solid line. If an appropriately selected model is fitted directly to the raw data using weighted nonlinear least-squares methods (the correct way of approaching the problem of determining physical constants from such data) the results, when plotted in the Scatchard domain, may not correspond to the manually drawn curve at all. The errors in the manually obtained results are due to the difficulty in estimating corrections (such as for nonspecific binding) which must be applied to the data prior to plotting the Scatchard, and to the severe statistical problems inherent in this graphic technique.

other forms than the concave upward curve of Fig. 5, and there are multiple explanations for all of these forms. Clearly the interpretation of Scatchard plots is highly ambiguous.

Another more severe problem arises from the statistical nature of the Scatchard plot. Not only is the error severely nonuniform over the curve, but there are also components of random error in both the x and the y directions. For radioassay data, the scatter in the points becomes more severe as the data reach the horizontal axis. Fitting such data manually is highly arbitrary and open to bias, and using least-squares methods is invalid due to the presence of error in the x direction. There are computer algorithms for fitting data with such errors, but they are either complex and possibly statistically invalid (O'Neill et al., 1961; Southwell, 1976) or limited to straight-line comparisons, as for correlation of different assay methods (Smith et al., 1980). The fitting of curves to data with errors in both directions is a complex and largely unresolved problem, so the reader should be wary of any seemingly easy methods.

Furthermore, such algorithms are unnecessary in this instance, as there is a better way to approach the problem. The proper method is to fit an appropriate equation by nonlinear least-squares directly to the raw data, using total ligand concentration as the independent variable. Scatchard analysis is only justifiable when the system being studied is well-described by Eq. (13-15), the errors are small, and the plots are being used only as crude indicators of gross changes in the values of K and q. Even then, there is evidence that the Scatchard plot may not be as reliable for such crude assessment as other methods, notably the Woolf plot and the direct linear plot (Cressie and Keightley, 1981; Woosley and Muldoon, 1977).

The literature abounds in papers that overinterpret Scatchard plots and related graphic procedures; Scatchard analysis is probably one of the most abused mathematical tools of contemporary biochemistry. Only recently have some misgivings crept into the major scientific journals (Klotz, 1982, 1983; Scheinberg, 1982; Munson and Rodbard, 1983) and even here confusion remains about the complex statistical issued involved. Unless extensive chemical information is available about the system producing a curve similar to that in Fig. 5, almost nothing of a qualitative or quantitative nature can be said on the basis of this plot.

X. CONCLUSIONS AND RECOMMENDATIONS

I have tried in this chapter to give a broad review of some of the important aspects of data processing and quality control as they apply to assay, at the expense of superficial treatment of individual topics. The bibliography should be a gateway for more detailed knowledge. I close with a few practical recommendations:

All assay results should be calculated automatically. Automation of assay calculations saves time and eliminates many potential errors (Challand, 1978; Jeffcoate and Das, 1977; Pegg and Miner, 1982). Properly implemented, it overcomes the increased variability of results observed when calculations are performed manually. It is increasingly difficult to muster valid objections to automation. Older excuses, such as "Our assay is being run on too small a scale" or "we can't afford a computer," are not tenable in the presence of a sound program for an inexpensive handheld calculator (Dudley et al., 1978; Dudley, 1981).

Perhaps in rare instances, an assay would yield calibration curves that are not adequately fitted by the model embodied in an available program. Many of the better packages are based on either the four- or the five-parameter logistic. The wide variety of assay methods to which the logistic has been applied is a clear testament to the versatility of this family of equations (see Fig. 1; Rodbard and McClean, 1977; Canellas and Karu, 1981). The burden of proof that a given assay can not be adequately described by one of these models lies with the assayist: in the rare event that such an assay were found, the developers of a program might wish to collaborate in improving their package to accommodate the assay.

In addition to increasing speed, decreasing cost, and improving accuracy, an automated package can analyze an assay in much greater depth than would be possible manually. Perhaps the greatest advantage is that a good package forces everyone concerned with the assay to think more systematically about sound assay design principles by presenting them with direct indices of assay performance, such as the imprecision profile, rather than nearly irrelevant and potentially misleading indices such as the slope of the calibration curve.

Don't write your own data processing program. The proliferation of programs only serves to dilute the impact of the good ones. Finney (1979, 1982) has outlined the requirements of a good assay data processing program; writing such a program is a major undertaking. The packages recommended below all represent multiple man-years of labor by skilled mathematicians and programmers. Your time will be better spent in helping to improve the documentation of an existing package, or in collaborating with its authors in improving the program.

Be wise in your choice of an automated package. Select a program that represents the current state of the art, such as the Edinburgh FORTRAN program (Raab, 1978; McKenzie and Thomson, 1982), the Middlesex-WHO BASIC program (Malan et al., 1978), the IAEA HP-41CV calculator program (Dudley, 1981), or the NIH FORTRAN program (Rodbard et al., 1975). Some published programs provide such limited assistance to the assayist as to not be worthwhile implementing (Brooks and Wittliff, 1973). Be cautious about programs that purport to be copies of well-established packages. For example, some of the many copies of the software of Rodbard are very good (Grotjan, 1978) while many others leave out crucial elements of the original software.

The choice may be partially dictated by the computing equipment available. It is also desirable to have a method for transferring raw data to the computer, so as to avoid tedious and error-prone manual data entry.

The broader requirements of an assay laboratory will often stipulate that the data processing package function within the framework of other software which allows the archiving and retrieval of a large number of results. This may require additional programming at the local site.

Be wary of commercial software. The best programs for assay analysis have uniformly come from academic or public institutions, and are available free or at minimal expense. Commercial firms seem to be more interested in glamorous options than statistical or computational validity. I have seen several commercial programs which offer a choice of calibration curve-fitting methods. Unfortunately, the methods selected are virtually identical, and the programs forego any statistical or quality control computations. Sometimes attractive features, such as optimized counting control for radioassays (see Interlude 1) are available within a program which is otherwise unacceptable (Challand, 1978). One commercial program which I recently saw under development showed a complete disregard for accepted statistical and computational practice. The analyte concentration was fit as a function of response, rather than vice versa, invalidating the assumptions underlying least-squares methods. A high-order polynomial was used, which is inherently unstable and which required the construction of clumsy constraints to prevent the fit from deviating wildly from the actual calibration curve. No consideration was given to the most basic statistical computations such as the estimation of confidence intervals.

The advent of the microprocessor has made it all too easy for equipment manufacturers to dash off a small program to sell with their machinery. Davis et al. (1980) have pointed out the deficiencies in these programs, as has Finney (1982), who adds that: "I do not believe we can at present countenance the uncritical acceptance of manufacturer's programs; I see reliance on these as part of standard (assay) practice to be improper, especially where patient care is involved."

Establish a comprehensive internal quality control program. This should include consideration of the requirements of the psychosocial and biophysical milieu as well as the mathematical. The stringency of the control rules employed in the program will be determined by the possible impact which an assay error can have: this may be more severe in a clinical setting. The use of a good automated data processing program is an inseparable part of a minimum acceptable quality control program.

Join a good external quality control program. Various schemes are sponsored by government, professional, and private organizations at the local, regional, national, and international level (Cresswell et al., 1978). The standardization of assay computational packages is as important a goal as the standardization of reagents and analytical methods which these organizations have pursued, in the effort to reduce unnecessary between-laboratory variation in results, and in attempting to facilitate the pooling of expertise among assayists.

Be very skeptical about results obtained from Scatchard analysis and related graphic procedures. This is especially important when medical decisions may be involved.

Never hesitate to call upon the services of an assay biostatistician. He or she may be able to offer helpful advice on the selection or implementation of a computer program, assist with assay design problems, or supervise modeling efforts.

ACKNOWLEDGMENTS

Prof. W. H. C. Walker provided the initial impetus which led to my writing this chapter. Dr. H. Feldman provided me with a preprint of his forthcoming review. Drs. D. Rodbard and R. Dudley have supplied me with copies of their assay software. Dr. D. Delacroix allowed me to use the assay data plotted in Fig. 1. The computations and text editing required by this chapter were made possible by Mr. P. Breesch and the staff of the Scientific Computing Center of the Agfa-Gevaert Company, Antwerp, Belgium, who gave me access to their excellent computing facilities. I thank the Bureau d'Etudes I. Ralston, Brussels, for assistance in the preparation of the figures.

Special gratitude is due to Profs. M. J. R. Healy and David Finney, and to Mrs. Gillian Raab. These individuals have made their software available, and have also criticized earlier drafts of this chapter. Prof. Finney and Mrs. Raab have contributed particularly detailed remarks, and the blame for any infelicities or errors remaining must fall to the author alone.

REFERENCES

1. Ackoff, R. L. (1968). Decision-making, In National Science Policy, Churchill Livingstone, London. Excerpted in: Mackay, A. L. (1977). The Harvest of a Quiet Eye, Institute of Physics, London.
2. Arrigucci, A., Forti, G., Fiorelli, G., Pazzagli, M., and Serio, M. (1973). Mathematical analysis of the results of competitive binding methods. In The Endocrine Function of the Human Testis, Vol. 1, V. H. T. James, M. Serio, and L. Martini (eds.), Academic Press, New York, pp. 73-90.
3. Ayers, G., Burnett, D., Griffiths, A., and Richens, A. (1981). Quality control of drug assays. Clin. Pharmacokinet. 6: 106-117.
4. Barnett, V., and Lewis, T. (1978). Outliers in Statistical Data. John Wiley, New York.
5. Baulieu, E., and Raynaud, J. P. (1970). A "proportion graph" method for measuring binding systems. Eur. J. Biochem. 13: 293.
6. Baxter, R. C. (1980). Simplified approach to confidence limits in radioimmunoassay. Clin. Chem. 26: 763-765.
7. Bennett, C. A., and Franklin, N. L. (1954). Statistical Analysis in Chemistry and the Chemical Industry. John Wiley, New York.
8. Besch, H. R., and Watanabe, A. M. (1975). Radioimmunoassay of digoxin and digitoxin. Clin. Chem. 21: 1815.
9. Bliss, C. I. (1970). Dose-response curves for radioimmunoassays. In Statistics in Endocrinology, J. W. McArthur, and T. Colton (eds.), MIT Press, Cambridge, Mass., pp. 432-47.
10. Boeynaems, J. M., and Dumont, J. E. (1980). Outlines of Receptor Theory. Elsevier-North Holland, New York.
11. Brooks, W. F., and Wittliff, J. L. (1973). Quantitation of radioligand binding data using a desk-top calculator in the program mode. Anal. Biochem. 54: 464-476.
12. Broughton, P. M. G., and Raine, D. N. (eds.). (1969). Quality control in clinical biochemistry. Ann. Clin. Biochem. 6: 88-148.

13. Brown, G. M., Boshans, R. L., and Schalch, D. S. (1970). Computer calculation of radioimmunoassays. Comput. Biomed. Res. 3: 212-217.
14. Burger, H. G., Lee, V. W. K., and Rennie, G. C. (1972). A generalized computer program for the treatment of data from competitive protein-binding assays including radioimmunoassays. J. Lab. Clin. Med. 80: 302-312.
15. Büttner, H., (ed.) (1968). International Symposium on Statistical Quality Control in the Analytical Laboratory, Z. Anal. Chem. 243: 751.
16. Büttner, H., Borth, R., Boutwell, J. H., and Broughton, P. M. G. (1975, 1976). Provisional recommendations on quality control in clinical chemistry (International Federation of Clinical Chemistry, Committee on Standards). Clin. Chim. Acta 63: F25-F38 or Clin. Chem. 22: 532-540.
17. Canellas, P. F., and Karu, A. E. (1981). Statistical package for analysis of competition ELISA results. J. Immunol. Methods 47: 375-385.
18. Catt, K. J., Tregear, G. W., and Burger, H. G. (1970). Radioimmunoassay of polypeptide hormones in antibody coated tubes. In In Vitro Procedures with Radioisotopes in Medicine, IAEA, Vienna, pp. 633-644.
19. Cekan, Z., Robertson, D. M., and Diczfalusy, E. (1973). Calculation of radioimmunoassay results on a Wang programmable electronic calculator. Acta Endocrinol. Suppl. 177: 101.
20. Cernosek, S. F., and Gutierrez-Cernosek, R. M. (1978). Use of a programmable desk-top calculator for the statistical quality control of radioimmunoassays. Clin. Chem. 24: 1121-1125.
21. Challand, G. S. (ed.) (1978). Automated calculation of radioimmunoassay results. Ann. Clin. Biochem. 15: 123-135.
22. Colquhoun, D. (1971). Lectures on Biostatistics. An Introduction to Statistics with Applications in Biology and Medicine, Clarenden Press, Oxford.
23. Computer programs in endocrinology, Endocrinology 91: 329-331.
24. Cook, B. (1975). Automation and data processing for radioimmunoassays. In Steroid Immunoassay. Proceedings of the Fifth Tenovus Workshop, Alpha Omega, Cardiff, pp. 293-310.
25. Copeland, B. E. (1973). Quality Control in Clinical Chemistry. American Society of Clinical Pathology, Chicago.
26. Cressie, N. A. C., and Keightley, D. D. (1981). Analyzing data from hormone-receptor assays. Biometrics 37: 235-249.
27. Cresswell, M. A., Hall, P. E., and Hurn, B. A. L. (1978). Quality control in RIA: a preliminary report of the results of the World Health Organization's programme for external quality control. In Radioimmunoassay and Related Procedures in Medicine, IAEA, Vienna, pp. 149-158.

28. Davidsohn, I., and Henry. F. B. (1972). Clinical Diagnosis by Laboratory Methods, 15th ed. Hafner, New York.
29. Davis, S. E., Jaffe, M. L., Munson, P. J., and Rodbard, D. (1980). RIA data processing with a small programmable calculator. J. Immunoassay 1: 15-25.
30. DeBoor, C. (1978). A Practical Guide to Splines. Applied Mathematics Series, no. 27. Springer Verlag, New York.
31. DeJong, R. H. (1973). Generalized analysis of quantal dose-response curves. Comput. Biomed. Res. 6: 588-595.
32. DeLean, A., Munson, P. J., and Rodbard, D. (1978). Dose-response curve analysis. Am J. Physiol. 235: E97-102.
33. Deming, W. E. (1967). What happened in Japan? Ind. Qual. Control 24: 89-93.
34. Deming, W. E. (1975). My view of quality control in Japan. Rep. Stat. Appl. Res. JUSE 22: 73-80.
35. Dixon, W. J. (1953). Processing data for outliers. Biometrics 9: 74-89.
35a. Dobbelaer, R., and Legrand, M. (1982). Probit analysis of multi-preparation bio-assays on a programmable calculator. Comput. Programs Biomed. 15: 103-110.
35b. Doetsch, K. (1983). Validating new methods (letter). Clin. Chem. 29: 581.
36. Duddleson, W. G., Midgley, A. R., and Niswender, G. D. (1973). Computer program sequence for analysis and summary of radioimmunoassay data. Comput. Biomed. Res. 5: 205-217.
37. Dudley, R. A. (1981). Programs for Data Processing in Radioimmunoassay using the HP-41C Programmable Calculator. IAEA-TECDOC-252 (plus supplements for on-line processing and quality control). IAEA, Vienna.
38. Dudley, R. A., Figdor, H. C., Keroe, E. A., Morris, A. C., and Mutz, O. J. (1978). Well scintillation counter with automatic sample changing and data processing: an inexpensive instrument incorporating consumer products. In Radioimmunoassay and Related Procedures in Medicine, vol. 1, IAEA, Vienna, pp. 457-467.
39. Duncan, A. J. (1974). Quality Control and Industrial Statistics, 4th ed. Irwin, Homewood, Ill.
39a. Edwards, G. C. (1983). How much quality control is enough? Clin. Chem. 29: 732-733.
40. Ekins, R. P. (1978). Quality control and assay design. In Radioimmunoassay and Related Procedures in Medicine, IAEA, Vienna, pp. 39-56.
41. Ekins, R. P. Newman, G. B., and O'Riordan, J. L. H. (1968). Theoretical aspects of "saturation" and radioimmunoassay. In Radioisotopes in Medicine. In Vitro Studies, Hayes, R. L., Goswitz, F. A., and B. E. P. Murphy (eds.), USAEC, Oak Ridge, Tenn., 1968, pp. 59-100.

42. Ekins, R. P., Sufi, S., and Malan, P. G. (1978). An "intelligent" approach to radioimmunoassay sample counting employing a microprocessor-controlled sample counter. In Radioimmunoassay and Related Procedures in Medicine, vol. 1 IAEA, Vienna, pp. 437-455.
43. Engel, G. (1974). Estimation of binding parameters of enzyme-ligand complex from fluorometric data by curve fitting procedure: seryl-tRNA synthetase-tRNA^{ser} complex. Anal. Biochem. 61: 184-191.
44. Faure, A., Nemoz, C., Claustrat, B., Paultre, C. Z., and Site, J. (1980a). Control of routine radioimmunoassays: a computer proggram for calculation of control charts for precision and accuracy. Comput. Biomed. Res. 12: 105-110.
45. Faure, A., Nemoz, C., Claustrat, B., and Site, J. (1980b). A computer program for determination of equilibrium constants and binding site concentrations in radioimmunoassay systems, Comput. Programs Biomed. 12: 19-26.
46. Feldman, H. A. (1983). Statistical analysis in radioligand assay: binding curves and the error around them. In In Vitro Nuclear Medicine, 2nd ed., B. Rothfeld (ed.), Lippincott, Philadelphia.
47. Fernandez, A. A., and Loeb, H. G. (1975). Practical applications of radioimmunoassay theory. Simple procedure yielding linear calibration curves. Clin. Chem. 21: 1113-20.
47a. Fernandez, A. A., Stevenson, G. W., Abraham, G. E., and Chiamori, N. Y. (1983). Interrelations of the various mathematical approaches to radioimmunoassay. Clin. Chem. 29: 284-289.
48. Finney, D. J. (1976). Radioligand assay. Biometrics 32: 721-740.
49. Finney, D. J. (1977). Bioassay. In Lecture Notes in Biomathematics no. 18: Mathematics and the Life Sciences, S. Levin (ed.), Springer, New York, pp. 67-151.
50. Finney, D. J. (1978). Statistical Method in Biological Assay, 3rd ed. Hafner, New York.
51. Finney, D. J. (1979). The computation of results from radioimmunoassay. Methods Inf. Med. 18: 164-171.
52. Finney, D. J. (1982). Software requirements for statistical analysis of immunoassays. In Immunoassays for Clinical Chemistry: A Workshop Meeting, W. M. Hunter and J. E. T. Corrie (eds.), Churchill Livingstone Edinburgh.
52a. Finney, D. J. (1983). Response curves for radioimmunoassay. Clin. Chem. 29: 1562-1566.
53. Finney, D. J., and Phillips, P. (1977). The form and estimation of a variance function, with particular reference to radioimmunoassay. Appl. Stat. 26: 312-320.
53a. Fischer, L. (1983). Logit-log radioimmunoassay data reduction: Weighted vs. unweighted (letter). Clin. Chem. 29: 321.
54. Flanagan, M. T., Tattam, F. G., and Green, N. M. (1978). The characterization of heterogeneous antibody-hapten interactions using non-linear regression analysis on fluorescence quenching curves. Immunochemistry 15: 261-267.

54a. Forget, G., and Sirois, P. (1983). Development of a radioimmunoassay data analysis pack (RIADAP) in level II BASIC for microcomputers. Comput. Biomed. Res. 16: 160-168.
55. Gilman, A. G. (1970). A protein binding assay for adenosine 3':5'-cyclic monophosphate. Proc. Natl. Acad. Sci. 67: 305-312.
56. Gomeni, R., and Gomeni, C. (1980). A conversational graphic program for the analysis of the sigmoid curve. Comput. Biomed. Res. 13: 489-499.
57. Griner, P. F., Mayewski, R. J., Mushlin, A. I., and Greenland, P. (1981). Selection and interpretation of diagnostic tests and procedures. Principles and applications. Ann. Intern. Med. 94 (Part 2): 553-600.
58. Groth, T., Falk, H., and Westgard, J. O. (1981). An interactive computer simulation program for the design of statistical quality control procedures in clinical chemistry. Comput. Programs Biomed. 13: 73-86.
59. Grotjan, H. E. (1978). WHAMBAM. Four Parameter Logistic Curve Fitting Program. Documentation and Program Listing. Dept. of Reproductive Medicine and Biology, Univ. of Texas Med. School, P. O. Box 20708, Houston, Texas 77025.
60. Grotjan, H. E., and Steinberger, E. (1977). Radioimmunoassay and bioassay data processing using a logistic curve fitting routine adapted to a desk top computer. Comput. Biol. Med. 7: 159-163.
61. Hainline, A., (ed.) (1982) Quality Control in Clinical Chemistry. USDHSS, PHS, CDC, Atlanta, Ga.
62. Hales, C. N. and Randel, P. J. (1963). Immunoassay of insulin with insulin-antibody precipitate. Biochem. J. 88: 137-46.
62a. Halfman, C. J. (1979). Concentrations of binding protein and labeled analyte for optimizing the response in immunoassays. Anal. Chem. 51: 2306-2311.
63. Harding, B. R., Thomson, R., and Curtis, A. R. (1973). A new mathematical model for fitting an HPL radioimmunoassay curve. J. Clin. Pathol. 26: 973-976.
64. Hatch, K. F., Coles, E., Busey, H., and Goldman, S. C. (1976). End-point parameter adjustment on a small desk-top programmable calculator for logit-log analysis of radioimmunoassay data. Clin. Chem. 22: 1383-1389.
65. Hawker, F. J., and Challand, G. S. (1981). Effect of outlying standard points on curve fitting in radioimmunoassay. Clin. Chem. 27: 14-17.
66. Healy, M. J. R. (1972). Statistical analysis of radioimmunoassay data. Biochem. J. 130: 207-210.
67. Healy, M. J. R. (1979). Outliers in clinical chemistry quality control schemes. Clin. Chem. 25: 675-677.

68. Healy, M. J. R., and Kimber, A. C. (1982). Robust estimation of variability in radioligand assays. In Immunoassays for Clinical Chemistry: A Workshop Meeting, W. M. Hunter and J. E. T. Corrie, (eds.), Churchill Livingstone, Edinburgh.
69. Henry, J. B., and Giegel, J. L. (1977). Quality Control in Laboratory Medicine. Transactions of the First Interamerican Symposium, Key Biscayne, Fla., April 1976. Masson, New York.
70. Herndl, R., and Marschner, I. (1975). Comparison of various mathematical methods for the calculation of radioimmunoassay data. Acta Endocrinol. Suppl. 193: 117.
71. Hidalgo, J. U., Maduell, C. R., Bloch, T., Spohrer, L. R., and Wooten, M. (1975). Precision of radioimmunoassay with emphasis on curve-fitting procedures. Semin. Nucl. Med. 5: 153-156.
72. Hosmer, D. W., Wang, C.-Y., Lin, I.-C., and Lemeshow, S. (1978). A computer program for stepwise logistic regression using maximum likelihood estimation. Comput. Programs Biomed. 8: 121-134.
73. Ichihara, K., Yamamoto, T., Kumahara, Y., and Miyai, K. (1977). An improved processing of radioimmunoassay data by means of a desk-top calculator. (1) Comparison of regression procedures applied to selected kinds of radioimmunoassay. Clin. Chim. Acta 79: 331-340.
74. Ichihara, K., Yamamoto, T., and Miyai, K. (1977). An improved processing of radioimmunoassay data by means of a programmable desk-top calculator. (2) A study of indices for discrimination of standard curve with large aberrance. Clin. Chim. Acta 80: 37-47.
75. Inhorn, S. L. (ed.) (1978). Quality Assurance Practices for Health Laboratories. American Public Health Association, Washington, D.C.
76. Jaarsma, D., Horne, E. J., and Spooner, R. L. (1978). Computer-assisted radioimmunoassay analysis. In Proceedings, The Second American Symposium on Computer Applications in Medical Care, Nov. 5-9, 1978, F. H. Orthner (ed.), pp. 613-616.
77. Jeffcoate, S. L. (1981). Efficiency and Effectiveness in the Endocrine Laboratory. Academic Press, London.
78. Jeffcoate, S. L., and Das, R. E. G. (1977). Interlaboratory comparison of radioimmunoassay results. Ann. Clin. Biochem. 14: 258-260.
79. Ketelslegers, J.-M., Knott, G. D., and Catt, K. J. (1975). Kinetics of gonadtrophin binding by receptors of the rat testis. Analysis by a non-linear curve-fitting method. Biochemistry 14: 3075-3083.
79a. Klotz, I. (1982). Numbers of receptor sites from Scatchard graphs: Facts and fantasies. Science 217: 1247-1249.
79b. Klotz, I. M. (1983). Reply to ref. 100a). Science 220: 979-981.
79c. Knudsen, L. F., and Curtis, J. M. (1947). The use of the angular transformation in biological assays. J. Am. Stat. Assoc. 42: 282-296.
80. Koshiver, J., and Moore, D. (1979). LOGIT: A program for dose-response analysis. Comput. Programs Biomed. 10: 61-65.

81. Leclercq, R., Taljedal, I.-B., and Wold, S. (1972). Evaluation of radio-isotope data in steroid assays based on competetive protein binding. Clin. Chim. Acta 36: 257-259.
82. Lee, E. Y. T., and Biggers, J. D. (1974). Abstract. Bull. Am. Phys. Soc., Ser. 2, 19: 170.
83. Livesey, J. H. (1974). Computation of radioimmunoassay data using segmentally linearized standard curves. Comput. Biomed. Res. 7: 7-20.
84. Lotz, A., Vogt, W., Popp, B., and Knedl, M. (1976). Polygonal interpolation, a simple, rapid, and versatile approximation method requiring minimal computing facilities. Comput. Biomed. Res. 9: 21-30.
85. Malan, P. G., Newman, G. B., and Ekins, R. P. (1973). Non-linear curve fitting to radioimmunoassay standard curves, Acta Endocrinol. Suppl. 177: 99.
86. Malan, P. G., Cox, M. G., Long, E. M. R., and Ekins, R. P. (1978). A multi-binding site model-based curve-fitting program for the computation of RIA data. In Radioimmunoassay and Related Procedures in Medicine. IAEA, Vienna, pp. 425-436.
87. Marbach, P., Gotz, U., Veteau, J. P., and Wagner, H. (1978). RIA analysis by means of non-linearized response functions: application to an automated rat-growth-hormone assay. In Radioimmunoassay and Related Procedures in Medicine, IAEA, Vienna, pp. 383-397.
87a. Marschner, I., Dobry, H., Erhardt, F., Landersdorfer, T., Popp, B., Ringel, C., and Scriba, P. C. (1974). Berechnung radioimmunologischer Messwerte mittels Spline-Funktionen. Artzl. Lab. 20: 184-191.
88. Marschner, I., Erhardt, F., and Scriba, P. C. (1974). Calculation of the radioimmunoassay standard curve by "spline function." In Radioimmunoassay and Related Procedures in Medicine, vol. 1. IAEA, Vienna, pp. 111-122.
89. Martinez-Alonso, J. R., Zunzunegui-Pastor, V., Marin-Rojas, M. C., Moreno-Gonzalez, J., Millan-Santos, I., Martinez-Sales, V., and Ortiz-Berrocal, J. (1976). Automatic treatment of assay data by Competetive protein binding and radioimmunoassay. Comput. Programs Biomed. 6: 249-262.
90. Massart, D. L., Dijkstra, A., and Kaufman, L. (1978). Evaluation and Optimization of Laboratory Methods and Analytical Procedures. A Survey of Statistical and Mathematical Techniques. Techniques and Instrumentation in Analytical Chemistry, vol. 1. Elsevier, Amsterdam.
91. Masson, P. L., Cambiaso, C. L., Collet-Cassart, D., Magnussen, C. G. M., and Sindic, C. J. M. (1981). Particle counting immunoassay (PACIA). In Methods in Enzymology, vol. 74, J. J. Langone, (ed.), Academic Press, New York, pp. 106-139.

92. McDonagh, B. F., Munson, P. J., and Rodbard, D. (1977). A computerized approach to statistical quality control for radioimmunoassays in the clinical chemistry laboratory. Comput. Programs Biomed. 7: 179-190.
93. McDonald, M. (1981). Calculator program for weighted logit-log radioimmunoassay data reduction. Clin. Chem. 27: 1946.
94. McIntosh, J. E. A., and McIntosh, R. P. (1980). Mathematical Modelling and Computers in Endocrinology. Springer-Verlag, New York, 1980.
95. McKenzie, G. M., and Thompson, R. C. H. (1982). Design and implementation of a software package for analysis of immunoassay data. In Immunoassays for Clinical Chemistry: A Workshop Meeting, W. M. Hunter and J. E. T. Corrie, (eds.), Churchill Livingstone, Edinburgh.
96. Meinert, C. L., and McHugh, R. B. (1968). The biometry of an isotope displacement immunologic microassay. Math. Biosci. 2: 319-338.
97. Morgan, C. R., Hardigg, J. B., and Fisher, D. D. (1967). A computer program for immunoassay of protein hormones, with specific reference to insulin and growth hormone. Diabetes 16: 734-737.
97a. Morrison, G. H. (1983). "Key" words (editorial). Anal. Chem. 55: 177.
98. Motta, M., and Esposti, A. D. (1981). A computer program for mathematical treatment of data in radioimmunoassay. Comput. Programs Biomed. 13: 121-129.
99. Munson, P., and Rodbard, D. (1978). An elementary components of variance analysis for multi-centre quality control. In Radioimmunoassay and Related Procedures in Medicine. IAEA, Vienna, pp. 105-25.
100. Munson, P. J., and Rodbard, D. (1980). LIGAND: A versatile computerized approach for characterization of ligand-binding systems. Anal. Biochem. 107: 220-239.
100a. Munson, P. J., and Rodbard, D. (1983). Number of receptor sites from Scatchard and Klotz graphs: A constructive critique. Science 220: 979-981.
101. Murphy, B. E. P. (1967). Some studies of the protein-binding of steroids and their application to the routine micro and ultramicro measurement of various steroids in body fluids by competetive protein-binding radioassay. J. Clin. Endocrin. Metab. 27: 973-990.
102. Natrella, M. G. (1963). Experimental Statistics. U. S. Bureau of Standards Handbook. 91, Washington, D. C. (Reprinted with corrections, 1966.)
103. Naus, A. J., Kuppens, P. S., and Borst, A. (1977). Calculation of radioimmunoassay standard curves. Clin. Chem. 23: 1624-1627.

104. Nemoz, C., Faure, A., Bizzollon, Ch., and Site, J. (1978). Treatment of radioimmunoassay data: a computer program for curve-fitting and dose interpolation using orthogonal polynomials. Comp. Programs Biomed. 8: 99-109.
104a. Nolte, H., Mühlen, A. v. Z., and Hesch, R. D. (1976). Auswertung radioimmunochemischer Bestimmungsmethoden durch "Spline-Approximation." J. Clin. Chem. Clin. Biochem. 14: 253-259.
105. O'Neill, M., Sinclair, I. G., and Smith, F. J. (1961). Polynomial curve fitting when abscissas and ordinates are both subject to error. Comput. J. 12: 52-56.
105a. Oppenheimer, L., Capizzi, T. P., Weppelman, R. M., and Mehta, H. (1983). Determining the lowest limit of reliable assay measurement. Anal. Chem. 55: 638-643.
106. Ottaviano, P. J., and DiSalvo, A. F. (1977). Quality Control in the Clinical Laboratory: A Procedural Text. University Park Press, Baltimore.
107. Pagano, M., and Tiede, J. J. (1977). The Application of Robust Regression to Radioimmunoassay. Technical report no. 52. State University of New York at Buffalo, Dept. of Computer Science.
107a. Pegg, P. J., and Miner, E. M. (1982). The effect of data reduction technic on ligand assay proficiency survey results. Am. J. Clin. Pathol. 77: 334-337.
108. Pekary, A. E. (1979). Parallel line and relative potency analysis of bioassay and radioimmunoassay data using a desk top computer. Comput. Biol. Med. 9: 355-362.
109. Peters, M., Clark, I. R., and Holder, R. L. (1982). A comprehensive data processing package for analyzing and reporting immunoassay data. Wolfson Research Laboratories, Dept. of Clinical Chemistry, Queen Elizabeth Medical Center, Birmingham, B15 2TH United Kingdom.
110. Pilo, A., and Zucchelli, G. C. (1975). Automatic treatment of radioimmunoassay data: an experimental validation of the results. Clin. Chim. Acta 64: 1-9.
111. Prentice, R. L. (1976). A generalization of the probit and logit methods for dose response curves. Biometrics 32: 761-768.
112. Raab, G. M. (1978). A modular program for the automated calculation of radioimmunoassay results. Ann. Clin. Biochem. 15: 129-130.
113. Raab, G. M. (1981a). Estimation of a variance function, with application to immunoassay. J. R. Stat. Soc. C 30: 32-40.
113a. Raab, G. M. (1981b). Robust calibration and radioimmunoassay (letter). Biometrics 37: 839-841.
114. Raab, G. (1982). Validity tests in the statistical analysis of immunoassay data. In Immunoassays for Clinical Chemistry: A Workshop Meeting, W. M. Hunter and J. E. T. Corrie (eds.). Churchill Livingstone, Edinburgh.
114a. Raab, G. M. (1983). Comparison of a logistic and a mass-action curve for radioimmunoassay data. Clin. Chem. 29: 1757-1761.

115. Raab, G. M., and McKenzie, I. G. M. (1982). A modular computer program for processing immunoassay data. In Quality Control in Clinical Endocrinology: Proceedings of the 8th Tenovus Workshop, D. W. Wilson, S. J. Gaskell, and K. Kemp (eds.), Alpha Omega, Cardiff.
116. Raab, G. M., Thompson, R., and McKenzie, I. (1980). Variance function estimation for immunoassays. Comput. Programs Biomed. 12: 111-120.
117. Rappoport, A. E. (ed.) (1971). Quality Control in Clinical Chemistry. Transactions of the IVth International Symposium. Hans Huber, Berne.
118. Rautela, G. S., Snee, R. D., and Miller, W. K. (1979). Response-surface co-optimization of reaction conditions in clinical chemical methods. Clin. Chem. 25: 1954-1964.
119. Reed, L. J., and Berkson, J. (1929). The application of the logistic function to experimental data. J. Phys. Chem. 33: 760-779.
120. Rodbard, D. (1971). Statistical aspects of radioimmunoassays. In Principles of Competetive Protein-Binding Assays, W. Odell and W. H. Daughaday (eds.), Lippincott, Philadelphia, pp. 204-259.
121. Rodbard, D. (1974). Statistical quality control and routine data processing for radioimmunoassays and immunoradiometric assays. Clin. Chem. 20: 1255-1270.
122. Rodbard, D. (1978a). Data processing for radioimmunoassays: an overview. In Clinical Immunochemistry: Chemical and Cellular Bases and Applications in Disease. Current Topics in Clinical Chemistry, vol. 3, S. J. Natelson, A. J. Pesce, and A. A. Dietz (eds.), American Association for Clinical Chemistry, Washington, D.C., pp. 477-494.
123. Rodbard, D. (1978b). Quality control for RIA; recommendations for a minimal program. In Radioimmunoassay and Related Procedures in Medicine. IAEA, Vienna, pp. 21-38.
124. Rodbard, D. (1978c). Statistical estimation of the minimal detectable concentration ("sensitivity") for radioligand assays. Anal. Biochem. 90: 1-12.
124a. Rodbard, D. (1983). Reply to ref. 53a. Clin. Chem. 29: 391-392.
125. Rodbard, D., Bridson, W., and Rayford, P. L. (1969). Rapid calculation of radioimmunoassay results. J. Lab. Clin. Med. 74: 770-781.
126. Rodbard, D., and Cooper, J. A. (1970). A model for prediction of confidence limits in radioimmunoassays and competitive protein binding assays. In Proceedings, Symposium on Radioisotopes in Medicine: in Vitro Studies. IAEA, Vienna, pp. 659-674.
127. Rodbard, D., Faden, V. B., Knisley, S., and Hutt, D. M. (1975). Radioimmunoassay Data Processing; Logit-log, Logistic Method, and Quality Control, 3rd ed., Reports nos. PB246222, PB246223, and PB246224, National Technical Information Service, Springfield, Va. 22161.

128. Rodbard, D., and Hutt, D. M. (1974). Statistical analysis of radioimmunoassays and immunoradiometric (labelled antibody) assays: a generalized, weighted, least squares method for logistic curve fitting. In Symposium on RIA and Related Procedures in Medicine. IAEA, Vienna, pp. 165-192.
129. Rodbard, D., Lenox, R. H., Wray, H. L., and Ramseth, D., (1976). Statistical characterization of the random errors in the radioimmunoassay dose-response variable. Clin. Chem. 22: 350-358.
130. Rodbard, D., and Lewald, J. E. (1970). Computer analysis of radioligand assay and radioimmunoassay data. In Steroid Assay by Protein Binding. Karolinska Symposium of Research Methods in Reproductive Biology, 2nd Symposium, pp. 79-103.
131. Rodbard, D., and McClean, S. W. (1977). Automated computer analysis for enzyme-multiplied immunological techniques. Clin. Chem. 23: 112-115.
132. Rodbard, D., Munson, P., and DeLean, A. (1978). Improved curve-fitting, parallelism testing, characterization of sensitivity and specificity, validation, and optimization for radioligand assays. In Radioimmunoassay and Related Procedures in Medicine, vol. 1. IAEA, Vienna, pp. 469-504.
133. Rodbard, D., Munson, P. J., and Thakur, A. K. (1980). Quantitative characterization of hormone receptors. Cancer 46: 2907-2918.
134. Rodbard, D., Rayford, P. L., Cooper, J. A., and Ross, G. T., (1968). Statistical quality control of radioimmunoassays. J. Clin. Encodrinol. Metab. 28: 1412-1418.
135. Rodbard, D., and Tacey, R. L. (1978). Radioimmunoassay dose interpolation based on the mass action law with antibody heterogeneity. Exact correction for variable mass of labelled ligand. Anal. Biochem. 90: 13-21.
136. Rodgers, R. C. (1976). Radioimmunoassay Theory for Health Care Professionals. Hewlett-Packard Company, Loveland, Colo., 2nd printing.
137. Rodgers, R. P. C. (1981). Data Analysis and Quality Control in Binder-Ligand Assay, vols. 1 and 2. Scientific Newsletters, Inc., Anaheim, Calif.
138. Rodgers, R. P. C., Elliot, S., and Ekins, R. P. (1983). A comprehensive theoretical model for binder-ligand assay: a guide to radioimmunoassay design principles (in preparation).
139. Rolleri, E., Novario, P. G., and Pagliano, B. (1973). Automated treatment of radioimmunoassay data. J. Nucl. Biol. Med. 17: 128-141.
140. Rubin, R. T. (1973). Program for routine RIA data analysis on the Hewlett-Packard HP 9100 desk-top programmable calculator, Harbor General Hospital, Torrance, Calif.
141. Russell, C. D., DeBlanc, H. J., and Wagner, H. N. (1974). Components of variance in laboratory quality control. Johns Hopkins Med. J. 135: 344-357.

142. Sandel, P., and Vogt, W. (1978). Performance of various mathematical methods for calculation of radioimmunoassay results. In Radioimmunoassay and Related Procedures in Medicine, vol. 2. IAEA, Vienna, pp. 373-381.
143. Sandel, P., Vogt, W., Popp, B., and Knedel, M. (1976). Smoothing-Spline und Polygonal-Interpolation, Logit-log- und Logit-Quadrat-Regression: eine Gegenuberstellung mathematischer Verfaren zur Ergebniswertberechnung von Radioimmunoassay. Z. Anal. Chem. 279: 126.
143a. Scheinberg, I. H. (1982). Scatchard plots. Science 215: 312-313.
144. Schöneshöfer, M. (1975). Suitability of four standard curve models for computer evaluation of steroid and peptide radioimmunoassays. Acta Endocrinol. Suppl. 193: 116.
145. Schöneshöfer, M. (1977). Computer program for evaluation, physicochemical characterization and optimization of competetive protein binding assays: Comparison of four curve-fitting models in peptide and steroid radioimmunoassays. Clin. Chim. Acta 77: 101-115.
146. Shaw, W., Smith, J., Spierto, F. W., and Agnese, S. T. (1977). Linearization of data for saturation-type competetive protein binding assay and radioimmunoassay. Clin. Chim. Acta 76: 15-24.
147. Sheiner, L. B. (1981). ELSFIT: A Program for the Extended Least Squares Fit to Individual Pharmacokinetic Data. User's Manual, Feb. 1981 (with corrections). A technical report of the Division of Clinical Pharmacology, University of California, San Francisco, Calif. 94143.
148. Sinterhauf, K., Müller, T., Spira, H. J., and Lommer, D. (1976). Mathematical models for the processing of data from binding radioassays; transformations and approximations. Z. Anal. Chem. 279: 127-128.
149. Sinterhauf, K. Müller, T., Spira, H. J., Müller, D., and Lommer, D. (1975). Comparison of mathematical transformations and approximations of radioassay dose-response curves. Acta Endocrinol. Suppl. 193: 115.
150. Smigel, M. D., and Lazar, J. D. (1981). Improved fitting of radioimmunoassay by Scatchard analysis. Int. J. Biomed. Comput. 12: 189-203.
151. Smith, D. S., Pourfarzaneh, M., and Kamel, R. S. (1980). Linear regression by Deming's method. Clin. Chem. 26: 1105-1106.
152. Southwell, W. H. (1976). Fitting data to nonlinear functions with uncertainty in all measurement variables. Comput. J. 19: 69-73.
153. Spona, J. (1974). Rapid assay for luteinizing hormone and evaluation of data by new computer program. In Radioimmunoassay and Related Procedures in Medicine, vol. 1. IAEA, Vienna, pp. 123-130.

154. Taljedal, I.-B., and Wold, S. (1970). Fit of some analytical functions to insulin radioimmunoassay standard curves. Biochem. J. 119: 139-143.
155. Tiede, J. J., and Pagano, M. (1979). The application of robust calibration to radioimmunoassay. Biometrics 35: 567-574.
156. Tonks, D. B. (1970). Quality Control in Clinical Laboratories. Warner-Chilcott Laboratories, Co., Canada.
157. Valleron, A.-J., and Rosselin, G.-E. (1971). Apport de l'informatique a l'analyse des hormones dans le plasma, Ann. Biol. Clin. 29: 145-152.
158. Vikelsoe, J. (1973). Spline-fits: An alternative route to computer treatment of data from competitive protein-binding assays. Acta Endocrinol. Suppl. 177: 100.
159. Vivian, S. R., and Labella, F. S. (1971). Classic bioassay statistical procedures applied to radioimmunoassay of bovine thyrotropin, growth hormone and prolactin. J. Clin. Endorcinol. Metab. 33: 225-233.
160. Weaver, C. K., and Cargille, C. M. (1971). Radioimmunoassay: A simplified procedure for dose interpolation. J. Lab. Clin. Med. 77: 661-664.
161. Vogt, W., Sandel, P., Langfelder, C., and Knedel, M. (1978). Performance of various mathematical methods for computer-aided processing of radioimmunoassay results. Clin. Chim. Acta 87: 101-111.
161a. Walker, R. O., and Uddin, Z. (1983). Immunoassay data handling with an Apple II + microcomputer system. Clin. Chem. 29: 727-728.
162. Weinstein, M. C., Fineberg, H. V., Elstein, A. S., Frazier, H. S., Neuhauser, D., Neutra, R. R., and McNeil, B. J. (1980). Clinical Decision Analysis. Saunders, Philadelphia.
163. Westgard, J. O., Groth, T., Aronsson, T., and de Verdier, C.-H. (1977). Combined Shewhart-cusum control chart for improved quality control in clinical chemistry. Clin. Chem. 23: 1881-1887.
164. Whitehead, T. P. (1977). Quality Control in Clinical Chemistry. Wiley, New York.
165. Widman, J. C., and Powsner, E. R. (1973). Radioimmunoassay data handling. Am. J. Clin. Pathol. 60: 480-486.
166. Wilkins, T. A., and Chadney, D. C. (1981). Effect of outlying standard points on curve fitting in radioimmunoassay. Clin. Chem. 27: 1770-1771.
167. Wilkins, T. A., Chadney, D. C., Bryant, J., Palmstrøm, S. H., and Winder, R. L. (1978). Non-linear least-squares curve-fitting of a simple theoretical model to radioimmunoassay dose-response data using a mini-computer. In Radioimmunoassay and Related Procedures in Medicine, vol. 1. IAEA, Vienna, pp. 399-423.

167a. Wilson, D. T., Maxwell, D., Carr, W. R. (1973). RIADP: Radioimmunoassay data processing program package. Animal Breeding Research Organization, West Mains Road, Edinburgh, United Kingdom.
168. Wilson, D., Sarfaty, G., Clarris, B., Douglas, M., and Cranshaw, K. (1971). The prediction of standard curves and errors for the assay of estradiol by competetive protein binding. Steroids 17: 77-90.
169. Woodward, R. H., and Goldsmith, P. L. (1964). Cumulative Sum Techniques. ICI Monograph no. 3. Oliver and Boyd, Edinburgh.
170. Woosley, J. T., and Muldoon, T. G. (1977). Comparison of the accuracy of the Scatchard, Lineweaver-Burk and direct linear plots for the analysis of steroid-protein interactions. J. Steroid Biochem. 8: 625-629.
171. Yalow, R. S., and Berson, S. A. (1968). General principles of radioimmunoassay. In Radioisotopes in Medicine: In Vitro Studies, R. L. Hayes, F. A. Goswitz, and B. E. P. Murphy (eds.), U. S. A. E. C., Oak Ridge, Tenn., pp. 7-41.
172. Youden, W. J., and Steiner, E. H. (1975). Statistical Manual of the Association of Official Analytical Chemists. Association of Official Analytical Chemists, Washington, D. C.
173. Zivitz, M., and Hidalgo, J. U. (1977). A linearization of the parameters in the logistic function; curve fitting radioimmunoassays. Comput. Programs Biomed. 7: 318.

Index

Accelerated precipitation (see Polyethylene glycol)
Accuracy, of immunoassays, 3, 57, 163-165, 168, 189, 293
Acetyl cholinesterase, 43, 44
Acridinium esters, 105-107, 111
Adenovirus, 54
Adrenal function, tests of, 13
Adrenocorticotrophin (ACTH), 183, 185, 188, 191-193
Affinity
 of antibodies, 79, 81, 185-187, 251, 256, 290
 McAb, 199, 208, 210, 211
 in nephelometry (passim) 121-167
 constant, 188, 189, 211
 of enzyme inhibitors, 43
 of labelled antigens, 20, 29, 31, 73
 purification (chromatography), 46, 49, 179-189, 194, 203-205, 212
Aflatoxin B1, 54, 55
African swine fever virus, 55
Albumin, as binding protein, 217, 224, 225, 229, 239, 240, 241
Alkaline phosphatase, 144, 147
Amebiasis, 55
Amersham International, fT_4 method, 239-241, 247
p-Aminobenzoic acid, 58
Amino-hexyl-ethyl-isoluminol (AHEI), 110-111
4-Amino-N^{10}-methylpteroic acid, 58
Aminopterin, 58
Analog, current measurements in luminometers, 107
Analysis of variance (ANOVA), in data analysis, 272-274, 279, 284
Androstenedione, 29

Angiotensin, 26, 180
Aniline, 27
Antibodies (see Antisera)
Antigen excess, in nephelometric assays, 130-133, 138-140, 149, 152, 154, 162-163, 170
Antisera
 isotypes, 208
 monoclonal (McAb), 2, 55, 103, 111-113, 130, 137, 165, 189, 194, 199-212
 in nephelometry, requirements, 127-130, 136-137, 162
 raising of, 1, 2, 3, 5, 6, 199-208, 211
 specificity of, 1, 3, 5-6, 8, 46, 57, 71, 97, 127, 181, 191, 194
 titer of, 1, 6, 9, 128, 162, 184, 185, 200, 210
α_1 Antitrypsin, 145
Arklone, P., 130
Ascitic fluid, in preparation of McAb, 203-204, 206
Association constants (see Equilibrium constants)
Automation
 of calculation, (passim) 254-294
 of immunoassays, 13, 58, 113, 117, 118, 129, 132, 143, 146-154, 167-170, 188-194, 209
Avidity (see Equilibrium constant)

Bacterial endotoxin, 169
Bacteriophages, as labels, 37
Baker Instruments (see Nephelometers)
Beckman Instruments (see Nephelometers)

Behringwerke Ag (see Nephelometers)
Bence Jones protein, 205, 206
Bias (see Error)
Bifunctional reagents, 88-92, 94, 95
Bile salts, 45
Binding proteins, 217-249
  cortisol binding globulin (CBG), 217, 224, 227
  progesterone binding protein (PBP), 217, 227
  sex-hormone binding globulin (SHBG), 217
  thyroxine binding globulin (TBG), 1, 166, 217-246, passim
  thyroxine binding prealbumin (TBPA), 217, 229
Biological activity, 1, 3, 14, 20, 24, 26, 27, 103, 191, 218-219, 227, 283
Bioluminat (see Luminometers)
Bioluminescence, 107
Biotin, 108
Blank signals (see Fluorescence, background
Bleaching, in fluorescence, 84
Bound fraction
  in circulation, 217-249
  in immunoassays, 3, 10, 12, 13, 50, 73-79, 258, 260, 265, 290
Bovine serum albumin (BSA), for linking to haptens, 3-5, 11, 29-31
16β-Bromoestradiol, 31
N-Bromosuccinimide, for iodinations, 20, 25
isoButylchlorocarbonate, 4-5

Calcitonin, 22, 180
Calibration curves, 256, 263-264, 268-270, 272, 275, 277, 279-286, 292
Carbodiimide reaction, 46, 184
Carboxylic ester hydroxylase, 43

O-Carboxymethyl oximes, 4, 29
Carcinoembryonic antigen, 13, 54
Cascade procedure, in production of McAb, 203
Catalase, 44
Cell surface antigen, 55
Cellulose particles, 6, 8, 9, 31, 32, 52, 112, 183-185, 187, 189, 192, 234
Ceruloplasmin, 145
Chaotropic ions, 124
Charcoal (see Dextran-coated charcoal)
Chemiluminescent methods, 1, 14, 37, 44, 83, 84, 104-114, 194, 254
Chlamydia trachomatis, 55
Chloramine-T, reagent for iodination, 20-23, 24-27, 29, 185, 212
Chlorine, use in iodinations, 20, 23-24
Chorionic gonadotropin (see Human chorionic gonadotropin)
Chromic chloride, 201
Cloning, in preparation of McAb, 199, 201-203, 207, 211
Clostridium botulinum toxin type A, 54
Cocktail (see Enhancement solution)
Coefficient of variation (CV), 56, 57, 145, 189, 258-260
Cofactors, and enzymic activity, 41
Competitive enzyme immunoassays, 38, 51
Competitive protein binding assays (CPBA), 1, 19, 73, 103, 254
Complement, 145
  fixation methods, 1, 208, 210
Concanavalin A, 169, 188
Conjugation labelling, 27-28
Corning "Immophase" method, 230, 242-244
Corticosteroids, 3
Cortisol, 10-12, 13, 54, 108, 109, 237, 246
  binding globulin (CBG) (see Binding proteins)

Index

Cyanogen bromide, 9, 10, 183
L-Cysteine, 22, 24, 29
Cytomegalovirus, 55
Curve fitting, 260-294
  least squares, 270, 273, 275, 276, 282, 291, 292, 294
  logistic, 264-269, 270, 272, 276, 280, 292
  manual methods, 270, 276, 277, 278, 292-293
  polygonal interpolation, 264, 278
  polynomial, 146, 150, 263-264, 268-272, 281
  spline, 264, 278

Damon "Liquisol" method, 244-246
Dansyl chloride (DNS-Cl), 86, 87
Data analysis, 253-295
Decay time (see Fluorescence lifetime)
Design, of assays, 287-290, 293
Dextran-coated charcoal, use in separation, 6, 10, 31, 109, 244
Dextran sulfate, 129
Dialysis, equilibrium, 228, 230, 231-233, 246, 247
2,6-Diamino-6-methylpteridine, 58
Diazonium compounds, 27, 110, 112, 184, 185
Diazoreaction, 9, 94
Digital photon counting, 107
Digoxin, 54, 169
β-Diketone, 88-94
Dimethyl sulfoxide (DMSO), 204-205
Dip sticks, 58
Direct sandwich assays (see Sandwich assays)
Disulfide bridges, 48
Drugs, immunoassays of, 3, 46, 51, 60, 72, 104, 149, 168, 211, 218
Donor-chelating chromophore, 88, 89

Double-antibody
  magnetic particles (DAMP), 8-10
  method, 7, 9, 38, 39, 51, 94, 189
  solid phase methods (DASP), 8
Double-stranded RNA, enzyme immunoassay of, 54

Electron donors, 89
Electrophoresis
  in gels, 1, 7, 22, 31, 163, 166
  Laurell rockets, 117, 128, 167
Emission wavelength, 89
Energy transfer, 88-92, 104
Enhancement solution, 88, 93, 95
Enzacryl (polymerized m-diaminobenzene), 9
Enzymes
  antibody conjugates, 47-50, 59
  channeling, 41-42
  electrodes, 43
  immunoassay methods (EIA), 1, 14, 37-60, 81-83, 104, 166
  inhibitors, 41, 43
  for labelling, 26-27
  ligand conjugates, 39-41, 45-46, 59
  linked immunosorbent assay (ELISA), 39, 52-53, 201, 210, 211-254
  multiplied immunoassay technique (EMIT), 39-41, 43, 118
Equilibrium constant, 6, 78, 79, 81, 82, 219, 236, 238, 239, 241, 270
Erythropoietin, 26
Error
  in assays, 3, 258, 260, 274, 278-294
  bias, 257, 263, 272-291
  counting, 258-260
  quadrature, 259-260
  random, 257, 260, 271-291
Estradiol, 14, 31, 46, 108, 110
Estriol-16α-glucuronide (E_3-16α-G), 12-13, 14, 108, 109, 110
Estrone, 14, 108, 110

Ethylenediaminetetracetic acid (EDTA), 94, 95
Europium, 87, 89-95, 97, 98
Excitation band, 87

Ferric oxide (black iron oxide, Fe_3O_4), for preparation of magnetic particles, 9, 10
Ferritin, 54, 112, 181, 184, 186, 187
Fetal calf serum (FCS), 204, 205, 207
α-Fetoprotein (AFP), immunoassay of, 13, 110, 111, 166, 181, 188-189
Fibrinogen, use as carrier, 3
Fibronectin, enzyme immunoassay of, 54
Fieller's theorem, 280, 281, 283
Fluorescein isothiocyanate (FITC), 86, 87, 104
Fluorescence
background, 84, 142
of enzyme labels, 43-45, 53, 82
lifetime, 85-89, 92
Fluorescent compounds, 84, 86-99
Fluoroimmunoassays, 1, 37, 72, 89-99, 104, 166, 169, 211
Fluorometers, 93, 95-97, 151, 152, 154
Folic acid, 58
Folinic acid, 51, 58
Follicle stimulating hormone (FSH), 5, 6, 7, 22, 24, 25, 26, 180, 181, 185
Foot and mouth virus, 54
Free fraction, in immunoassay, 3, 10, 12, 13, 14, 50, 73, 74, 78, 79, 109, 256, 260, 290
Free hormones, in blood, 217-249
Fusion, in raising McAb, 200, 202, 205-207

β-Galactosidase, use of as label in EIA, 40-53, 56
"Gated" detection system, in fluorometers, 85
Gelatin, use of, as carrier, 32
Gentamicin, 41, 169, 207
Glial fibrillary acid protein, 182-183
Glucoamylase, 44
Glucose oxidase, use of
as enzyme label, 44
in iodinations, 26
Glucose-6-phosphate
dehydrogenase, 44
isomerase, 44
β-Glucuronidase, 12-13
Glucuronides, of steroids, 12-13, 14, 29
Glutamic acid, 58
Glutaraldehyde, 48, 201
Glycine, 27
Glycoproteins, as antigens, 5-6, 24, 26, 169 (see also FSH, HCG, LH, TSH)
Gonadotropins, 5, 6, 24, 181 (see also FSH, HCG, LH)
Growth hormone (see Human growth hormone)

Haptens, 58, 149, 155, 168-169, 185
conjugates, 3-5
labelled, 29-31, 39-42, 45-46, 108-109
Haptoglobin, 145, 165, 166
Hemagglutination, 2, 37, 201, 206 210, 211
Hemisuccinates, 4, 29, 31
Hemopexin, 145
Hemophilic factor, 185
Hepatitis B surface antigen (HBS), 94, 97, 98, 110, 111, 181
Hepes buffer, 201, 207

Herpes virus, 55
Heterogeneous immunoassays, 38-39, 41, 43, 44, 50, 60, 108
Heterologous immunoassays, 29-31
Hexokinase, 44
High-dosage hook, in immunoradiometric assays, 187
Histidine, 20, 184
Histone, 55
Homogeneous immunoassays, 38, 39-42, 43, 44, 55, 60, 108, 110
Homologous immunoassays, 29-31
Horseradish peroxidase, 26, 44, 47, 105
Human chorionic gonadotropin (HCG), 2, 5, 6, 54, 112, 211-212
Human chorionic somatomammotropin (HCS) (see HPL)
Human growth hormone (HGH), 20, 25, 32, 52-53, 180, 181, 211
Human placental lactogen (HPL) (human chorionic somatomammotropin, HCS), 166, 211
Hydrogen peroxide, 26, 104
4-Hydroxy-3-nitro-methyl benzimidate hydrochloride, 48
11α-Hydroxyprogesterone, 29
Hyland (see Nephelometers)
Hypochlorite, sodium, use of as iodinating agent, 20, 24-25
Hypoxanthine-aminopterin-thymidine (HAT) medium, 200, 202, 204-205, 206-207

2-Iminothiolane/4,4'-dithiodipyridine, 48
Immunization, 200-201 (see also Antisera) multisite intradermal method, 6
Immunoadsorbents, 48, 179-188, 193 (see also Affinity chromatography)

Immunochemiluminometric assays (ICLA) (see Chemiluminescent methods)
Immunoenzymetric methods (see Enzymes, immunoassay methods (EIA))
Immunoglobulins (see also Double antibody method)
assays of
 A (IgA), 165, 267
 E (IgE), 54, 58-59
 G (IgG), 40, 50, 54, 95-96, 110-111, 129, 145, 158-167, 180-185, 192, 200, 208-212
 M (IgM), 131, 137, 145, 149, 162, 165, 166, 184, 208
Immunohistology, 211
Immunoinhibition (see Nephelometric methods)
Immunoprecipitation (see also Polyethylene glycol, (PEG)), 1, 209-210
Immunoradiometric assays (IRMA), 2, 10, 13, 19, 71-83, 103, 179-194, 212, 241, 254
Indirect sandwich assays (see Sandwich assays)
Influenza virus, 55
Instrumentation Laboratory (see Nephelometers)
Insulin, 54, 179, 180, 184, 186
Interferon, 199, 203
Iodination (see Radioiodination)
Iodo-Gen (1,3,4,6-tetrachloro-3α, 6α-diphenylglycoluril), reagent for iodinations, 20, 25-26
Iron oxide (see Ferric oxide)

Jablonski diagram, 105

Kemtek 3000, for automated immunoassay, 188-190
Keyhole limpet hemocyanin, 3, 21

Labeled antibodies (see Immunoradiometric assays)
Labeled antigens (see Chemiluminescent methods; Enzyme immunoassays; Fluoroimmunoassays; Radioiodination; Specific activity, etc.)
Lactate dehydrogenase, 44
Lactoperoxidase, reagent for iodination, 20, 26, 185, 193
Lanthanide chelate fluorescence, 89-91
Lanthanide chelates (see Rare earth chelates)
Lasers, 118, 120, 132, 134, 143-153, 155, 210
Latex particles, in immunoassays, 2, 37, 169, 267
Laurell rocket electrophoresis (see Electrophoresis)
Least squares (see Curve fitting)
Leischmania donovani soluble antigen, 54
Lepitit, kit for fT_4, 233
Ligand differentiation immunoassay (LIDIA), 12
Ligand modulator conjugates, 41
Light scattering (see Nephelometry)
Lipoamide-dehydrogenase, 44
β-Lipoprotein, 121, 125, 128, 130, 137, 161
Logistic technique (see Curve fitting)
Logit transformation, 265, 269, 277, 278
Lower dilution limit (LDL), also Lower limit of reliable assay measurement (LLORAM), Minimum detection concentration (MDC), 283, 286
Luciferin-luciferase, 104
Lucigenin, 111
Luminol, 104-111
Luminometers, 107, 108, 111
  Biolumat, 950 and 9520, 107, 108, 109

Luteinizing hormone (LH), 5, 6, 25, 26, 31, 180, 185, 212
Luteinizing hormone-releasing hormone (LH-RH), 20
Lysozyme, 43, 44

$α_2$-Macroglobulin, 145
Magnetized particles, 6-7, 8, 51, 185
Malate dehydrogenase, 39, 44
Maleimide groups, 48, 49-50
m-Maleimidobenzoyl-N-hydroxysuccinimide (MBS), 48, 49
Mancini technique, 209
Mass spectrometry, 3
Menstrual cycle, hormonal changes, 14
2-Mercaptoethanol (2-ME), 45, 49, 51, 202, 205-207
Metabisulfite, sodium, 22, 24
Methionine, 24, 204
Methotrexate, 45-46, 51-52, 54, 57, 58, 59, 206-207
10-Methylacridone, 106, 111
N^{10}-Methylfolic acid, 58
Methyl p-hydroxybenzimidate, 27
4-Methyl mercaptobutyrimidate, 48
6-Methyl-prednisolone, 45
4-Methylumbelliferyl-β-galactopyranoside, 53
$β_2$-Microglobulin, 54, 130, 131, 163, 168
Microperoxidase, 105
Microtiter plates, 58, 107, 111, 201, 202, 206, 207
Misclassification (see Nonspecific binding)
Mixed anhydride reaction, 4-5, 45
Monoclonal antibodies (McAb) (see Antisera)
Mononuclear cells, 55
Morphine, 169

Index 315

Mouse mammary tumor virus, 54
Mucopolysaccharide, nephelometric assay of, 169
Multisite intradermal immunization method (see Immunization)
Mycobacterium tuberculosis, 55
Mycoplasma hyperpneumoniae, 54
Myelin, 55

α-Naphthoyltrifluoroacetone (NTA), 93
Neonatal hypothyroidism, 189-190
Nephelometers
 Baker Instruments, 140, 143-147, 150, 157, 169
 Beckman Instruments, 136, 138 140, 144, 147-149, 153, 169
 Behringwerke, Ag, 135, 144, 149-151, 153, 156, 159
 Hyland, 126, 135, 144, 153-154, 157, 158, 162-164, 169
 Instrumentation Lab., IL Multistat III, 144, 151
 Technicon Instruments, 118, 125,
Nephelometric methods (light scattering), 1, 8, 117-170, 209-210, 267
 immunoinhibition, 168
Neural tube defects, 111, 188
O-Nitrophenyl-β-D-galactopyranoside, 51
Nomenclature, of immunoassay, 73-75
Noncovalent coupling, in enzyme-antibody conjugates, 48
Nonparametric statistics, 254, 274, 275
Nonradioactive labels (see Chemiluminescent methods; Enzyme immunoassay; Fluoroimmunoassays)
Nonspecific binding, 10, 25, 75, 78, 79, 81, 82, 183, 270, 290

Nortryptyline, 54, 57, 59

Optimal concentration, in immunoassays, 73-77, 79, 81
Optimization, in design of immunoassays, 208, 287-290, 293
Oral bacteria, 55
Ornithine-aminotransferase, 45
Orosomucoid, 145
Orthophenanthroline isothiocyanate, 91, 92
Outliers, in assays, 275-276, 277, 287
Ovalbumin, 3
Ovulation, 13-14
Oxytocin, 184

Parametric statistics, 254, 270
Paraproteins, 130, 131, 140, 163
Parathyroid hormone (PTH), 180, 183, 184, 185, 186
Particle-counting immunoassays (PACIA), 254, 267
Penicillinase, 44
Periodate oxidation, 48
Phenobarbital, 54, 169
Phenylenedimaleimide, 48
bis-Phenyl oxalate, 104
Phenytoin, 169
Phosphorescence, 91, 104
Photonic emission, 104, 111
π-Electron delocation, 89
Plasmacytoma cells, in raising McAb, 199, 200, 201-202, 208
Plasmodium falciparum, 55
Platelets, 55
Pneumoccocal capsular, polysaccharide antigen, 55
Polarizability of molecules (α), 118-119
Polyethylene glycol (PEG), accelerated precipitation with

[Polyethylene glycol]
application, in preparing McAbs, 201, 207, 209, 211
nephelometric methods (passim), 125-167
Radioimmunoassay, 7, 8
Poly(L-glutamic acid), as carrier, 3
Polygonal interpolation (see Curve fitting)
Poly(L-lysine), as carrier, 3
Polymer enhancement, in nephelometry, 159-161
Polymers, and precipitin reactions, 125-126, 128, 156
Polynomials (see Curve fitting)
Porous glass, as solid phase, 185
Prealbumin, nephelometric assay of, 157-159
Precipitation methods (see Polyethylene glycol, (PEG))
Precipitin curve, 121-125, 135-136, 156
Precision (Imprecision) profiles, 113, 281-283, 287, 289, 292
Pregnanediol glucuronide, 109
Pristane (2,6,10,14-tetramethylpentadecane), 202, 203
Progesterone, 3, 13, 29, 31, 54, 108, 109, 169
 binding protein (PBP) (see Binding proteins)
Proinsulin, 182, 183, 185
Prolactin, 8, 9, 22, 24
Prosthetic groups and enzyme activity, 41
Protamine sulfate, 129
Pulsed-light time-resolving fluorometer, 95
Pulsed nitrogen lasers, 97
Pulsed xenon light source, 91, 93, 97
N-3-Pyrene maleimide (NPM), 86, 87

Quadrature, in assay error (see Error)
Quality control, 2, 253-294, passim
Quantum efficiency, 92,
 yield, 105, 106, 110
Quenching effects, 84, 91, 92, 110

Radial immunodiffusion, 117, 129, 166, 167
Radiation damage, 19, 24, 27, 104, 185, 186, 193
Radioiodination, with ^{125}I, methods of, 19-32, 80-83, 185-186
Radioreceptor assays (see Competitive protein binding assays (CPBA))
Rare earth chelates, 86-89, 94, 104
Resin uptake, assay for free T_4 fraction, 229, 233, 243
Response-error relationships (RER), 281-282, 283, 284, 287
 variance relationships (RVR), 281-282
Retinol binding protein (RBP), 218
Rheumatoid factors, 165
Rhodamine, 104
RPMI-1640 medium, use in preparation of McAb, 201, 204-207
Rubella virus, 54, 55

Saliva, 13-14
Salmonella virus, 55
Sandwich assays (two-site assays) (passim), 39, 77, 182, 184-194
 direct, 94, 98
 indirect, 94, 97

Saturation assays (see Competitive protein binding assays (CPBA))
Scatchard plot, 290-292, 294
Schistosoma mansoni, 55
Screening procedure, in raising McAb, 199-200, 201, 211
Second antibody methods (see Double antibody method)
Sendai virus, 201
Sensitivity
 of chemiluminescent assays, 83-84, 105, 108, 111-113
 of enzyme immunoassays, 41, 46, 51, 59, 201
 of fluoroimmunoassays, 82-84, 87-88, 91, 97
 of immunoradiometric assays, 71-84, 103, 182-194
 of nephelometric methods, 151, 153, 161-162, 167, 170
 of radioimmunoassays, 1, 3, 13, 29, 71-84, 103-104, 188, 201, 211, 212, 231, 286
Sex-hormone-binding globulin (SHBG) (see Binding proteins)
Sialic acid, 24
Simplex methods, for optimizing assays, 288-289
Sodium hypochlorite (see Hypochlorite, sodium)
Sodium metabisulphite (see Metabisulfite, sodium)
Solid phase methods, 6, 26, 109, 234, 236, (see also Double antibody magnetic particle (DAMP); Double antibody solid phase (DASP); Enzyme linked immunosorbent assay (ELISA); Immunoradiometric assays (IRMA))
Specific activity, of labelled antigens, 19-25, 75, 78, 80, 81, 84, 110, 113, 186, 187, 194, 243

Specificity (see Antisera)
Spectrophotometry, 43, 44, 53, 58
Spleen cells, in raising McAb, 201, 205
Splines (see Curve fitting)
Standard deviation (S.D.), in assays, 254, 258-259, 273, 279, 280
Standard error (S.E.), in assays, 254
Staphylococcal protein A (SpA), 208, 210
Steroid conjugates, 4-5, 12-13, 29
Steroid hormones (see under individual steroids)
Stokes' shift, 87, 88
Streptococcal antigen, 54
Subunits, of glycoproteins, 5, 6
Succinic anhydride, 4
N-Succinimidyl 3-(4-hydroxy 5-[^{125}I]iodophenyl) propionate, 27
Sucrose layering separation technique, 10-12
Sulfhydryl groups, in proteins, 21, 47, 48, 48-50
Synergistic agents, 91, 92

Talc, as solid phase reagent, 31
Technicon Instruments (see Nephelometers)
Temporal decay (see Fluorescence lifetime)
Terbium, use of, in fluoroimmunoassay, 89
Testosterone, immunoassay of, 3, 13, 54, 108, 246
Thenoyltrifluoroacetone (TTA), 91, 92, 94
Theophylline, 169
Thermometry, and enzyme activity, 43

Thioguanine, use of, in preparing McAb, 200, 207
Thyroglobulin, 3, 191
Thyroid binding globulin (TBG) (see Binding proteins)
Thyroid hormones, 3, 8, 217-249, passim (see also Thyroxine; Triiodothyronine)
Thyroid stimulating hormone (TSH)
antibodies to, 5, 6
binding immunoglobulin, 190
chemiluminescent immunoassays, 112, 113
enzyme immunoassays, 54
immunoradiometric assays, 13, 181, 187-191
iodination of, 24, 25
radioimmunoassay of, 8, 212
Thyrotoxicosis, 113
Thyroxine (T_4) (passim), 8, 39, 54, 108, 190, 211, 217-249
analogue of, 239-241
binding prealbumin (TBPA) (see Binding proteins)
free (fT_4), in blood, 217-249, passim
free index (FTI), 228-230, 232, 244, 245, 249
Time resolution, in fluoroimmunoassay, 86
Titer (see Antisera)
Tolerance, 6

Tomato ring spot virus, 54
Transferrin, 145
Transition metal cations, 105
Tri-N-butylamine, 4-5
Trichinella spiralis, 55
Trifluoroacetylacetone (TFAC), 90, 91
Triiodothyronine (T_3), 8, 54, 57, 59, 211, 217, 222, 228-229, 231
free, in blood (FT_3), 233, 237
Trioctylphosphine oxide (TOPO), 91-93
Triplet excited state, 89, 90
Tryptophan, 24
Turbimetry, 43, 117, 118, 151, 155, 166, 168, 170, 209
Two-site assays (see Sandwich assays)
Tyramine, 29
Tyrosine, 20, 21, 25, 27, 179, 184

"Ultra"-sensitive immunometric assays, 81-84
Umbelliferone, 41

Variance, 254, 258, 259, 273, 281-282
Vitamin, A, 218